Optimization of Sustainable Process Systems

Optimization of Sustainable Process Systems

Multiscale Models and Uncertainties

Edited by
Can Li

WILEY

Published by John Wiley & Sons, Inc., Hoboken, New Jersey.

Library of Congress Cataloging-in-Publication Data Applied for

Hardback ISBN: 9781394205578

Cover Design: Wiley
Cover Image: © akindo/Getty Images

Set in 9.5/12.5pt STIXTwoText by Straive, Chennai, India

Contents

List of Contributors

Andrew Allman
Department of Chemical Engineering
University of Michigan
Ann Arbor
MI, USA

Styliani Avraamidou
Department of Chemical and Biological Engineering
University of Wisconsin-Madison
Madison
WI, USA

Michael Baldea
McKetta Department of Chemical Engineering
The University of Texas at Austin
Austin
TX, USA

Debangsu Bhattacharyya
Department of Chemical and Biomedical Engineering
West Virginia University
Morgantown
WV, USA

Rahul Bindlish
Technical Expertise and Support
Technology Center
The Dow Chemical Company
Houston
TX, USA

Seolhee Cho
Department of Chemical Engineering
Carnegie Mellon University
Pittsburgh
PA, USA

Prodromos Daoutidis
Department of Chemical Engineering and Materials Science
University of Minnesota
Minneapolis
MN, USA

Valeria González
Instituto de Ingeniería Química
Facultad de Ingeniería
Universidad de la República
Montevideo, Uruguay

Ignacio E. Grossmann
Department of Chemical Engineering
Carnegie Mellon University
Pittsburgh
PA, USA

Rishabh Gupta
Department of Chemical Engineering and Materials Science
University of Minnesota,
Minneapolis
MN, USA

Soledad Gutiérrez
Instituto de Ingeniería Química
Facultad de Ingeniería
Universidad de la República
Montevideo, Uruguay

Jnana S. Jagana
Department of Chemical Engineering and Materials Science
University of Minnesota
Minneapolis
MN, USA

Zheyu Jiang
School of Chemical Engineering
Oklahoma State University
Stillwater
OK, USA

Tylee Kareck
School of Chemical Engineering
Oklahoma State University
Stillwater
OK, USA

Roberto Kreimerman
Departamento de Inserción Social del Ingeniero
Facultad de Ingeniería
Universidad de la República
Montevideo, Uruguay

Akshay Kudva
Department of Chemical and Biomolecular Engineering
The Ohio State University
Columbus
OH, USA

Can Li
Davidson School of Chemical Engineering
Purdue University
West Lafayette
IN, USA

Mariano Martín
Department of Chemical Engineering
University of Salamanca
Salamanca, Spain

Kenneth Martinez
Department of Chemical and Biological Engineering
University of Wisconsin-Madison
Madison
WI, USA

Ilias Mitrai
McKetta Department of Chemical Engineering
The University of Texas at Austin
Austin
TX, USA

Ana Flávia Monteiro
Department of Chemical and Biomedical Engineering
West Virginia University
Morgantown
WV, USA

Paola Munoz-Briones
Department of Chemical and
Biological Engineering
University of Wisconsin-Madison
Madison
WI, USA

Saba Ghasemi Naraghi
School of Chemical Engineering
Oklahoma State University
Stillwater
OK, USA

Joel A. Paulson
Department of Chemical and
Biomolecular Engineering
The Ohio State University
Columbus
OH, USA

Alejandro Pedezert
Refinación
Ancap
Montevideo, Uruguay

Efstratios N. Pistikopoulos
Artie McFerrin Department of
Chemical Engineering,
Texas A&M University
College Station
TX, USA

and

Texas A&M Energy Institute
Texas A&M University
College Station
TX, USA

Lucia Pittaluga
Instituto de Economía
Facultad de Ciencias Económicas y
Administración
Universidad de la República
Montevideo, Uruguay

and

Instituto Tecnológico Regional
Centro-Sur
Universidad Tecnológica
Durazno, Uruguay

Paritosh Ramanan
School of Industrial Engineering and
Management
Oklahoma State University
Stillwater
OK, USA

Asha Ramanujam
Davidson School of Chemical
Engineering
Purdue University
West Lafayette
IN, USA

Tushar Rathi
Department of Chemical Engineering
and Materials Science
University of Minnesota
Minneapolis
MN, USA

Richard Reed
School of Industrial Engineering and
Management
Oklahoma State University
Stillwater
OK, USA

Antonio Sánchez
Department of Chemical Engineering
University of Salamanca
Salamanca, Spain

Diego Santamaría
Department of Chemical Engineering
University of Salamanca
Salamanca, Spain

Kyeongjun Seo
McKetta Department of Chemical Engineering
The University of Texas at Austin
Austin
TX, USA

and

Department of Chemical Engineering and Biotechnology
Tech University of Korea
Siheung-Si
Republic of Korea

Mark A. Stadtherr
McKetta Department of Chemical Engineering
The University of Texas at Austin
Austin
TX, USA

Wei-Ting Tang
Department of Chemical and Biomolecular Engineering
The Ohio State University
Columbus
OH, USA

Yuhe Tian
Department of Chemical and Biomedical Engineering
West Virginia University
Morgantown
WV, USA

Ana I. Torres
Department of Chemical Engineering
Carnegie Mellon University
Pittsburgh
PA, USA

Javiera Vergara-Zambrano
Department of Chemical and Biological Engineering
University of Wisconsin-Madison
Madison
WI, USA

Hongxuan Wang
Department of Chemical Engineering
University of Michigan
Ann Arbor
MI, USA

Lingyun Xiao
Operations Research and Industrial Engineering
University of Texas
Austin
TX, USA

Qi Zhang
Department of Chemical Engineering and Materials Science
University of Minnesota
Minneapolis
MN, USA

Preface

The world is at a pivotal moment in its energy transition. Major economies have all pledged to achieve net-zero carbon emissions within the coming decades. Reaching this ambitious goal requires a profound transformation of the way we design, operate, and integrate process systems. Chemical engineers, at the intersection of energy, technology, and sustainability, are uniquely positioned to lead this transformation.

A critical part of this journey is the adoption of renewable energy sources such as solar and wind. While they hold tremendous promise, their inherent variability, driven by weather patterns and seasonal cycles, poses significant challenges. Designing reliable and sustainable energy systems, therefore, requires confronting two intertwined issues: uncertainty and decision-making across multiple spatial and temporal scales. Uncertainty arises in the availability of renewables, market conditions, and technological performance, while multi-scale decisions span from short-term plant operations to long-term infrastructure planning. Despite growing research efforts, there is still no book publication that unifies optimization under uncertainty with multi-scale decision-making for process systems. This book is written to help close that gap.

The text is organized into two complementary parts. The first part (Chapters 1–6) establishes the theoretical foundation, introducing the mathematical frameworks and algorithmic tools that underpin optimization under uncertainty and multi-scale modeling. Readers will find discussions on Bayesian optimization, multi-objective and bilevel optimization, decomposition algorithms, generalized disjunctive programming, and the solution of problems that span multiple spatial and temporal scales.

The second part (Chapters 7–13) brings these methods to real-world process systems. Case studies and applications include biomass processing, carbon capture, chemical energy storage, food packaging, and power systems. The book also emphasizes forward-looking concepts such as the circular economy and process intensification, illustrating how they can be integrated into sustainable process systems. In particular, attention is given to the role of renewable-based energy systems in reshaping the chemical industry, and to the methodologies required for achieving both sustainability and resilience in practice.

This book is intended as both a reference and a guide for graduate students, researchers, and practitioners alike, who wish to gain a deeper understanding of how uncertainty and optimization can be harnessed to design the sustainable processes and energy systems of the future. My hope is that it serves as a bridge between theory and practice, providing both the intellectual foundation and the practical tools needed to contribute to the global effort toward decarbonization.

Chapter 1 by Gupta et al. introduces bilevel optimization as a framework for hierarchical decision-making, where a leader anticipates a follower's response in complex systems. The authors present its mathematical formulation, solution approaches, and applications in sustainable systems engineering, machine learning, and robust optimization.

Chapter 2 by Mitrai and Daoutidis demonstrates how stochastic block modeling and Bayesian inference can uncover the structure of optimization problems. They further show how this structure enables distributed and hierarchical decomposition algorithms, with a case study on transitioning conventional ammonia supply chains to green ammonia.

Chapter 3 by Kudva, Tang, and Paulson presents a unifying approach to gray-box multi-objective optimization by representing complex systems as networks of functions. The proposed MOBONS algorithm leverages Bayesian optimization principles to handle cyclic dependencies, constraints, and parallel evaluations, with applications to sustainable process design.

Chapter 4 by Ramanujam and Li provides a tutorial on multi-time scale optimization models, where long-term and short-term decisions must be solved in an integrated manner. The chapter reviews algorithms, introduces the metric Value of the Multi-scale Model (VMM), and illustrates its use through a capacity expansion planning case study.

Chapter 5 by Allman and Wang examines many-objective optimization problems (MaOPs), where four or more competing objectives make both solution and interpretation challenging. The authors discuss objective selection, dimensionality reduction methods, and showcase case studies on demand response operations and planetary boundary-constrained energy supply chains.

Chapter 6 by Cho and Grossmann reviews discrete and continuous optimization models, including LP, NLP, MIP, GDP, and SP, and the decomposition algorithms that enable large-scale solutions. The authors also discuss algebraic modeling languages such as GAMS, JuMP, and Pyomo, and illustrate applications to sustainable energy systems involving hydrogen, biomass, methanol, and power systems.

Chapter 7 by Seo, Stadtherr, and Baldea highlights recent advances in multi-scale modeling and optimization of point-source carbon capture processes, with an emphasis on solvent-based systems. The chapter reviews phenomena across spatial and temporal scales, outlines design and operational optimization problems, and surveys methods for addressing uncertainty in carbon capture.

Chapter 8 by Tian, Bindlish, and Pistikopoulos introduces process intensification as a means of achieving step-change improvements in cost, energy efficiency, and sustainability. The authors present an integrated synthesis and operability optimization framework, supported by case studies in dividing wall column purification and reaction–separation integration.

Chapter 9 by Munoz-Briones et al. examines circular economy assessment frameworks for industrial process systems, focusing on chemical, food, and energy applications. The authors compare existing approaches, discuss limitations of current circularity metrics, and highlight the need for more holistic methods that integrate economic and social dimensions alongside environmental metrics.

Chapter 10 by Ghasemi Naraghi et al. addresses the decarbonization of steam cracking for olefins production through microgrid-based electrification strategies. A stochastic optimization framework is developed to manage renewable intermittency and electricity price uncertainty, with case studies showing the trade-offs between process economics and decarbonization levels.

Chapter 11 by Santamaría, Sánchez, and Martín explores the role of chemicals as energy storage carriers, focusing on hydrogen, methane, methanol, and ammonia. The chapter provides a comprehensive review of process design, operation, control, and supply chain strategies for Power-to-X-to-Power systems, emphasizing their potential to support renewable integration.

Chapter 12 by González et al. presents a techno-environmental evaluation of repurposing oil refineries for biomass-based aviation fuel production. Two process designs are developed and assessed with EPA-GREENSCOPE sustainability metrics, demonstrating promising yields and properties compared to ASTM D7566 standards.

Chapter 13 by Monteiro and Bhattacharyya discusses uncertainty quantification in solid sorbent-based CO_2 capture processes, where reaction, transport, and heat transfer mechanisms remain highly uncertain. The authors present Bayesian UQ methods for model discrepancy and parameter estimation, with case studies demonstrating reduced uncertainty in key performance metrics for sorbent-based contactors.

I would like to thank all the authors of the chapters for their thoughtful and creative contributions, which make this book both comprehensive and distinctive. Their expertise and dedication have shaped a resource that I hope will inspire and guide future research and practice in sustainable process systems.

Can Li
West Lafayette
1 December 2025

1

An Introduction to Bilevel Optimization and Its Application to Sustainable Systems Engineering

Rishabh Gupta, Jnana S. Jagana, Tushar Rathi, and Qi Zhang

Department of Chemical Engineering and Materials Science, University of Minnesota, Minneapolis, MN, USA

1.1 Introduction

Bilevel optimization refers to a class of optimization problems that involves a hierarchy of two interdependent decision-making levels. At the upper level, the "leader" makes decisions while anticipating the reaction of the "follower," who solves their own optimization problem at the lower level based on the leader's decisions. The leader's problem can then be formulated as a bilevel optimization problem (also called a bilevel program), which considers the follower's optimization problem in its constraints.

The leader–follower terminology originates from the leader–follower game conceptualized by von Stackelberg [1, 2]. Also known as the Stackelberg game, it is a strategic sequential game in which one player, the leader, makes the first move, followed by the move made by the second player, the follower. The two players are assumed to be noncooperative, that is, they both try to maximize their own utilities. Bracken and McGill [3] were the first to formulate such a Stackelberg game, in their case for a military application, as an optimization problem, which marked the beginning of bilevel optimization.

Many problems can be interpreted as Stackelberg games and hence be modeled as bilevel optimization problems. This is especially true in sustainable systems engineering as these systems often involve conflicts between upper-level policy goals, for example, sustainability and equity, and lower-level goals of individual stakeholders, who typically focus on economic performance. For example, a city government may want to assign tolls to some roads in an effort to reduce traffic, and thereby air pollution, in certain parts of the city [4]. With a toll in place, a driver now has to consider not only the travel time but also the toll amount when choosing their route; here, the driver's route choice will only be different from the case with no tolls if the imposed toll is sufficiently high. In this setting, the city government is the leader and each driver is a follower. To design effective tolls, the city government needs to anticipate the paths the drivers take as a function of the toll amount. Assuming that drivers make their decisions by solving an optimization problem (e.g. the shortest-path problem), the city government's toll design problem can be formulated as a bilevel program.

In this chapter, we provide an introductory overview of bilevel optimization and its applications. In Section 1.2, the fundamentals of bilevel optimization are presented, including its formal mathematical formulation and the most commonly applied solution approaches. Section 1.3 provides a brief overview of common bilevel optimization problems in sustainable systems engineering. In Sections 1.4 and 1.5, we discuss the use of bilevel optimization in machine learning and robust optimization, respectively, two more general areas that have gained significant relevance in sustainable systems engineering. Finally, we close with some concluding remarks in Section 1.6.

1.2 Fundamentals of Bilevel Optimization

In the following, we provide an introduction to the mathematical fundamentals of bilevel optimization. For a more comprehensive and in-depth discussion, we refer the reader to the book by Dempe et al. [5]. For a "gentle" (but perhaps not as gentle as this book chapter) introduction to the topic, we recommend the lecture notes from Beck and Schmidt [6]. Recent review papers on bilevel optimization include Colson et al. [7], Kleinert et al. [8], and Beck et al. [9].

1.2.1 Mathematical Formulation

Let x and y denote the leader's and the follower's decision variables, respectively. Given the leader's decision x, the follower's problem is as follows:

$$\begin{aligned} \underset{y \in \mathcal{Y}}{\text{minimize}} \quad & f(x,y) \\ \text{subject to} \quad & g(x,y) \leq 0, \end{aligned} \tag{1.1}$$

where $\mathcal{Y} \subseteq \mathbb{R}^q$, and the objective function f and the constraint functions g generally depend on both x and y. Now let $S(x)$ denote the set of optimal solutions to problem (1.1), also referred to as the *rational reaction set*, that is,

$$S(x) = \underset{y \in \mathcal{Y}}{\arg\min} \left\{ f(x,y) \,:\, g(x,y) \leq 0 \right\}. \tag{1.2}$$

Then, the leader's problem, which is the bilevel program, can be formulated as follows:

$$\begin{aligned} \underset{x \in \mathcal{X}, y}{\text{minimize}} \quad & F(x, y) \\ \text{subject to} \quad & G(x, y) \leq 0 \\ & y \in \mathcal{S}(x), \end{aligned} \tag{1.3}$$

where $\mathcal{X} \subseteq \mathbb{R}^p$, and F and G can similarly be functions of both x and y. Note that the last constraint forces y to be an optimal solution to the follower's optimization problem (1.1), which is what makes problem (1.3) a bilevel program. Acknowledging the given bilevel decision-making hierarchy, problems (1.1) and (1.3) are also referred to as the lower- and upper-level problems, respectively.

1.2.1.1 Optimistic Versus Pessimistic Bilevel Optimization

When $\mathcal{S}(x)$ is a singleton, that is, the follower's problem has a unique optimal solution given any feasible x, there is no ambiguity in the definition of the bilevel optimization problem. If, however, there are multiple optimal solutions to the follower's problem, it is not always clear which one the follower will choose. In that case, the leader needs to make some assumptions in terms of how they expect the follower to react.

One option is to assume that the follower is "leader-friendly," that is, that they will choose among the optimal solutions the one that is best for the leader. This applies if the follower is, for example, a colleague or a friendly neighbor, who still primarily cares about their own objective but is willing to help you out given that it does not negatively affect their own performance. This is the most optimistic assumption from the leader's perspective; hence, the associated bilevel optimization problem is called an *optimistic* bilevel problem. Indeed, formulation (1.3) does model the optimistic case, where y is chosen from $\mathcal{S}(x)$ such that $F(x, y)$ is minimized.

Another option is to assume that the follower tries to work against the leader by choosing the solution from their rational reaction set that is the least beneficial to the leader. This assumption is more appropriate when the follower is an adversary or competitor. The bilevel optimization problem that considers this worst case for the leader is called a *pessimistic* bilevel problem and can be formulated as follows:

$$\underset{x \in \mathcal{X}}{\text{minimize}} \; \max_{y \in \mathcal{S}(x)} \; \{F(x, y) : G(x, y) \leq 0\}, \tag{1.4}$$

where for a given x, $y \in \mathcal{S}(x)$ is chosen such that $F(x, y)$ is maximized. Implicit in this formulation is the use of the convention that a y that violates the constraints $G(x, y) \leq 0$ leads to a value of $F(x, y)$ of $+\infty$. Alternatively, we apply the following formulation, which explicitly states that the upper-level constraints $G(x, y) \leq 0$

must be feasible for all $y \in S(x)$:

$$\begin{aligned} \underset{x \in \mathcal{X}}{\text{minimize}} \quad & \max_{y \in S(x)} F(x, y) \\ \text{subject to} \quad & G(x, y) \leq 0 \quad \forall y \in S(x). \end{aligned} \tag{1.5}$$

The optimistic and pessimistic bilevel optimization problems represent the two extreme cases; in reality, the solution could also be anywhere in-between, that is, the follower chooses a y that satisfies neither of these two assumptions. However, in that case, it is more difficult to clearly define the problem; as a result, the literature mainly considers the optimistic and pessimistic cases. For the remainder of this chapter, we will restrict our discussion to the optimistic bilevel problem, which is the more commonly used one, in part, because it is significantly easier to solve than the pessimistic case.

1.2.1.2 High-Point Relaxation

A straightforward and useful relaxation of the bilevel program (1.3) is the high-point relaxation, which is formulated as follows:

$$\begin{aligned} \underset{x \in \mathcal{X},\, y \in \mathcal{Y}}{\text{minimize}} \quad & F(x, y) \\ \text{subject to} \quad & G(x, y) \leq 0 \\ & g(x, y) \leq 0, \end{aligned} \tag{1.6}$$

where the constraint $y \in S(x)$ from the bilevel problem has been replaced with $y \in \mathcal{Y}$ and $g(x, y) \leq 0$. This means that in problem (1.6), y can be chosen from the set of all feasible instead of only the set of optimal solutions of the lower-level problem; as a result, problem (1.6) is clearly a relaxation of problem (1.3). Importantly, problem (1.6) is a single-level problem and hence much easier to solve than problem (1.3).

1.2.1.3 When to Not Use Bilevel Optimization

There are sometimes problems that have a seemingly bilevel decision-making structure but upon closer inspection, one realizes that their solution does not really require the use of bilevel optimization. In the following, we discuss two such situations.

i) *The problem does not involve two noncooperative decision-makers.* For example, if the two-decision makers are from different parts of the same company but share the same overall business goals, they will be willing to cooperate to achieve the best overall outcome. In that case, instead of a bilevel problem, the problem should be an *integrated* optimization problem where the decisions of the two decision-makers are optimized jointly, which can be formulated as follows:

$$\begin{aligned} \underset{x\in\mathcal{X},\, y\in\mathcal{Y}}{\text{minimize}} \quad & F(x,y)+f(x,y) \\ \text{subject to} \quad & G(x,y) \leq 0 \\ & g(x,y) \leq 0. \end{aligned} \tag{1.7}$$

Note that here we assume that the overall objective function is $F(x,y)+f(x,y)$, but it could also be a differently weighted sum of F and f. An example for such a problem is production planning and scheduling where in practice, the planning and scheduling decisions are often made in a sequential manner, that is, production targets are first determined by solving a long-term planning problem after which short-term scheduling problems are solved to meet these targets. We are often restricted to such a sequential decision-making process due to a stringent hierarchical organizational structure or computational limitations, but given that these challenges can be overcome, there are significant benefits in integrating planning and scheduling decisions in a single optimization problem [10].

ii) *The leader and follower have nonconflicting objectives.* Let Z^* be the optimal value of the high-point relaxation, that is,

$$Z^* = \min_{(x,y)\in\Omega} F(x,y), \tag{1.8}$$

where

$$\Omega = \{(x,y) : G(x,y) \leq 0, \quad g(x,y) \leq 0, \quad x \in \mathcal{X}, \quad y \in \mathcal{Y}\}. \tag{1.9}$$

Similarly, let z^* be the optimal value of the same problem but with $F(x,y)$ replaced with the follower's objective function $f(x,y)$, that is,

$$z^* = \min_{(x,y)\in\Omega} f(x,y). \tag{1.10}$$

We say that the leader and follower have nonconflicting objectives if there exists a solution $(x^*,y^*) \in \Omega$ for which $F(x^*,y^*) = Z^*$ and $f(x^*,y^*) = z^*$. Such a solution can be directly obtained by solving the integrated problem (1.7). A corollary of this is that in such a case, an optimal solution to problem (1.7) is also an optimal solution to the bilevel problem (1.3) since y^* will automatically be in the rational reaction set $\mathcal{S}(x^*)$; note that the converse may not be true. Hence, there is no need to solve a bilevel optimization problem in this case. Consider the example of a regional transmission organization (RTO) that performs power transmission expansion planning with the sole goal of maximizing the penetration of wind energy in the grid and a wind energy company that aims to maximize its profit by building and operating wind farms. This is clearly a bilevel decision-making problem where the RTO is the leader and the wind energy company is the follower. However, if generating energy from wind is already cheaper than all alternatives at a given location, the optimal solution for both the RTO and the wind energy company may be to expand transmission to that location as much as possible so that the wind farm can sell the maximum amount of energy to the grid. In that case, the leader's and follower's objectives are nonconflicting, and we can find the solution to the bilevel problem by simply solving the integrated single-level problem.

1.2.2 KKT Reformulation

Most methods for solving bilevel programs are based on either a Karush-Kuhn-Tucker (KKT) or value-function reformulation. The KKT reformulation was introduced in the pioneering work of Fortuny-Amat and McCarl [11], which considered convex-quadratic bilevel problems. In the following, we present the KKT reformulation and its most common variants as well as address some of the intricacies in applying it.

1.2.2.1 "Naive" KKT Reformulation

The basic idea here is to obtain a single-level reformulation by replacing the lower-level problem with its optimality conditions. Applying the KKT optimality conditions, one arrives at the following formulation:

$$\underset{x\in\mathcal{X},\, y\in\mathbb{R}^q,\, \lambda\in\mathbb{R}^m}{\text{minimize}} \quad F(x,y) \tag{1.11a}$$

$$\text{subject to} \quad G(x,y) \leq 0 \tag{1.11b}$$

$$\nabla_y f(x,y) + \sum_{i=1}^{m} \lambda_i \nabla_y g_i(x,y) = 0 \tag{1.11c}$$

$$g(x,y) \leq 0 \tag{1.11d}$$

$$\lambda \geq 0 \tag{1.11e}$$

$$\lambda_i\, g_i(x,y) = 0 \quad \forall\, i = 1, \dots, m, \tag{1.11f}$$

where m is the number of lower-level constraints, and λ_i denotes the dual variable associated with inequality $g_i(x,y) \leq 0$. Constraints (1.11c), (1.11d), (11.11e), and (1.11f) are the stationarity, primal feasibility, dual feasibility, and complementarity conditions, respectively, for problem (1.1).

The KKT reformulation (1.11) and the bilevel problem (1.3) are equivalent if the lower-level problem (1.1) is convex and Slater's constraint qualification is satisfied for problem (1.1) for all feasible x [12]. In that case, if (x^*, y^*, λ^*) is a global optimal solution of problem (1.11), (x^*, y^*) is a global optimal solution of problem (1.3).

1.2.2.2 Mixed-Integer Programming Reformulation

While problem (1.11) is conceptually clear and easy to formulate, we call it the "naive" KKT reformulation because it cannot be directly solved by standard nonlinear programming (NLP) solvers. This is because of the complementarity conditions (1.11f) due to which classic constraint qualifications such as the linear independence and Mangasarian-Fromowitz constraint qualifications are violated at every feasible point of problem (1.11) [13]. One way to avoid this issue is by modeling the complementarity constraints as disjunctions, that is, for each $i = 1, \dots, m$, we have:

$$[\lambda_i = 0] \ \vee \ [g_i(x, y) = 0], \tag{1.12}$$

which can be written as mixed-integer constraints. This results in the following mixed-integer programming (MIP) formulation of the KKT reformulation [11]:

$$\underset{x \in \mathcal{X},\, y \in \mathbb{R}^q,\, \lambda \in \mathbb{R}^m,\, z \in \{0,1\}^m}{\text{minimize}} \quad F(x, y) \tag{1.13a}$$

$$\text{subject to} \quad \text{Constraints (1.11b)–(1.11e)}$$

$$\lambda_i \leq M z_i \quad \forall i = 1, \dots, m \tag{1.13b}$$

$$g_i(x, y) \geq -M(1 - z_i) \quad \forall i = 1, \dots, m, \tag{1.13c}$$

where z_i is a binary variable that equals 1 if $g_i(x, y) = 0$ and 0 if $\lambda_i = 0$ as enforced by constraints (1.13b) and (1.13c), where M is a sufficiently large big-M parameter. Although problem (1.13) is now a formulation that includes discrete variables, it can be directly solved using off-the-shelf MIP solvers.

1.2.2.3 Branching on Complementarity Constraints

A major challenge in using the MIP reformulation (1.13) is the proper choice of the big-M parameter, which must be sufficiently large for the formulation to be valid but should not be too large as that can lead to numerical issues or slow progress in the branch-and-bound algorithm. While one can quite easily derive bounds on the primal constraint functions in most practical applications, bounding the dual variables can be problematic. In particular, if the lower-level problem is a linear program (LP) with a bounded primal feasible set, the corresponding dual feasible set is unbounded [14]; this typically means that one needs to choose a rather large big-M value.

When a standard MIP solver is used to solve problem (1.13), the classic branch-and-bound method is applied, which branches on the binary variables z associated with the complementarity constraints. Alternatively, one can directly branch on the complementarity constraints where branching on a node in the branch-and-bound tree is achieved by creating two child nodes, one in which we set $\lambda_i = 0$ for a chosen i and another in which we add the constraint $g_i(x, y) = 0$ for the same i. At each node, we solve a problem that corresponds to problem (1.11) without constraints (1.11f) but with the added constraints from branching up to that node. Branching directly on the complementarity constraints is equivalent to branching on the binary variables but has the advantage of not requiring big-M parameters.

Branching on complementarity constraints can be conveniently implemented by using special ordered sets of type 1 (SOS1). To do so, we reformulate constraints (1.11f) as follows:

$$s_i = g_i(x, y) \quad \forall i = 1, \dots, m \tag{1.14a}$$

$$\text{SOS1}(s_i, \lambda_i) \quad \forall i = 1, \dots, m, \tag{1.14b}$$

where we introduce a new auxiliary variable s_i that equals the constraint function $g_i(x, y)$. The SOS1 constraints (1.14b) state that for each i, either s_i or λ_i must be zero. Modern MIP solvers handle these SOS1 constraints by reformulating them using big-M constraints if provably correct bounds on s_i and λ_i are available or, otherwise, directly branching on the variables analogous to what we described above. The SOS1 approach may not lead to reduced solution times compared to the big-M approach; in fact, results from Kleinert and Schmidt [15] indicate that the big-M formulation is more efficient if good big-M parameters are available. However, the SOS1 approach eliminates the need to find valid big-M parameters, which is often challenging. Kleinert and Schmidt [15] also find that in the case of linear bilevel programs, one can derive valid inequalities that can help significantly speed up the SOS1 approach.

1.2.2.4 Penalty-Based Reformulation

Another way to address the issues related to the complementarity constraints is to move these constraints to the objective function using a penalty reformulation [16]:

$$\begin{aligned} \underset{x\in\mathcal{X},\, y\in\mathbb{R}^q,\, \lambda\in\mathbb{R}^m}{\text{minimize}} \quad & F(x, y) + \rho \sum_{i=1}^{m} |\lambda_i g_i(x, y)| \\ \text{subject to} \quad & \text{Constraints (1.11b)–(1.11d),} \end{aligned} \tag{1.15}$$

where the added term in the objective function penalizes violations of the complementarity constraints. This formulation uses an exact penalty function based on the ℓ_1-norm. It is exact in the sense that if the penalty parameter ρ is sufficiently large, a local optimal solution of problem (1.15) that is feasible for the KKT reformulation (1.11) is also a local optimal solution of problem (1.11). Generally, one does not know in advance how large ρ needs to be. Therefore, typically, problem (1.15) is solved iteratively where the value of ρ is increased from one iteration to the next until the obtained solution satisfies all complementarity constraints.

1.2.3 Value-Function Reformulation

The KKT reformulation can only be applied if the lower-level problem is convex. We now present another reformulation, the value-function reformulation, which applies much more broadly; however, solving the resulting problem often requires exploiting some special structure of the problem.

1.2.3.1 Reformulation Using the Optimal-Value Function

The value-function reformulation of the bilevel problem (1.3) is based on the most general optimality condition for the lower-level problem, namely that the chosen solution y must result in the minimum objective function value. The reformulation can be written as follows:

$$
\begin{aligned}
\underset{x \in \mathcal{X},\, y}{\text{minimize}} \quad & F(x, y) && (1.16a)\\
\text{subject to} \quad & G(x, y) \leq 0 && (1.16b)\\
& g(x, y) \leq 0 && (1.16c)\\
& f(x, y) = \phi(x), && (1.16d)
\end{aligned}
$$

where $\phi(x)$ denotes the optimal-value function defined as

$$
\phi(x) = \min_{\hat{y} \in \mathcal{Y}} \{f(x, \hat{y}) \,:\, g(x, \hat{y}) \leq 0\}. \qquad (1.17)
$$

At first glance, the value-function reformulation does not seem to be a major improvement to the original bilevel formulation since it still contains an embedded optimization problem in the form of the optimal-value function. However, given this reformulation, the tractability of the problem now depends on the properties of the optimal-value function, which may be different for different classes of bilevel programs and hence give rise to different solution methods.

1.2.3.2 *K*th-Best Algorithm

Consider the special case of a linear bilevel program (i.e. $\mathcal{X} = \mathbb{R}^p$, $\mathcal{Y} = \mathbb{R}^q$, and all upper- and lower-level objective and constraint functions are linear) of the following form:

$$
\begin{aligned}
\underset{x \in \mathbb{R}^p,\, y}{\text{minimize}} \quad & c^\top x + \hat{c}^\top y && (1.18a)\\
\text{subject to} \quad & Ax \geq b && (1.18b)\\
& y \in \arg\min_{\hat{y} \in \mathbb{R}^q} \{\hat{h}^\top \hat{y} \,:\, Ex + \hat{E}\hat{y} \geq d\}, && (1.18c)
\end{aligned}
$$

where $c \in \mathbb{R}^p$, $\hat{c} \in \mathbb{R}^q$, $A \in \mathbb{R}^{\ell \times p}$, F, $b \in \mathbb{R}^\ell$, $h \in \mathbb{R}^p$, $F \in \mathbb{R}^{m \times p}$, $G \in \mathbb{R}^{m \times q}$, and $d \in \mathbb{R}^m$. This problem does not contain any upper-level coupling constraints, that is, constraints (1.19b) do not depend on y; otherwise, the bilevel-feasible set may be disconnected. Also note that the lower-level objective function is assumed to not depend on x; this does not cause any loss of generality since from the lower-level LP's perspective, such a dependence on x would only add a constant cost term.

The value-function reformulation of problem (1.18) is as follows:

$$\begin{aligned}
\underset{x\in\mathbb{R}^p,\, y\in\mathbb{R}^q}{\text{minimize}} \quad & c^\top x + \hat{c}^\top y && (1.19a)\\
\text{subject to} \quad & Ax \geq b && (1.19b)\\
& Ex + \hat{E}y \geq d && (1.19c)\\
& \hat{h}^\top y = \phi(x) = \min_{\hat{y}\in\mathbb{R}^q} \{\hat{h}^\top \hat{y} : Ex + \hat{E}\hat{y} \geq d\}. && (1.19d)
\end{aligned}$$

Due to strong LP duality, we can also rewrite the optimal-value function as

$$\phi(x) = \max_{\lambda\in\mathbb{R}^m} \{(d - Ex)^\top \lambda : \hat{E}^\top \lambda = \hat{h},\ \lambda \geq 0\}, \tag{1.20}$$

where λ are the dual variables. Note that the feasible set of the dual does not depend on x. From classic LP theory, we know that there is always an optimal solution to an LP that is attained at a vertex of the polyhedral feasible region. As a result, given that $\hat{\lambda}^1, \dots, \hat{\lambda}^r$ are all the vertices of the polyhedron defined by the constraints in (1.20), we can further rewrite the optimal-value function as follows:

$$\phi(x) = \max_{\lambda} \{(d - Ex)^\top \lambda : \lambda \in \{\hat{\lambda}^1, \dots, \hat{\lambda}^r\}\}. \tag{1.21}$$

Equation (1.21) indicates that $\phi(x)$ is a convex piecewise-linear function. This is a key property that allows one to show that an optimal solution of the bilevel problem (1.18) always lies at a vertex of the set

$$\Omega = \{(x, y) \in \mathbb{R}^{p+q} : Ax \geq b,\ Ex + \hat{E}y \geq d\}. \tag{1.22}$$

This insight [6] has led to the development of the Kth-best algorithm [17], one of the first solution methods for bilevel programs. The Kth-best algorithm searches over the vertices of Ω, similar to the simplex algorithm for solving LPs. The following is an outline of the algorithm, where we assume that the lower-level LP has a unique solution for every feasible x:

- **Step 1:** Solve the high-point relaxation of the linear bilevel program (1.18) to obtain (x^1, y^1). Set $\mathcal{V} \leftarrow \emptyset$ and $k \leftarrow 1$.
- **Step 2:** Solve the lower-level LP with $x = x^k$ to obtain y^*.
- **Step 3:** If $y^* = y^k$, stop with optimal solution (x^k, y^k). Otherwise, find all vertices of Ω adjacent to (x^k, y^k); let $\overline{\mathcal{V}}^k$ denote this set of vertices. Set $\mathcal{V} \leftarrow \mathcal{V} \cup \overline{\mathcal{V}}^k \setminus \{(x^1, y^1), \dots, (x^k, y^k)\}$ and obtain

  $$(x^{k+1}, y^{k+1}) \in \arg\min_{x,y}\{c^\top x + \hat{c}^\top y : (x, y) \in \mathcal{V}\}.$$

  Set $k \leftarrow k + 1$ and go to Step 2.

Note that finding all adjacent vertices in Step 3 is the computationally most intensive part of the algorithm.

1.2.3.3 Cutting-Plane Approach

For the special case where the lower-level constraints are not affected by the upper-level decisions, that is, g only depends on y, we can reformulate problem (1.16) as follows:

$$\underset{x\in\mathcal{X},\,y\in\mathcal{Y}}{\text{minimize}} \quad F(x,y) \tag{1.23a}$$

$$\text{subject to} \quad G(x,y) \le 0 \tag{1.23b}$$

$$g(y) \le 0 \tag{1.23c}$$

$$f(x,y) \le f(x,\hat{y}) \quad \forall \hat{y} \in \hat{\mathcal{Y}} \tag{1.23d}$$

where $\hat{\mathcal{Y}} := \{y \in \mathcal{Y} : g(y) \le 0\}$. Here, constraint (1.16d) has been replaced with (1.23d), where the rationale is that if $f(x,y)$ takes its minimum value for a given x, then $f(x,y) \le f(x,\hat{y})$ for all feasible $\hat{y}$.

Problem (1.23) is generally a semi-infinite program (SIP) since we may have an infinite number of constraints depending on the nature of the set $\hat{\mathcal{Y}}$. To solve the SIP, we can use a cutting-plane approach [18], where we replace $\hat{\mathcal{Y}}$ in (1.23d) with a finite subset $\overline{\mathcal{Y}}_k \subseteq \hat{\mathcal{Y}}$ that gets updated at each iteration k with additional points obtained from solving a separate cut-generation problem. The outline of such a cutting-plane algorithm is as follows:

- **Step 1:** Initialize $\overline{\mathcal{Y}}_1$ with a set of feasible points or set $\overline{\mathcal{Y}}_1 = \leftarrow \emptyset$. Set $k \leftarrow 1$.
- **Step 2:** Solve the following problem:

$$\begin{aligned} \underset{x\in\mathcal{X},\,y\in\mathcal{Y}}{\text{minimize}} \quad & F(x,y) \\ \text{subject to} \quad & G(x,y) \le 0 \\ & g(y) \le 0 \\ & f(x,y) \le f(x,\hat{y}) \quad \forall \hat{y} \in \overline{\mathcal{Y}}_k, \end{aligned} \tag{1.24}$$

  which is a relaxation of problem (1.23). Obtain x^* and y^*.
- **Step 3:** Solve the following cut-generation problem:

$$\begin{aligned} \underset{y\in\mathcal{Y}}{\text{minimize}} \quad & f(x^*,y) \\ \text{subject to} \quad & g(y) \le 0, \end{aligned} \tag{1.25}$$

  which is the lower-level problem for the upper-level decision x^*. Obtain $\hat{y}$.
- **Step 4:** If $f(x^*,y^*) \le f(x^*,\hat{y})$, stop with optimal solution (x^*,y^*). Otherwise, set $\overline{\mathcal{Y}}_{k+1} \leftarrow \overline{\mathcal{Y}}_k \cup \{\hat{y}\}$, set $k \leftarrow k+1$, and go to Step 2.

For certain bilevel programs, for example, where the lower-level problem is an LP or mixed-integer linear program (MILP) with a bounded feasible region, the cutting-plane algorithm is guaranteed to converge in a finite number of iterations.

1.3 Some Applications in Sustainable Systems Engineering

Achieving improved sustainability often requires the interaction between multiple decision-makers. These decision-makers typically follow a hierarchical structure similar to a leader and follower(s), such as a government imposing limits on carbon emissions and individual actors (e.g. power plants) optimizing their operations accordingly. For this reason, the bilevel optimization framework is commonly used in decision-making for sustainable systems, with several applications discussed below.

Supply chain planning, which is one of the most studied classes of optimization problems, is commonly formulated as a centralized decision-making problem. However, in reality, it almost always involves multiple decision-makers, who could be different companies providing products and services to each other at different levels of the supply chain, or government-controlled agencies, such as the Environmental Protection Agency (EPA), that regulate or influence some aspects of the supply chain through emission targets, incentives/subsidies, and other means. Some of the relevant work on sustainable supply chains include the following: Yue and You [19] deal with a decentralized biofuel supply chain problem, with the biorefinery investor (leader) making decisions on the location, capacities, and technology of the facilities, and the suppliers/consumers (followers) making decisions on the input/output material transactions. Yue and You [20] consider an integrated forestry and biofuel supply chain problem, with a forestry company (leader) dictating the biorefinery investment decisions, while the pulp company (follower) decides the pulp production plans and recipe. Chalmardi and Camacho-Vallejo [21] present a bilevel model that allows the EPA to offer financial incentives to supply chain managers for choosing cleaner production technologies, integrating both the environmental and economic aspects in a hierarchical decision-making setting. Avval et al. [22] propose a cap-and-trade scheme that enables the government to set an optimal emission allowance and supply chains to adapt their operations and trade allowances among each other accordingly. Lastly, recent works [23, 24] have used bilevel optimization as a solution framework to solve bi-objective hydrogen supply chain planning problems (minimizing economic costs and green house gas (GHG) emissions), with the upper-level problem determining the location and size of the facility, and the lower-level problem determining production and transportation decisions.

Energy systems planning requires coordination of policies and regulations among decision-makers, such as those operating at different levels of energy distribution. This is a complex problem, particularly as renewable energy increasingly penetrates the energy sector, further complicating decentralized decision-making processes due to factors such as emission caps, consumer preferences, advancements in storage technologies, and more. In Martelli et al. [25], the government optimizes carbon tax and incentives for renewable energy to achieve emission objectives, while operators of multi-energy systems (MESs) optimize their design and operations accordingly. In Cheng et al. [26], a bilevel expansion planning model is proposed, with the upper level optimizing integrated electricity and natural gas network planning in a multi-regional MES and the lower level optimizing the planning of a district MES modeled as an energy hub under carbon emission constraints. Additionally, the bilevel framework has been used to model coordination and exploit synergies between transmission- and distribution-level MESs under different carbon pricing strategies [27, 28]. Lastly, Ma et al. [29] propose an optimal sizing and operations strategy for shared energy storage (SES) systems involving multiple power generation stakeholders. Specifically, the upper-level problem determines SES capacity, while the lower-level problem addresses day-ahead power dispatching strategies.

The maritime industry is the backbone of global trade, which, unfortunately, also makes it a major contributor to GHG emissions. These emissions result not only from the ships traveling between distant ports but also from the burning of fuel, due to the auxiliary engines operating, while the ships are docked at the ports for loading/unloading. Decarbonizing the maritime industry requires the cooperation of several entities, including international organizations, various governments, port authorities, port operators, and fuel producers, among others [30]. Tan et al. [31] utilize bilevel programming to model the government's goal of setting emission targets, violation penalties, and deciding subsidies for various mitigation technologies, mostly involving vessel retrofitting, while allowing the industry to respond by selecting the most cost-effective option. In Molavi et al. [32], the authors investigate the effectiveness of regulatory authorities imposing carbon taxes and providing incentive policies on various sustainable options available to ports, such as onshore power supply (OPS), trucks and locomotives using liquefied natural gas, and hybrid electric tugboats. Wang et al. [33] extend the work by considering a hybrid economic approach involving a regulatory agency and a port entity to promote the adoption of OPS for the reduction of at-berth emissions. Lastly, Wen et al. [34] use bilevel programming to model the collaborative decision-making between a seaport microgrid, which establishes the energy prices, and the ship operators, who decide their charging and service plans.

The transformation of the transportation sector is crucial for minimizing rising GHG emissions. Implementing mitigation measures and new infrastructure requires interaction between regulatory agencies, stakeholders, and the population, making bilevel optimization an appropriate framework for their design and implementation. In Qiu et al. [35], a bilevel framework is proposed to model a pollution routing problem involving interactions between a regulatory authority that sets carbon tax and subsidy prices to minimize emissions, and a freight company that minimizes costs by determining its optimal freight schedule. Similarly, Qiu et al. [36] focus on the aviation industry, with the government implementing a carbon tax incentive scheme and airlines altering their air transport plans accordingly. Zhao et al. [37] propose a bilevel optimization model to assist a policymaker (leader) in determining a mode-of-transport-dependent tolling scheme that encourages travelers (followers) to choose a greener mode of transportation. Finally, in the context of electric vehicle recharging, Piedra-de-la Cuadra and Ortega [38] utilize a bilevel model to determine the minimum number of charging stations required to ensure reinforced service coverage for all users while maximizing demand satisfaction under budgetary constraints, addressing both administrative and private sector objectives.

Sustainability encompasses the environmental, social, and economic aspects of society, leading to numerous other applications in the literature where bilevel optimization has been used as a decision-making tool. A comprehensive review, including a discussion of applications in green facility location, eco-industrial parks, waste and water management, agriculture, and disaster prevention and response, can be found in Caselli et al. [39].

1.4 Bilevel Optimization for Machine Learning

The hierarchical modeling framework offered by bilevel optimization has proven valuable in several key applications within machine learning. In this section, we highlight two prominent applications: *Inverse optimization*, which aims to infer a data-driven model for a decision-making process as a mathematical optimization problem, and *hyperparameter optimization*, which involves tuning model hyperparameters to achieve optimal learning performance. We conclude by discussing *first-order solution methods for bilevel optimization*, which are especially suited for efficiently solving high-dimensional problems in machine learning.

1.4.1 Data-Driven Inverse Optimization

Inverse optimization is an emerging paradigm in operations research that refers to the problem of inferring the underlying model of a decision-making process, framed as a mathematical optimization problem. This is motivated by the idea that optimization problems are natural models for decision-making processes.

Inverse optimization has been applied in various fields, with early applications proposed by Burton and Toint [40] in traffic modeling and seismic topography. In the context of inverse shortest-path optimization, Burton and Toint [40, 41] developed a framework to estimate arc costs using observed shortest-path data. In traffic modeling, their approach uses actual traffic flow data to infer travel times on road segments as *perceived* by the users. In seismic topography, they use seismic wave arrival times to estimate travel times through subsurface layers, revealing the Earth's structure. Since then, numerous applications of inverse optimization have emerged, in areas such as healthcare [42, 43], energy systems [44, 45], and finance [46–48] among others; a comprehensive overview of these applications can be found in Chan et al. [49].

Inverse optimization has also been applied to design incentive mechanisms for achieving sustainability goals across various sectors. For example, in the aviation sector, Rathore and Jakhar [50] consider inverse optimization to design a differential carbon tax policy that guides airlines to operate on routes aligning with environmental and social sustainability objectives. In renewable energy systems, Zhou et al. [51] apply inverse optimization to determine the level of policy intervention required to boost investment in renewable energy production. Lastly, Tan and Aviso [52] study the estimation of appropriate incentive or disincentive required to form an optimal resource conservation network among a set of chemical plants in an industrial park. In all these examples, the desired sustainable outcome serves as the input, and inverse optimization identifies the appropriate policy intervention needed to achieve it as a solution to the forward problem.

Formulating the inverse optimization problem as a bilevel optimization problem is natural. The lower-level problem represents the agent's assumed decision process, whereas the upper-level problem seeks to "invert" this decision-making process by adjusting the model parameters to minimize discrepancies between predicted and observed decisions. Formally, consider the parameterized form of the agent's assumed optimization problem:

$$\begin{aligned} \underset{x \in \mathcal{X}}{\text{minimize}} \quad & f(x, u; \theta) \\ \text{subject to} \quad & g(x, u; \theta) \leq 0, \end{aligned} \tag{1.26}$$

where x are the decision variables, u denote inputs to the agent based on which it makes a decision, and θ are the model parameters, some or all of which may be unknown. In this formulation, f is the agent's objective function, reflecting its preferences or goals, and g are its constraints. Given inputs u and corresponding observed decisions x, the inverse optimization problem aims to infer the unknown parameters θ. This is formulated as the following bilevel optimization problem:

$$\begin{aligned} \underset{\hat{\theta} \in \Theta, \hat{x}}{\text{minimize}} \quad & \sum_{i \in \mathcal{I}} \ell(x_i, \hat{x}_i) \\ \text{subject to} \quad & \hat{x}_i \in \underset{\tilde{x} \in \mathcal{X}}{\arg\min} \left\{ f(\tilde{x}, u_i; \hat{\theta}) \, : \, g(\tilde{x}, u_i; \hat{\theta}) \leq 0 \right\} \quad \forall i \in \mathcal{I}, \end{aligned} \tag{IOP}$$

where ℓ is a loss function providing a measure of the discrepancy between the observed decisions x and their estimates $\hat{x}$. Problem (IOP) seeks to find parameter estimates $\hat{\theta}$ from the set Θ that minimize the loss function, ensuring that $\hat{x}_i$ are optimal solutions to (1.26) for the given input u_i across all $i \in \mathcal{I}$.

Problem (IOP) is a generalized formulation of the data-driven inverse optimization problem, accommodating noisy observed data and unknown parameters in both the objective and constraint functions. Due to its complexity, only a few works, such as a recent Bayesian optimization-based approach by Lu et al. [53], directly address the solution of this general form of (IOP), while most focus on special cases. Early studies on the solution of (IOP) considered cases with deterministic (non-noisy) observed decisions and linear forward problems [54, 55]. More recent works have tackled noisy data, mainly for convex lower-level problems with unknown parameters in the objective function [56–58]. A typical solution method for these problems involves reformulating (IOP) into a single-level problem by replacing the lower-level problems with their optimality conditions. Some authors have focused exclusively on linear forward problems leveraging polyhedral insights for specialized solutions strategies [59–61]. Others propose efficient approximate algorithms for joint estimation of objective and constraint parameters, assuming convex forward problems [62–64]. Additionally, a few works consider inverse mixed-integer optimization, employing a value-function reformulation to transform (IOP) into a single-level problem [65]. In this case, the resulting semi-infinite optimization problem is often solved with a cutting-plane algorithm [66].

A key feature of inverse optimization that sets it apart from traditional black-box learning methods like decision trees or deep learning is the modeling flexibility offered by its optimization framework. This flexibility enables the direct incorporation of domain knowledge through constraints. For example, in traffic modeling, constraints can explicitly capture the structure of the road network and impose limits on traffic flow for specific segments. By embedding such domain-specific knowledge, inverse optimization models become inherently interpretable, more generalizable, and less dependent on extensive training data, enhancing their practicality and effectiveness in real-world applications.

More recently, inverse optimization has also been applied to decision-focused learning [67, 68] and the construction of optimization proxies [69–71]. Decision-focused learning integrates the prediction of model parameters that are part of an optimization model and the solving of that optimization model to make decisions. The key idea is to train the predictive models such that the decision error is minimized, which is defined as the difference between the optimal solution obtained using the actual model parameter values and the optimal solution obtained with the predicted parameter values. Similarly, an optimization proxy is a trained model that predict solutions that are close to the true optimal solutions of the original complex optimization model; here, the primary objective is to obtain a surrogate model that is computationally efficient yet leads to the same or close to the same optimal solutions (i.e., an optimization proxy). It turns out that these problems can be formulated as data-driven inverse optimization problems, which can benefit from the ability of inverse optimization to explicitly incorporate constraints.

1.4.2 Hyperparameter Tuning

Hyperparameter optimization has become one of the quintessential applications of bilevel optimization in machine learning. Training machine learning models often involves tuning hyperparameters – parameters that are not directly learned from data but instead control various aspects of the learning process, such as learning rates, regularization factors, or neural network architectures. Unlike model parameters, which are learned during the training process, hyperparameters are set prior to training and can significantly influence the model's performance.

Typically, hyperparameters are optimized using methods such as grid search, random search, and Bayesian optimization:

1) **Grid Search:** A predefined set of hyperparameter values is specified, and the model is trained and evaluated for each combination. Cross-validation is often used in conjunction with grid search to assess the model's performance across different subsets of the data, ensuring that the chosen hyperparameters generalize well to unseen data.
2) **Random Search:** Instead of exhaustively searching a fixed grid, random search samples hyperparameters from predefined distributions, offering a more efficient exploration of the search space.
3) **Bayesian Optimization:** This method models the objective function as a probabilistic surrogate (e.g. a Gaussian process) and sequentially selects promising hyperparameter configurations based on an acquisition function, balancing exploration and exploitation.

However, these methods have limitations. Grid search and random search can be computationally expensive, especially for high-dimensional hyperparameter spaces, as they require training the model multiple times. Additionally, grid search may miss optimal configurations that lie between predefined grid points. Bayesian optimization mitigates some inefficiencies but still requires repeated model evaluations and does not scale well with complex learning tasks.

An alternative and more principled approach is to frame hyperparameter optimization as a bilevel optimization problem. The upper-level problem optimizes the hyperparameters to minimize a validation error, while the lower-level problem corresponds to the standard training procedure that minimizes the training loss for a fixed set of hyperparameters. Formally, this can be expressed as:

$$\begin{aligned} \underset{\lambda}{\text{minimize}} \quad & L_{\text{validation}}(\omega(\lambda)) \\ \text{subject to} \quad & \omega(\lambda) \in \underset{\omega}{\arg\min}\ L_{\text{training}}(\omega, \lambda), \end{aligned} \tag{1.27}$$

where L_{training} and $L_{\text{validation}}$ denote the training and validation losses, respectively. The lower-level optimization problem represents training a model with fixed hyperparameters λ, yielding optimal weights $\omega^*(\lambda)$. The upper-level problem then seeks the hyperparameters that minimize the validation loss. This formulation generalizes naturally to scenarios involving cross-validation, where the validation loss aggregates errors across multiple data folds:

$$L_{\text{validation}}(\omega(\lambda)) = \frac{1}{K}\sum_{k=1}^{K} L^{(k)}_{\text{validation}}(\omega^{(k)}(\lambda)), \tag{1.28}$$

where K is the number of folds, and $\omega^{(k)}(\lambda)$ are the optimal model parameters for the kth fold.

Example 1.1 Support vector machines (SVMs) are widely used for classification tasks. Given a training dataset $\{x_i, y_i\}_{i=1}^n$, an SVM classifier aims to find a hyperplane that maximizes the margin between different classes. The standard SVM training problem with a nonlinear kernel is formulated as:

$$\begin{aligned} \underset{\omega,b,\xi}{\text{minimize}} \quad & \frac{1}{2}|\omega|^2 + C\sum_{i=1}^{n}\xi_i \\ \text{subject to} \quad & y_i(\omega^T\phi_\gamma(x_i) + b) \geq 1 - \xi_i \quad \forall i \in \{1, \dots, n\}, \\ & \xi_i \geq 0, \quad \forall i \in \{1, \dots, n\}, \end{aligned} \tag{1.29}$$

where $\phi_\gamma(x)$ is a feature mapping induced by the kernel function with a mapping parameter $\gamma > 0$, C is a regularization parameter, and ξ_i are slack variables allowing for misclassification.

To illustrate bilevel hyperparameter optimization, consider tuning the regularization parameter C and the kernel parameter γ for the SVM. The bilevel problem is formulated as:

$$\begin{aligned} \underset{C,\gamma}{\text{minimize}} \quad & \sum_{i\in\mathcal{D}_{\text{validation}}} \ell(y_i, f(x_i; \omega(C,\gamma))) \\ \text{subject to} \quad & \omega(C,\gamma) \in \underset{\omega}{\arg\min}\left[\sum_{i\in\mathcal{D}_{\text{training}}} \ell(y_i, f(x_i;\omega)) + CR(\omega)\right], \end{aligned} \tag{1.30}$$

where $\ell(y, f(x;\omega))$ is the loss function and $R(\omega)$ is the regularization term. The sets $\mathcal{D}_{\text{training}}$ and $\mathcal{D}_{\text{validation}}$ contain the training and validation samples, respectively. Gradient-based methods such as implicit differentiation or gradient unrolling are commonly used to solve this bilevel problem efficiently.

Unlike grid search and random search, which require training multiple models independently, bilevel optimization directly leverages the structure of the problem, leading to better generalization and reduced computational overhead. Bayesian optimization, while more efficient than grid search, still struggles with high-dimensional search spaces and requires expensive function evaluations. In contrast, bilevel optimization integrates hyperparameter tuning with the training process itself, allowing for a more natural and adaptive exploration of the hyperparameter space.

From a sustainability perspective, inefficient hyperparameter tuning exacerbates the already significant carbon footprint of machine learning training. Large-scale deep learning models require substantial computational resources, leading to high-energy consumption and carbon emissions. Bilevel optimization offers a more efficient approach to hyperparameter tuning by optimizing hyperparameters within a structured mathematical framework, reducing unnecessary training runs and computational overhead. By improving the efficiency of model training, bilevel optimization not only enhances performance but also contributes to more sustainable AI development by lowering energy consumption and mitigating the environmental impact of large-scale machine learning.

While bilevel optimization is inherently difficult to solve, recent advances in gradient-based optimization have enabled fast solutions for large-scale instances. Hypergradient-based methods compute the gradient of the validation loss with respect to hyperparameters using techniques such as implicit differentiation or unrolling strategies, significantly reducing the number of function evaluations required. These approaches make bilevel hyperparameter optimization practical for large-scale machine learning applications. We provide an overview of these methods in the next subsection.

Recent studies have established bilevel optimization as a powerful framework for hyperparameter tuning in machine learning. Bengio [72] pioneered this direction by demonstrating how reverse-mode differentiation can compute gradients of the validation loss with respect to hyperparameters, effectively formulating hyperparameter tuning as a bilevel optimization problem. Expanding on this foundation, Baydin and Pearlmutter [73] reviewed automatic differentiation techniques and highlighted their role in efficiently computing hypergradients for such problems. Maclaurin et al. [74] further advanced these ideas by introducing a reversible learning method that enables efficient hypergradient computation, making it feasible to optimize thousands of hyperparameters – including detailed learning-rate schedules and weight initialization schemes – without excessive memory costs. More recently, Franceschi et al. [75] provided a formalized bilevel optimization framework for hyperparameter tuning and meta-learning, demonstrating improved model performance and generalization compared to traditional search methods. In neural architecture search, Liu et al. [76] leveraged bilevel optimization in the Differentiable Architecture Search (DARTS) framework, optimizing architecture hyperparameters for neural networks with tens of millions of parameters, yielding strong results on datasets such as CIFAR-10 and ImageNet. Collectively, these works illustrate the progression of bilevel optimization from theoretical foundations to scalable, high-impact techniques for large-scale hyperparameter tuning.

1.4.3 Algorithms for Large-Scale Bilevel Optimization

Solving bilevel optimization problems exactly is often computationally intractable, particularly in large-scale settings such as those encountered in machine learning, where models involve millions of parameters. The inner-level optimization itself can be expensive, making traditional approaches, such as reformulating the bilevel problem using optimality conditions, impractical for large problems. This challenge has motivated the development of gradient-based methods, which offer a scalable alternative by leveraging the gradient (often computed approximately) to iteratively update the upper-level variables. These methods efficiently handle high-dimensional problems by avoiding the need to fully solve the inner optimization at each step, instead relying on implicit differentiation, unrolling, or approximation techniques to compute gradients. The key advantage of these approaches lies in their ability to balance computational efficiency and solution quality, making bilevel optimization feasible for modern applications such as hyperparameter tuning, meta-learning, and reinforcement learning.

Mathematically, assuming a general bilevel optimization problem of the following form:

$$\begin{aligned} \underset{\theta\in\Theta}{\text{minimize}} \quad & F(\theta, x^*(\theta)) \\ \text{subject to} \quad & x^*(\theta) \in \arg\min_{x\in\mathcal{X}(\theta)} f(x,\theta), \end{aligned} \tag{1.31}$$

where F and f are, respectively, the upper and lower-level objective functions. The goal of (1.31) is to find $\theta \in \Theta$ that minimize F. In case of the hyperparameter optimization problem (1.27), θ are the hyperparameters for the machine learning model and x^* are the optimal model parameters for a given θ.

Gradient-based methods iteratively update θ by approximating the gradient of the upper-level objective function with respect to θ, while ensuring that $x^*(\theta)$ remains sufficiently close to an optimal solution. These methods typically follow an update rule of the form:

$$\theta^{k+1} = \theta^k - \eta\nabla_\theta F(\theta^k, x^*(\theta^k)), \tag{1.32}$$

where $\eta > 0$ is a step size (learning rate) and $\nabla_\theta F(\theta, x^*(\theta))$ is the total derivative of the upper-level objective with respect to θ. Using the chain rule, this gradient can be expressed as:

$$\nabla_\theta F(\theta, x^*(\theta)) = \frac{\partial F}{\partial \theta} + \frac{\partial F}{\partial x}\frac{dx^*}{d\theta}. \tag{1.33}$$

Here, the term $\frac{dx^*}{d\theta}$ accounts for the sensitivity of the optimal solution of the lower-level problem, $x^*(\theta)$, to changes in θ. Computing this term directly is often infeasible in large-scale settings, so various approximation techniques are employed to estimate it efficiently. In the following, we provide a brief overview of implicit and explicit methods developed for estimating these gradients and discuss their trade-offs in terms of accuracy and computational cost. For simplicity, we assume that in problem (1.31), the lower-level problem is differentiable, is unconstrained, and has a unique solution.

1.4.3.1 Implicit Estimation

Implicit estimation leverages the optimality conditions of the lower-level problem to compute gradients with respect to upper-level variables without explicitly solving the lower-level problem at each iteration. If the lower-level problem is differentiable and satisfies optimality conditions, its solution $x^*(\theta)$ can be implicitly characterized using the first-order optimality conditions.

More specifically, assuming the lower-level problem is unconstrained, the necessary optimality condition for $x^*(\theta)$ is given by the stationarity condition,

$$\nabla_x f(x^*, \theta) = 0. \tag{1.34}$$

Differentiating both sides with respect to θ, we obtain,

$$\nabla_{xx} f(x^*, \theta)\frac{dx^*}{d\theta} + \nabla_{x\theta} f(x^*, \theta) = 0, \tag{1.35}$$

where solving for $\frac{dx^*}{d\theta}$, we obtain,

$$\frac{dx^*}{d\theta} = -\left(\nabla_{xx} f(x^*, \theta)\right)^{-1}\nabla_{x\theta} f(x^*, \theta). \tag{1.36}$$

This equation implicitly characterizes how the lower-level solution changes with θ. Recall, the gradient of the upper-level objective with respect to θ is,

$$\nabla_\theta F(\theta, x^*(\theta)) = \frac{\partial F}{\partial \theta} + \frac{\partial F}{\partial x}\frac{dx^*}{d\theta}, \tag{1.37}$$

where substituting for $\frac{dx^*}{d\theta}$ yields:

$$\nabla_\theta F(\theta, x^*(\theta)) = \frac{\partial F}{\partial \theta} - \frac{\partial F}{\partial x}\left(\nabla_{xx} f(x^*, \theta)\right)^{-1}\nabla_{x\theta} f(x^*, \theta). \tag{1.38}$$

This formulation enables the computation of the gradient of the upper-level objective without explicitly re-solving the lower-level optimization problem at each iteration.

Computational Considerations Direct computation of the inverse Hessian $\left(\nabla_{xx} f(x^*, \theta)\right)^{-1}$ is often intractable for large-scale problems due to computational and storage constraints. Instead, iterative methods such as the conjugate gradient algorithm are employed to solve the linear system:

$$Hz = v, \tag{1.39}$$

where $H = \nabla^2_{xx} f(x^*, \theta)$ and $v = \nabla^2_{x\theta} f(x^*, \theta)$. These methods compute the product $z = H^{-1}v$ without explicitly forming the Hessian matrix, leveraging Hessian-vector products that can be efficiently computed using automatic differentiation tools. This approach is particularly advantageous in high-dimensional settings where forming and inverting the Hessian is impractical [77].

Some Key Extensions

- **Constrained Lower-Level Problems:** When the lower-level problem includes constraints, implicit differentiation can be extended by incorporating the KKT conditions, which provide necessary optimality conditions for constrained optimization problems. For example, Amos and Kolter [78] demonstrate this approach when the lower-level optimization problems are quadratic programs. Additionally, Yao et al. [79] develop algorithms capable of handling more general lower-level constraints.
- **Nonconvex Lower-Level Problems:** For nonconvex lower-level problems, the implicit function theorem may not directly apply due to the lack of a unique solution. Recent studies have explored algorithms that approximate implicit differentiation using iterative methods [80]. Zucchet and Sacramento [81] review gradient-based techniques for bilevel optimization, including those applicable to nonconvex settings.
- **Nondifferentiable Lower-Level Problems:** In cases where the lower-level problem is nondifferentiable, smoothing techniques or subgradient methods can be employed. These approaches approximate or handle nondifferentiability to enable the application of bilevel optimization algorithms [82].
- **Singular Hessian:** When the Hessian is singular or ill-conditioned, computing its inverse is problematic. Regularization techniques or generalized inverses (e.g. the Moore-Penrose pseudo-inverse) can be applied to stabilize the computations. Additionally, alternative formulations that do not rely on the explicit inversion of the Hessian have been proposed. For a comprehensive discussion, see the work by Arbel and Mairal [83].

1.4.3.2 Explicit Estimation

Explicit differentiation provides an alternative approach to computing gradients in bilevel optimization by directly differentiating through the iterative optimization process used to solve the lower-level problem. Unlike implicit differentiation, which relies on optimality conditions, explicit methods track the sequence of updates in the lower-level solver and differentiate through them to compute the upper-level gradient. This approach is particularly useful when the lower-level problem is solved using an iterative method such as gradient descent.

Instead of assuming that the lower-level problem has reached full optimality, explicit methods approximate the gradient by differentiating through the sequence of updates. In this approach, the lower-level solution $x^*(\theta)$ is approximated by the final iterate x^T after T steps of an iterative algorithm following which we compute the gradient through unrolling the iterative process.

Let Ψ represent the iterative process that updates the estimates of $x^*(\theta)$ as follows:

$$x_{t+1}(\theta) = \Psi(x_t, \theta), \quad t = 0, 1, \ldots, T-1, \tag{1.40}$$

with initial parameters x_0. After T iterations, the final parameters are

$$x_T(\theta) = \Psi(x_{T-1}(\theta), \theta) \circ \cdots \circ \Psi(x_0, \theta). \tag{1.41}$$

The objective is to minimize the upper-level objective given by

$$F(\theta, x_T(\theta)). \tag{1.42}$$

Since $x_T(\theta)$ depends on θ through the entire training process, the total derivative of F with respect to θ is

$$\nabla_\theta F(\theta, x_T(\theta)) = \nabla_\theta F(\theta, x_T(\theta)) + \nabla_x F(\theta, x_T(\theta)) \frac{\partial x_T(\theta)}{\partial \theta}. \tag{1.43}$$

Recursive Jacobian Computation Define the Jacobians at each iteration as follows:

$$A_t = \frac{\partial \Psi(x_t, \theta)}{\partial x}, \quad B_t = \frac{\partial \Psi(x_t, \theta)}{\partial \theta}.$$

Then, by recursively applying the chain rule through the T iterations, we obtain,

$$\frac{\partial x_T(\theta)}{\partial \theta} = \sum_{t=0}^{T-1} \left(\prod_{j=t+1}^{T-1} A_j \right) B_t, \tag{1.44}$$

with the convention that $\prod_{j=T}^{T-1} A_j = I$ (the identity matrix).

Thus, the explicit hypergradient is given by,

$$\nabla_\theta F(\theta, x_T(\theta)) = \nabla_x F(\theta, x_T(\theta)) \cdot \left[\sum_{t=0}^{T-1} \left(\prod_{j=t+1}^{T-1} A_j \right) B_t \right]. \tag{1.45}$$

In practice, computing the full Jacobian product over all iterations is computationally expensive. A common solution is to use *truncated backpropagation*, where differentiation is applied only over a fixed number of iterations rather than the entire trajectory. Further, modern deep learning frameworks such as PyTorch, TensorFlow, and JAX provide automatic differentiation (AD) tools that can compute explicit gradients efficiently. AD methods handle differentiation through iterative optimization steps using two primary modes:

- **Forward-Mode Differentiation:** Computes derivatives sequentially along the optimization process. This is efficient for problems with a small number of parameters but is not commonly used in large-scale bilevel optimization.
- **Reverse-Mode Differentiation (Backpropagation):** Tracks all operations in the forward pass and computes gradients efficiently in the backward pass. This is well-suited for problems where the number of parameters is large but the output dimension is small.

Explicit gradient estimation methods offer a versatile approach to computing gradients in bilevel optimization, particularly when the iterative nature of the lower-level problem is well understood and differentiable. This flexibility allows explicit methods to be applied across a broad spectrum of iterative algorithms, even in the absence of explicit stationarity conditions. A notable example of applying explicit gradient estimation to complex iterative algorithms is presented in the work of Tan et al. [63], where they unroll an interior point solver within a deep learning framework.

Finally, we close this section by presenting a comparative analysis of implicit and explicit gradient estimation methods used in bilevel optimization. A summary of these differences is also available in Table 1.1.

Table 1.1 Comparison of explicit and implicit hypergradient estimation methods in bilevel optimization.

Feature	Explicit estimation (Unrolling)	Implicit estimation
Memory cost	Requires storing all intermediate states (high memory usage)	Memory efficient; avoids storing the entire training trajectory
Computational cost	Computationally expensive, especially for long unrolls	More scalable for long training processes, but involves solving a linear system
Exactness of gradients	Computes exact gradients via backpropagation through the unrolled process	May introduce approximation errors when solving the Hessian system
Implementation	Straightforward to implement using automatic differentiation tools (e.g. TensorFlow, PyTorch, JAX)	More complex; requires handling Hessian-vector products and matrix inversion
Scalability	Limited to short or truncated unrolls due to memory and time constraints	Better suited for large-scale or long-horizon training procedures
Numerical stability	Prone to issues like vanishing or exploding gradients over long unrolls	More robust over long training horizons, assuming a well-conditioned Hessian

- **Memory and Computational Costs:** Explicit estimation methods involve unrolling the iterative process of the lower-level problem, which necessitates storing all intermediate states. This approach leads to high memory consumption and can be computationally expensive, especially when dealing with long unrolls. In contrast, implicit estimation methods are more memory-efficient, as they avoid storing the entire training trajectory. However, they may require solving a linear system, which can introduce additional computational overhead.
- **Gradient Accuracy and Implementation Complexity:** Explicit methods provide exact gradients through backpropagation across the unrolled iterations, ensuring precise gradient information. This exactness simplifies implementation, making explicit methods straightforward to apply using automatic differentiation tools such as TensorFlow, PyTorch, or JAX. On the other hand, implicit methods might involve approximations when solving the Hessian system, potentially affecting gradient accuracy. Implementing implicit methods is generally more complex, as it requires handling Hessian-vector products and matrix inversion.
- **Scalability and Numerical Stability:** Due to memory and time constraints, explicit estimation methods are typically limited to shorter or truncated unrolls. They can also face numerical stability issues, such as vanishing or exploding gradients, especially over extended unrolling. Implicit methods, however, are better suited for large-scale or long-horizon training procedures. They tend to be numerically more stable over longer training horizons, provided that the Hessian is well-conditioned.
- **Practical Considerations:** The choice between explicit and implicit gradient estimation methods depends on the specific requirements of the task at hand. Explicit methods are preferable when exact gradients are necessary, and computational resources allow for storing intermediate states. In contrast, implicit methods are suitable for large-scale problems where memory efficiency and numerical stability are paramount, despite their increased implementation complexity.

1.5 Robust Optimization

From a structural point of view, bilevel optimization problems can also be observed in the realm of optimization under uncertainty. Uncertainty in the parameters of an optimization problem is quite prevalent in real-life applications due to perturbations in data measurements or simply a lack of knowledge of the parameters ahead of time. Stochastic and robust optimization are the two main approaches used to solve optimization problems under uncertainty. Stochastic optimization uses a probability distribution to model uncertainty [84]. Robust optimization (RO), on the other hand, does not require prior knowledge of the probability distribution but instead assumes that the uncertain data vary within the limits of the *uncertainty set* and the optimal solution to the problem, if it exists, is robust against all possible uncertainty realizations within this set.

In this section, we provide an introduction to single-stage (or static) RO problems and show how they can be interpreted as bilevel optimization problems. We show the two popular solution methodologies for static RO problems and briefly show extensions to adjustable RO. Then, we highlight applications of RO from the sustainable systems engineering literature and provide a detailed discussion on the application of RO for scheduling in a manufacturing process providing ancillary services to the power grid.

1.5.1 Mathematical Formulation

Consider the following general formulation of a single-stage RO problem:

$$\underset{x}{\text{minimize}} \quad f(x) \tag{1.46a}$$

$$\text{subject to} \quad g(x, \xi) \leq 0 \quad \forall \xi \in \Xi, \tag{1.46b}$$

where x denotes the decision variables and ξ denotes the uncertain parameters. Without loss of generality, we do not consider uncertainty in the objective function $f(x)$. If in a problem the objective function is affected by uncertainty, i.e. we have $\tilde{f}(x, \xi)$, then an epigraph reformulation using the constraint $t \geq \tilde{f}(x, \xi)$ with auxiliary variable t can convert the problem into the general form of problem (1.46). Constraints (1.46b) require that the solution x be feasible for all possible realizations of the uncertain parameters that reside in the uncertainty set. Problem (1.46) is also referred to as a static RO problem because all the decision variables are *here-and-now* decisions, that is, their values are fixed before the uncertainty is realized.

The uncertainty set is assumed to be nonempty and compact. Also, replacing the uncertainty set, Ξ, with its convex hull does not affect the result [85]. This is because a robust feasible solution of problem (1.46) will remain robust feasible for convex combinations of possible uncertain parameter values. Therefore, assuming that the uncertainty set is convex is not restrictive. Further, it is assumed that uncertainty can be modeled constraint-wise. That is, problem (1.46) is equivalent to the following formulation:

$$\underset{x}{\text{minimize}} \quad f(x)$$

$$\text{subject to} \quad g_n(x, \xi) \leq 0 \quad \forall \xi \in \Xi, \quad n \in \mathcal{N},$$

where $\mathcal{N}$ is the set of constraints. It turns out that this assumption is also not restrictive in the case of single-stage RO problems [85]. The $\forall$ quantifier on the uncertainty parameter space can be equivalently modeled using the worst-case reformulation as follows:

$$\begin{aligned} &\underset{x}{\text{minimize}} && f(x) && (1.47a)\\ &\text{subject to} && \max_{\xi\in\Xi} g_n(x,\xi) \leq 0 \quad \forall n \in \mathcal{N}, && (1.47b)\end{aligned}$$

where the maximization problems on the left-hand sides of inequalities (1.47b) find the worst possible realizations of ξ for a given x and therefore, ensure that the constraints are satisfied for all possible values of ξ.

Problem (1.47) has a bilevel structure where $|\mathcal{N}|$ independent *followers* corresponding to the embedded maximization problems work against the *leader* who is concerned with minimizing $f(x)$ and controls the choice of x. Interestingly, solution strategies developed independently for RO are similar to those shown in Section 1.2.3 for bilevel optimization. Here, either (i) the lower-level maximization problems are replaced by their dual problems to reduce problem (1.47) to a single-level formulation or (ii) a cutting-plane approach is used where the problem is iteratively solved until the worst-case scenario for ξ is found. In the following, we briefly explain these reformulation and cutting-plane methods in Sections 1.5.1.1 and 1.5.1.2, respectively.

1.5.1.1 Reformulation

For simplicity, we focus on linear RO problems under polyhedral uncertainty sets for which tractable reformulations exist. However, a similar derivation can also be used to find tractable reformulations for problems under other uncertainty sets [86]. Consider the following linear RO problem with uncertain parameter dependence on both left-hand and right-hand side of the constraints:

$$\begin{aligned} &\underset{x}{\text{minimize}} && c^\top x\\ &\text{subject to} && a_n^\top x + \xi^\top \bar{A}_n x \leq b_n + \bar{b}_n^\top \xi \quad \forall n \in \mathcal{N}, \quad \xi \in \Xi,\end{aligned} \tag{1.48}$$

where the uncertainty set, $\Xi = \{\xi \in \mathbb{R}^m : W\xi \leq u\}$, is polyhedral and x can be continuous or discrete. As uncertainty can be modeled constraint-wise, we can focus on a single constraint for our derivation. The worst-case reformulation of the nth constraint is given by

$$a_n^\top x + \max_{\xi\in\Xi} (\bar{A}_n x - \bar{b}_n)^\top \xi \leq b_n. \tag{1.49}$$

The embedded maximization problem is an LP; hence, by strong duality, constraint (1.49) can be reformulated as follows:

$$a_n^\top x + \min_{\zeta_n \geq 0} u^\top \zeta_n \leq b_n \tag{1.50a}$$

$$W^\top \zeta_n = \bar{A}_n x - \bar{b}_n, \tag{1.50b}$$

where $\zeta_n \in \mathfrak{R}^{|u|}$ are the dual variables. If constraint (1.50a) holds for one feasible ζ_n, then it holds for the optimal ζ_n for the minimization problem as well.

Therefore, the min operator in (1.50a) can be dropped, which results in the following single-level formulation:

$$\begin{aligned}
\underset{x,\zeta_n\geq 0}{\text{minimize}} \quad & c^\top x \\
\text{subject to} \quad & a_n^\top x + u^\top \zeta_n \leq b_n \quad \forall n \in \mathcal{N} \\
& W^\top \zeta_n = \bar{A}_n x - \bar{b}_n \quad \forall n \in \mathcal{N}.
\end{aligned} \tag{1.51}$$

The above formulation is an LP or an MILP depending on whether x contains discrete variables and can be solved using off-the-shelf MIP solvers.

1.5.1.2 Cutting-Plane Approach

Adversarial or cutting-plane approaches developed to solve RO problems [87] are quite similar to the cutting-plane approach developed for bilevel optimization shown in Section 1.2.3. These adversarial approaches are based on the fact that only a subset of realizations of uncertain parameters are actually relevant to the RO problem [88, 89]. That is, optimal solutions of the lower-level maximization problems in (1.47b) or, equivalently, the worst-case uncertainty parameter realizations form a subset of the uncertainty set, and hence optimizing the original problem over this subset should provide a robust optimal solution.

However, finding this subset is not straightforward as the lower-level maximization problems depend on the decision variables x as well. Therefore, an iterative procedure can be employed, where the master problem is a relaxation of problem (1.46) with the uncertainty set Ξ replaced by a subset $\overline{\Xi}_k \subseteq \Xi$ that is updated in each iteration k with points in the uncertainty set obtained from the cut-generation problem. The cut-generation problem involves solving the lower-level maximization problem for $|\mathcal{N}|$ constraints (or followers) to check if $g_n(x^*, \hat{\xi}) \leq 0$ for all $n \in \mathcal{N}$ and setting $\overline{\Xi}_{k+1} = \overline{\Xi}_k \cup \hat{\xi}$ if any constraint violations are found. The master problem has the same complexity as the original problem without uncertainty in its parameters, and the complexity of evaluating the subproblem depends on the nature of the uncertainty set, making the method applicable to a broad class of problems.

1.5.2 Adjustable Robust Optimization

As mentioned above, all the decision variables in static RO are here-and-now decisions, that is, these decisions must be made ahead of uncertainty realization, which can lead to conservative solutions if in reality there is an opportunity to adjust decisions at a later time. Adjustable (or adaptive) robust optimization (ARO) was developed to also incorporate recourse (or wait-and-see) decisions. Consider the following two-stage RO problem:

$$\underset{x,y(\cdot)}{\text{minimize}} \quad c_1^\top x + \max_{\xi\in\Xi} c_2^\top y(\xi) \tag{1.52a}$$

$$\text{subject to} \quad (a_n^\top + \xi^\top \bar{A}_n)x + d_n^\top y(\xi) \leq b_n + \bar{b}_n^\top \xi \quad \forall n \in \mathcal{N}, \quad \xi \in \Xi, \tag{1.52b}$$

where the decision variables x are here-and-now decisions and y are wait-and-see decisions. Wait-and-see decisions, by definition, are fixed after the uncertain parameter ξ is realized, which is why y are expressed as functions of ξ. Note that the coefficients of y, c_2 and d_n, are assumed to be independent of ξ (this is referred to as fixed recourse).

Problem (1.52) is generally intractable as one would be optimizing over the set of all possible functions for y. One common solution approach is to restrict $y(\xi)$ to some classes of parameterized functions [90, 91]. These parameterized functions are also called decision rules and often, the decision rules are chosen to be affine functions of ξ:

$$y(\xi) = \overline{y} + \tilde{Y}\xi. \tag{1.53}$$

The parameters $\overline{y}$ and $\tilde{Y}$ become here-and-now decisions when equations (1.53) are substituted into problem (1.52) as follows:

$$\begin{aligned} \underset{x,\overline{y},\tilde{Y}}{\text{minimize}} \quad & c_1^\top x + t \\ \text{subject to} \quad & t \geq c_2^\top(\overline{y} + \tilde{Y}\xi) \quad \forall \xi \in \Xi \\ & (a_n^\top + \xi^\top \bar{A}_n)x + d_n^\top(\overline{y} + \tilde{Y}\xi) \leq b_n + \overline{b}_n^\top \xi \quad \forall n \in \mathcal{N}, \quad \xi \in \Xi. \end{aligned} \tag{1.54}$$

Note that, problem (1.54) is a single-stage RO problem and can be solved using the solution methods described in Sections 1.5.1.1 and 1.5.1.2. ARO with affine decision rules provides the same tractability as single-stage RO for problems of the form (1.52). However, this tractability comes at the cost of suboptimality as a better mapping from ξ to y could exist. This inspired the development of nonlinear decision rules such as polynomial decision rules [92] and piecewise affine decision rules [93] as well as alternative solution approaches such as column-and-constraint generation for two-stage RO [94] and the K-adaptability approach [95].

1.5.3 Applications

Many applications in the sustainable systems engineering are inherently optimization problems under uncertainty. This is due to various factors such as the increasing use of intermittent renewable energy sources, volatility in market conditions, and unknown future cost trajectories for new low-carbon technologies, which require development of techniques to handle uncertainty in the most economic, safe, and reliable way. RO has been applied extensively to such problems in recent years and has proven to be particularly useful when probability distributions of uncertain parameters are unknown or difficult to determine. In this section, we provide a few examples from the sustainable systems engineering literature to give a glimpse into the applicability of RO. Note that these examples are in no way exhaustive.

For scheduling and planning in power systems, Conejo and Wu [96] apply RO and ARO techniques to handle vulnerabilities in transmission lines. Lorca and Sun [97] consider uncertainty in the availability of wind energy for the economic dispatch of power systems using ARO and dynamic uncertainty sets to capture temporal and spatial correlations of wind power. Later, Lorca and Sun [98] extend their work to also consider uncertainty in solar power and include operation of energy storage devices. Apostolopoulou et al. [99] consider scheduling in a hydroelectric system with uncertainty in inflows to the system and load.

On the electricity consumer side, RO techniques are used to incorporate uncertainty in real-time electricity prices [100]. Demand response (DR), where electricity consumers alter their consumption in response to price signals from the electricity market, often introduces uncertainty into the consumer's operations due to volatility in prices or due to participation in the financially lucrative reserve market. In the latter, consumers commit to providing operating reserves to the grid operator within a short notice period in exchange for monetary rewards. Such DR activities help in maintaining grid reliability and also lower costs for the consumer. Therefore, various works have considered DR participation for residential [100], commercial [101], and industrial consumers [102]. One such example is also explored in Section 1.5.4.

With the increasing penetration of renewable energy sources, distributed generation resources, which are placed close to consumers, are growing in popularity. RO has been used to find the optimal placement of these distributed generators under uncertainty of the energy availability of the generators [103]. Rezvan et al. [104] find the optimal investment decisions for distributed generators in buildings under demand uncertainty.

At the supply chain level, Pishvaee et al. [105] develop a general MILP to design a closed-loop supply chain network with uncertainty in its input parameters such as demand, and capacities. Jabbarzadeh et al. [106] consider the design of a green supply chain network for electricity comprising suppliers and consumers along with DR activities.

1.5.4 Case Study

The case study is taken from Jagana et al. [107], where scheduling is considered for a power-intensive manufacturing facility that participates in DR programs. In general, DR programs can be categorized into nondispatchable and dispatchable DR. This case study considers DR programs from both categories, and we show results taken from Jagana et al. [107] that compare participation in only nondispatchable DR and participation in both nondispatchable and dispatchable DR. The latter, as we will see, is an optimization problem under uncertainty, and the authors use an ARO approach to solve it.

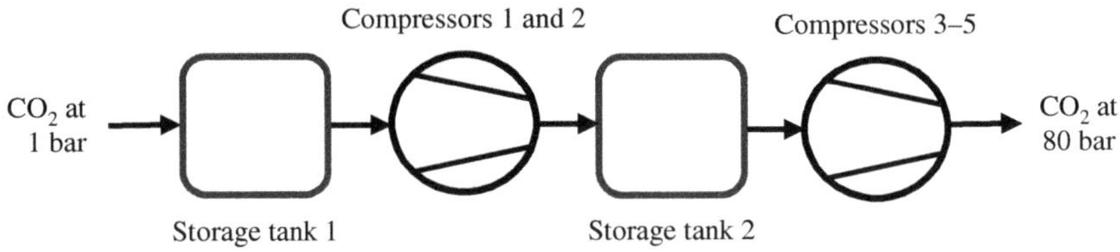

Figure 1.1 A train of five compressors with intermediate storage, which allows for operational flexibility. CO_2 flows in at a constant flowrate and needs to be compressed by the end of the day.

The base case considers participation in nondispatchable DR programs, where consumers schedule their operations in response to varying electricity prices. Consider, for example, scheduling in a train of five compressors with intermediate storage as shown in Figure 1.1. CO_2 flows at a constant input rate and has to be compressed by the end of the operating day. Compressors have given minimum and maximum flowrates and require an hour of startup time. Storage tank 1 is sized such that compressors 1 and 2 cannot be shutdown given the CO_2 flowrate. Storage tank 2, however, is sized such that compressors 3–5 can be shut down flexibly.

We assume that the compressor train participates in the day-ahead electricity market operated by the Electricity Reliability Council of Texas (ERCOT). Hourly electricity prices are assumed to be known ahead of the scheduling horizon and a deterministic optimization problem is solved to find the optimal production schedule. The results for one operating day with hourly resolution are shown in Figure 1.2. Here, the electricity prices are shown on the right *y*-axis and the

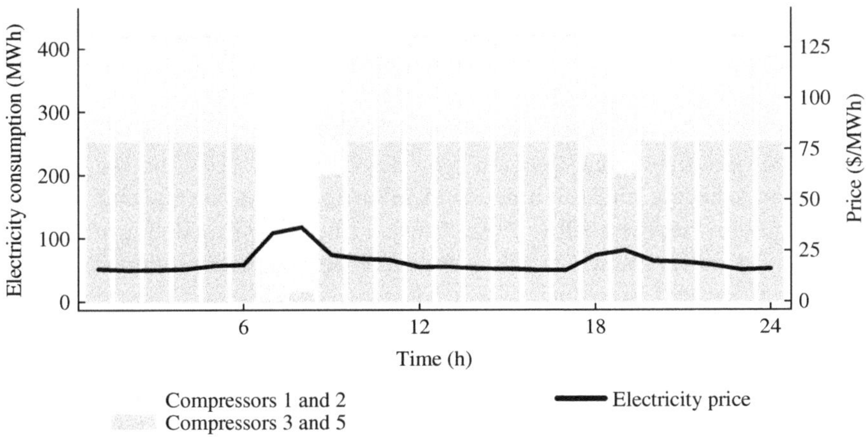

Figure 1.2 Electricity consumption in the base case, where the compressor train only participates in a nondispatchable DR program.

compressors' electricity consumption is shown on the left y-axis. As electricity price peaks in time period 7, the solution recommends shutting down compressors 3–5 with an hour of startup time allotted in time period 8. However, compressors 1 and 2 are not turned off during this peak, as storage tank 1 is not large enough to flexibly operate the two compressors. This base case has a total operating cost of $166,787.

Next, the model is extended to consider participation in dispatchable DR programs, where electricity consumers can adjust their power consumption when requested by the grid operator in exchange for financial incentives. The power grid operator utilizes the so-called *operating reserves* to correct supply-demand imbalances, maintain grid frequencies, handle contingencies, and so on. Operating reserves that are provided by dispatchable DR is often referred to as *interruptible load*.

The electricity consumer willing to participate is required to commit to the amount of interruptible load they are willing to provide ahead of the operating day. As the day progresses, the grid operator can request for load reduction up to the committed amount from the consumer with a short notice. Naturally, this brings in uncertainty into the production schedule as we do not know ahead of time when and how much load reduction will be requested. Quantifying this uncertainty as a probability distribution can be difficult because historical data on contingency events during which interruptible load is typically needed is not readily available, and even if it were available, the information would be given at the grid level, which does not allow a direct prediction of load reduction requests to individual interruptible load providers. Further, terms of participation in the ancillary services market require that the load reduction requests be satisfied without fail; noncompliance can result in heavy penalties and in some cases even removal from the ancillary services market. These requirements motivate a robust optimization approach to schedule operations under uncertainty due to interruptible load provision.

The details of the formulation can be found in Jagana et al. [107], where ARO with piecewise affine linear decision rules has been applied to find the optimal production schedule and the amount of interruptible to be provided in each operating hour. Piecewise affine linear decision rules enable modeling of discrete wait-and-see decisions, which allow for recourse actions like shutting down processes to satisfy load reduction requests, whereas affine decision rules (1.53) are restricted to continuous wait-and-see decisions like varying production rates.

When the compressor train also participates in the dispatchable DR program of providing interruptible load in the ancillary services market of ERCOT, the total operating cost decreases to $146,954. The 11.9% cost savings can be attributed to the revenue from providing interruptible load as seen in Figure 1.3. Here, the electricity prices we pay and interruptible load prices we are paid are shown on

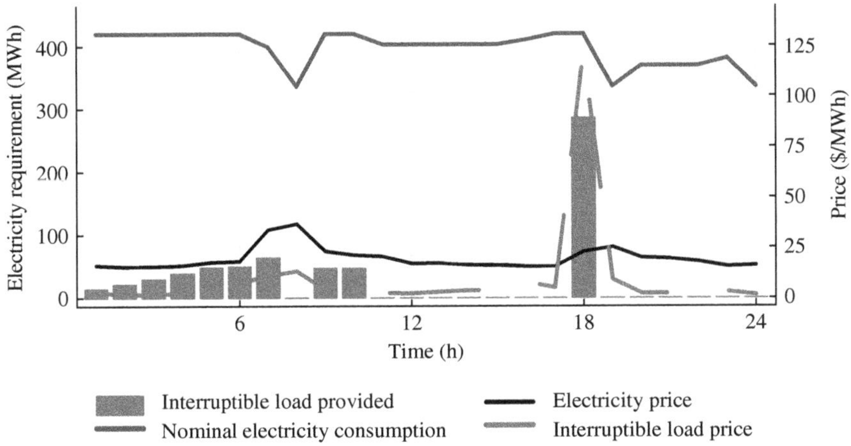

Figure 1.3 Electricity consumption and interruptible load provided when the compressor train participates in nondispatchable and dispatchable DR programs.

the right *y*-axis and the nominal electricity consumption and interruptible load provided are shown on the left *y*-axis. Nominal electricity consumption refers to the consumption profile when no load reduction is requested. Note that, unlike the base case, the electricity consumption does not have a drastic dip in value in time periods 7 and 8. This result makes sense when the decisions made for the interruptible load provided are also considered. Payment for interruptible load provided in time period 18 is very high, and consequently, the model allocates a significant fraction of the nominal electricity requirement as interruptible load. However, possible load reduction requests in this case can lead to shutdown of compressors 3–5. To shield production requirements from this uncertainty, additional CO_2 compression occurs in the previous time periods.

1.6 Conclusions

In this book chapter, we provided an introduction to bilevel optimization and its application in sustainable systems engineering as well as more broadly in machine learning and robust optimization. The objective was to present the mathematical fundamentals in an accessible way, assuming no prior knowledge in this area from the reader, and to demonstrate the relevance and broad applicability of the bilevel optimization framework.

Once considered a rather specialized optimization approach, bilevel optimization has gained significant attention and its use has increased extensively in the last two decades, precisely due to emerging applications in areas like sustainable systems engineering, machine learning, and robust optimization. Optimization researchers and practitioners have realized that many important problems can and should be formulated as bilevel programs, and that we now have the computational tools to solve a large fraction of these problems. The increasing use of bilevel optimization is also being facilitated by available software packages, such as YALMIP in MATLAB [108], PAO in Python [109], and BilevelJuMP.jl in Julia [110], that considerably ease the implementation of certain classes of bilevel programs.

While the tractability of bilevel optimization has improved significantly, major challenges remain. Some classes of bilevel programs are still notoriously difficult to solve, including those involving nonconvex lower-level problems, discrete decisions, multiple interdependent followers, or uncertainty. As a result, bilevel optimization continues to be an active area of research.

References

1 von Stackelberg, H. (1934). *Marktform und Gleichgewicht*. Springer.

2 von Stackelberg, H. (1952). *The Theory of the Market Economy*. Oxford University Press.

3 Bracken, J. and McGill, J.T. (1973). Mathematical programs with optimization problems in the constraints. *Operations Research* 21 (1): 37–44.

4 Wang, J.Y.T., Ehrgott, M., Dirks, K.N., and Gupta, A. (2014a). A bilevel multi-objective road pricing model for economic, environmental and health sustainability. *Transportation Research Procedia* 3: 393–402.

5 Dempe, S., Kalashnikov, V., Pérez-Valdés, G.A., and Kalashnykova, N. (2015). Bilevel programming problems. In: *Energy Systems*, vol. 10, 978–973. Berlin: Springer.

6 Beck, Y. and Schmidt, M. (2021). A gentle and incomplete introduction to bilevel optimization.

7 Colson, B., Marcotte, P., and Savard, G. (2007). An overview of bilevel optimization. *Annals of Operations Research* 153: 235–256.

8 Kleinert, T., Labbé, M., Ljubić, I., and Schmidt, M. (2021). A survey on mixed-integer programming techniques in bilevel optimization. *EURO Journal on Computational Optimization* 9: 100007.

9 Beck, Y., Ljubić, I., and Schmidt, M. (2023). A survey on bilevel optimization under uncertainty. *European Journal of Operational Research* 311 (2): 401–426.

10 Maravelias, C.T. and Sung, C. (2009). Integration of production planning and scheduling: overview, challenges and opportunities. *Computers & Chemical Engineering* 33 (12): 1919–1930.

11 Fortuny-Amat, J. and McCarl, B. (1981). A representation and economic interpretation of a two-level programming problem. *Journal of the Operational Research Society* 32 (9): 783–792.

12 Dempe, S. and Dutta, J. (2012). Is bilevel programming a special case of a mathematical program with complementarity constraints? *Mathematical Programming* 131: 37–48.

13 Ye, J.J. and Zhu, D.L. (1995). Optimality conditions for bilevel programming problems. *Optimization* 33 (1): 9–27.

14 Williams, A.C. (1970). Boundedness relations for linear constraint sets. *Linear Algebra and its Applications* 3 (2): 129–141.

15 Kleinert, T. and Schmidt, M. (2023). Why there is no need to use a big-m in linear bilevel optimization: a computational study of two ready-to-use approaches. *Computational Management Science* 20 (1): 3.

16 Baumrucker, B.T., Renfro, J.G., and Biegler, L.T. (2008). Mpec problem formulations and solution strategies with chemical engineering applications. *Computers & Chemical Engineering* 32 (12): 2903–2913.

17 Bialas, W.F. and Karwan, M.H. (1984). Two-level linear programming. *Management Science* 30 (8): 1004–1020.

18 Blankenship, J.W. and Falk, J.E. (1976). Infinitely constrained optimization problems. *Journal of Optimization Theory and Applications* 19: 261–281.

19 Yue, D. and You, F. (2014). Game-theoretic modeling and optimization of multi-echelon supply chain design and operation under stackelberg game and market equilibrium. *Computers & Chemical Engineering* 71: 347–361.

20 Yue, D. and You, F. (2017). Stackelberg-game-based modeling and optimization for supply chain design and operations: a mixed integer bilevel programming framework. *Computers & Chemical Engineering* 102: 81–95.

21 Chalmardi, M.K. and Camacho-Vallejo, J.-F. (2019). A bi-level programming model for sustainable supply chain network design that considers incentives for using cleaner technologies. *Journal of Cleaner Production* 213: 1035–1050.

22 Avval, A.E., Dehghanian, F., and Pirayesh, M. (2023). Joint optimization of multiple supply chains under cap-and-trade regulation: a bi-level programming model and solution algorithm. *Asia-Pacific Journal of Operational Research* 40 (2): 2250019.

23 Cantú, V.H., Azzaro-Pantel, C., and Ponsich, A. (2021). A novel matheuristic based on bi-level optimization for the multi-objective design of hydrogen supply chains. *Computers & Chemical Engineering* 152: 107370.

24 Cantú, V.H., Ponsich, A., Azzaro-Pantel, C., and Carrera, E. (2023). Capturing spatial, time-wise and technological detail in hydrogen supply chains: a bi-level multi-objective optimization approach. *Applied Energy* 344: 121159.

25 Martelli, E., Freschini, M., and Zatti, M. (2020). Optimization of renewable energy subsidy and carbon tax for multi energy systems using bilevel programming. *Applied Energy* 267: 115089.

26 Cheng, Y., Zhang, N., Zongxiang, L., and Kang, C. (2018). Planning multiple energy systems toward low-carbon society: a decentralized approach. *IEEE Transactions on Smart Grid* 10 (5): 4859–4869.

27 Cheng, Y., Zhang, N., Zhang, B. et al. (2019). Low-carbon operation of multiple energy systems based on energy-carbon integrated prices. *IEEE Transactions on Smart Grid* 11 (2): 1307–1318.

28 Dongge, W., Chang, X., Xue, Y. et al. (2024). Bilevel low-carbon coordinated operation of integrated energy systems considering dynamic tiered carbon pricing methodology. *Energy* 310: 133251.

29 Ma, M., Huang, H., Song, X. et al. (2022). Optimal sizing and operations of shared energy storage systems in distribution networks: a bi-level programming approach. *Applied Energy* 307: 118170.

30 Wang, H., Daoutidis, P., and Zhang, Q. (2023). Ammonia-based green corridors for sustainable maritime transportation. *Digital Chemical Engineering* 6: 100082.

31 Tan, R.R., Ivan Henderson, V., Gue, J.F. et al. (2023). Bilevel optimization model for maritime emissions reduction. *Journal of Cleaner Production* 398: 136589.

32 Molavi, A., Lim, G.J., and Shi, J. (2020). Stimulating sustainable energy at maritime ports by hybrid economic incentives: a bilevel optimization approach. *Applied Energy* 272: 115188.

33 Wang, L., Liang, C., Shi, J. et al. (2021). A bilevel hybrid economic approach for optimal deployment of onshore power supply in maritime ports. *Applied Energy* 292: 116892.

34 Wen, S., Dong, Z., Zhao, T. et al. (2024). Hierarchical game theory-based optimal power management for both seaports and ships. *Energy* 313: 133748.

35 Qiu, R., Jiuping, X., Ke, R. et al. (2020a). Carbon pricing initiatives-based bi-level pollution routing problem. *European Journal of Operational Research* 286 (1): 203–217.

36 Qiu, R., Jiuping, X., Xie, H. et al. (2020b). Carbon tax incentive policy towards air passenger transport carbon emissions reduction. *Transportation Research Part D: Transport and Environment* 85: 102441.

37 Zhao, L., Xiang, X., Oliver Gao, H. et al. (2016). A bi-level model for ghg emission charge based on a continuous distribution of travelers' value of time (vot). *Transportation Research Part D: Transport and Environment* 47: 371–382.

38 Cuadra, R.P.-d.-l. and Ortega, F.A. (2024). Bilevel optimization for the deployment of refuelling stations for electric vehicles on road networks. *Computers & Operations Research* 162: 106460.

39 Caselli, G., Iori, M., and Ljubić, I. (2024). Bilevel optimization with sustainability perspective: a survey on applications. *arXiv preprint arXiv:2406.07184.*

40 Burton, D. and Toint, P.L. (1992). On an instance of the inverse shortest paths problem. *Mathematical Programming* 53: 45–61.

41 Burton, D. and Toint, P.L. (1994). On the use of an inverse shortest paths algorithm for recovering linearly correlated costs. *Mathematical Programming* 63: 1–22.

42 Ajayi, T., Lee, T., and Schaefer, A.J. (2022). Objective selection for cancer treatment: an inverse optimization approach. *Operations Research* 70 (3): 1717–1738.

43 Chan, T.C.Y., Eberg, M., Forster, K. et al. (2022). An inverse optimization approach to measuring clinical pathway concordance. *Management Science* 68 (3): 1882–1903.

44 Birge, J.R., Hortaçsu, A., Michael, J., and Pavlin. (2017). Inverse optimization for the recovery of market structure from market outcomes: an application to the miso electricity market. *Operations Research* 65 (4): 837–855.

45 Konstantakopoulos, I.C., Ratliff, L.J., Ming Jin, S. et al. (2017). A robust utility learning framework via inverse optimization. *IEEE Transactions on Control Systems Technology* 26 (3): 954–970.

46 Bertsimas, D., Gupta, V., and Paschalidis, I.C. (2012). Inverse optimization: a new perspective on the black-litterman model. *Operations Research* 60 (6): 1389–1403.

47 Li, J.Y.-M. (2021). Inverse optimization of convex risk functions. *Management Science* 67 (11): 7113–7141.

48 Yu, S., Wang, H., and Dong, C. (2023). Learning risk preferences from investment portfolios using inverse optimization. *Research in International Business and Finance* 64: 101879.

49 Chan, T.C.Y., Mahmood, R., and Zhu, I.Y. (2023). Inverse optimization: theory and applications. *Operations Research* 73 (2): 1046–1074.

50 Rathore, H. and Jakhar, S.K. (2021). Differential carbon tax policy in aviation: one stone that kills two birds? *Journal of Cleaner Production* 296: 126479.

51 Zhou, Y., Wang, L., and McCalley, J.D. (2010). Effective incentives design for renewable energy generation expansion planning: an inverse optimization approach. In IEEE PES General Meeting, 1–7. IEEE.

52 Tan, R.R. and Aviso, K.B. (2012). An inverse optimization approach to inducing resource conservation in eco-industrial parks. In: *Computer Aided Chemical Engineering*, vol. 31, 775–779. Elsevier.

53 Lu, Y.-A., Hu, W.-S., Paulson, J.A., and Zhang, Q. (2025). Bo4io: a Bayesian optimization approach to inverse optimization with uncertainty quantification. *Computers & Chemical Engineering* 192: 108859.

54 Ahuja, R.K. and Orlin, J.B. (2001). Inverse optimization. *Operations Research* 49 (5): 771–783.

55 Heuberger, C. (2004). Inverse combinatorial optimization: a survey on problems, methods, and results. *Journal of Combinatorial Optimization* 8: 329–361.

56 Aswani, A., Shen, Z.-J., and Siddiq, A. (2018). Inverse optimization with noisy data. *Operations Research* 66 (3): 870–892.

57 Bertsimas, D., Gupta, V., and Paschalidis, I.C. (2015). Data-driven estimation in equilibrium using inverse optimization. *Mathematical Programming* 153: 595–633.

58 Esfahani, P.M., Shafieezadeh-Abadeh, S., Hanasusanto, G.A., and Kuhn, D. (2018). Data-driven inverse optimization with imperfect information. *Mathematical Programming* 167: 191–234.

59 Babier, A., Chan, T.C.Y., Lee, T. et al. (2021). An ensemble learning framework for model fitting and evaluation in inverse linear optimization. *Informs Journal on Optimization* 3 (2): 119–138.

60 Gupta, R. and Zhang, Q. (2022). Decomposition and adaptive sampling for data-driven inverse linear optimization. *INFORMS Journal on Computing* 34 (5): 2720–2735.

61 Shahmoradi, Z. and Lee, T. (2022). Quantile inverse optimization: improving stability in inverse linear programming. *Operations Research* 70 (4): 2538–2562.

62 Gupta, R. and Zhang, Q. (2023a). Efficient learning of decision-making models: a penalty block coordinate descent algorithm for data-driven inverse optimization. *Computers & Chemical Engineering* 170: 108123.

63 Tan, Y., Delong, A., and Terekhov, D. (2019). Deep inverse optimization. In: *Integration of Constraint Programming, Artificial Intelligence, and Operations Research: 16th International Conference, CPAIOR 2019, Thessaloniki, Greece, June 4–7, 2019, Proceedings 16*, 540–556. Springer.

64 Tan, Y., Terekhov, D., and Delong, A. (2020). Learning linear programs from optimal decisions. *Advances in Neural Information Processing Systems* 33: 19738–19749.

65 Wang, L. (2009). Cutting plane algorithms for the inverse mixed integer linear programming problem. *Operations Research Letters* 37 (2): 114–116.

66 Bodur, M., Chan, T.C.Y., and Zhu, I.Y. (2022). Inverse mixed integer optimization: polyhedral insights and trust region methods. *INFORMS Journal on Computing* 34 (3): 1471–1488.

67 Gupta, R. and Zhang, Q. (2023b). An inverse optimization approach to decision-focused learning. In: *Computer Aided Chemical Engineering*, vol. 52, 1359–1365. Elsevier.

68 Mandi, J., Kotary, J., Berden, S. et al. (2024). Decision-focused learning: foundations, state of the art, benchmark and future opportunities. *Journal of Artificial Intelligence Research* 80: 1623–1701.

69 Dixit, S., Gupta, R., and Zhang, Q. (2025). Decision-focused surrogate modeling for mixed-integer linear optimization. Transactions on Machine Learning Research. https://openreview.net/forum?id=A6tOXkkE4Z (accessed 1 December 2025).

70 Gupta, R. and Zhang, Q. (2024). Data-driven decision-focused surrogate modeling. *AIChE Journal* 70 (4): e18338.

71 Van Hentenryck, P. (2025). Optimization learning. *arXiv preprint arXiv:2501.03443.*

72 Bengio, Y. (2000). Gradient-based optimization of hyperparameters. *Neural Computation* 12 (8): 1889–1900.

73 Baydin, A. G. and Pearlmutter, B. A. (2014). Automatic differentiation of algorithms for machine learning. *arXiv preprint arXiv:1404.7456.*

74 Maclaurin, D., Duvenaud, D., and Adams, R. (2015). Gradient-based hyperparameter optimization through reversible learning. In International conference on machine learning, 2113–2122. PMLR.

75 Franceschi, L., Frasconi, P., Salzo, S., et al. (2018). Bilevel programming for hyperparameter optimization and meta-learning. In International conference on machine learning, 1568–1577. PMLR.

76 Liu, H., Simonyan, K., and Yang, Y. (2018). Darts: differentiable architecture search. *arXiv preprint arXiv:1806.09055.*

77 Blondel, M., Berthet, Q., Cuturi, M. et al. (2022). Efficient and modular implicit differentiation. *Advances in Neural Information Processing Systems* 35: 5230–5242.

78 Amos, B. and Kolter, J. Z. (2017). Optnet: differentiable optimization as a layer in neural networks. In International conference on machine learning, 136–145. PMLR.

79 Yao, W., Yin, H., Zeng, S., and Zhang, J. (2024). Overcoming lower-level constraints in bilevel optimization: a novel approach with regularized gap functions. *arXiv preprint arXiv:2406.01992.*

80 Giovannelli, T., Kent, G. D., and Vicente, L. N. (2021). Inexact bilevel stochastic gradient methods for constrained and unconstrained lower-level problems. *arXiv preprint arXiv:2110.00604.*

81 Zucchet, N. and Sacramento, J. (2022). Beyond backpropagation: bilevel optimization through implicit differentiation and equilibrium propagation. *Neural Computation* 34 (12): 2309–2346.

82 Bolte, J., Le, T., Pauwels, E., and Silveti-Falls, T. (2021). Nonsmooth implicit differentiation for machine-learning and optimization. *Advances in Neural Information Processing Systems* 34: 13537–13549.

83 Arbel, M. and Mairal, J. (2022). Non-convex bilevel games with critical point selection maps. *Advances in Neural Information Processing Systems* 35: 8013–8026.

84 Birge, J.R. and Louveaux, F. (2011). *Introduction to Stochastic Programming*. Springer Science & Business Media.

85 Ben-Tal, A., Nemirovski, A., and El Ghaoui, L. (2009). *Robust Optimization*. Princeton University Press.

86 Ben-Tal, A., Hertog, D.D., and Vial, J.-P. (2015). Deriving robust counterparts of nonlinear uncertain inequalities. *Mathematical Programming* 149 (1): 265–299.

87 Bertsimas, D., Dunning, I., and Lubin, M. (2016). Reformulation versus cutting-planes for robust optimization: a computational study. *Computational Management Science* 13: 195–217.

88 Bienstock, D. and Özbay, N. (2008). Computing robust basestock levels. *Discrete Optimization* 5 (2): 389–414.

89 Mutapcic, A. and Boyd, S. (2009). Cutting-set methods for robust convex optimization with pessimizing oracles. *Optimization Methods and Software* 24: 381–406.

90 Georghiou, A., Kuhn, D., and Wiesemann, W. (2019). The decision rule approach to optimization under uncertainty: methodology and applications. *Computational Management Science* 16 (4): 545–576.

91 Gorissen, B.L., Yanıkoğlu, İ., and Den Hertog, D. (2015). A practical guide to robust optimization. *Omega* 53: 124–137.

92 Bertsimas, D., Iancu, D.A., and Parrilo, P.A. (2011). A hierarchy of near-optimal policies for multistage adaptive optimization. *IEEE Transactions on Automatic Control* 56 (12): 2809–2824.

93 Bertsimas, D. and Georghiou, A. (2015). Design of near optimal decision rules in multistage adaptive mixed-integer optimization. *Operations Research* 63 (3): 610–627.

94 Zeng, B. and Zhao, L. (2013). Solving two-stage robust optimization problems using a column-and-constraint generation method. *Operations Research Letters* 41 (5): 457–461.

95 Subramanyam, A., Gounaris, C.E., and Wiesemann, W. (2020). K-adaptability in two-stage mixed-integer robust optimization. *Mathematical Programming Computation* 12: 193–224.

96 Conejo, A.J. and Wu, X. (2022). Robust optimization in power systems: a tutorial overview. *Optimization and Engineering* 23 (4): 2051–2073.

97 Lorca, A. and Sun, X.A. (2014). Adaptive robust optimization with dynamic uncertainty sets for multi-period economic dispatch under significant wind. *IEEE Transactions on Power Systems* 30 (4): 1702–1713.

98 Lorca, A. and Sun, X.A. (2016). Multistage robust unit commitment with dynamic uncertainty sets and energy storage. *IEEE Transactions on Power Systems* 32 (3): 1678–1688.

99 Apostolopoulou, D., De Grève, Z., and McCulloch, M. (2018). Robust optimization for hydroelectric system operation under uncertainty. *IEEE Transactions on Power Systems* 33 (3): 3337–3348.

100 Chen, Z., Lei, W., and Yong, F. (2012). Real-time price-based demand response management for residential appliances via stochastic optimization and robust optimization. *IEEE Transactions on Smart Grid* 3 (4): 1822–1831.

101 Hosseini, S. M., Carli, R., and Dotoli, M. (2022). Robust optimal demand response of energy-efficient commercial buildings. In *2022 European Control Conference (ECC)*, 1–6. IEEE.

102 Zhang, Q., Morari, M.F., Grossmann, I.E. et al. (2016). An adjustable robust optimization approach to scheduling of continuous industrial processes providing interruptible load. *Computers & Chemical Engineering* 86: 106–119.

103 Wang, Z., Chen, B., Wang, J. et al. (2014b). Robust optimization based optimal dg placement in microgrids. *IEEE Transactions on Smart Grid* 5 (5): 2173–2182.

104 Rezvan, A.T., Shams Gharneh, N., and Gharehpetian, G.B. (2012). Robust optimization of distributed generation investment in buildings. *Energy* 48: 455–463.

105 Pishvaee, M.S., Rabbani, M., and Torabi, S.A. (2011). A robust optimization approach to closed-loop supply chain network design under uncertainty. *Applied Mathematical Modelling* 35 (2): 637–649.

106 Jabbarzadeh, A., Fahimnia, B., and Rastegar, S. (2017). Green and resilient design of electricity supply chain networks: a multiobjective robust optimization approach. *IEEE Transactions on Engineering Management* 66 (1): 52–72.

107 Jagana, J.S., Amaran, S., and Zhang, Q. (2025). Multistage robust mixed-integer optimization for industrial demand response with interruptible load. *Computers & Chemical Engineering* 194: 108974.

108 Löfberg, J. (2004). Yalmip: a toolbox for modeling and optimization in Matlab. In *In Proceedings of the CACSD Conference*, Taipei, Taiwan.

109 Hart, W. and Castillo, A. (2019). Pao [Computer Software]. https://doi.org/10.11578/dc.20201105.1.

110 Garcia, J.D., Bodin, G., and Street, A. (2024). Bileveljump. jl: modeling and solving bilevel optimization problems in julia. *INFORMS Journal on Computing* 36 (2): 327–335.

2

Exploiting the Multiscale Structure of Sustainable Engineering Problems via Network-Based Decomposition

Ilias Mitrai[1] *and Prodromos Daoutidis*[2]

[1]*McKetta Department of Chemical Engineering, The University of Texas at Austin, Austin, TX, USA*
[2]*Department of Chemical Engineering and Materials Science, University of Minnesota, Minneapolis, MN, USA*

2.1 Introduction

The design of efficient production systems has always been a key objective in process engineering. Traditionally, increased efficiency has been achieved by optimizing individual processes, production facilities, and supply chain networks. However, novel approaches have recently been proposed to achieve a sustainable manufacturing paradigm. Typical examples include the decarbonization of the chemical sector by coupling chemical manufacturing with the power grid and renewable electricity resources [1–4], incorporation of new energy storage technologies such as ammonia and hydrogen [5–7], and the design of circular and sustainable supply chain networks [8–11]. The design of such systems requires new operational paradigms and poses unique challenges. For example, chemical processes in such systems typically operate in a dynamic mode rather than a steady-state one. Therefore, the dynamic behavior of the process, which is usually nonlinear, must be considered. Furthermore, the design of such systems must consider simultaneously decisions across multiple scales and is affected by uncertainties at the operational (hours or days) and strategic (years or decades) planning scale. This requires the simultaneous consideration of decisions across a wide range of temporal and spatial scales, leading to large and multiscale decision-making problems.

Mathematical modeling and optimization have been widely used to model and solve such decision-making problems across a wide range of scales, from the molecular to the enterprise-wide one. Significant advances have been made in optimization theory and algorithms for the solution of a broad class of optimization problems such as Mixed Integer Linear (MILP), Nonlinear (NLP) [12], Mixed Integer Nonlinear (MINLP) [13, 14], stochastic [15, 16], robust [17], and disjunctive [18] optimization problems. Despite these advances, the solution of large-scale and nonconvex optimization problems remains a challenging task.

Decomposition-based optimization algorithms have been extensively used to improve the tractability of large-scale optimization problems by exploiting the underlying structure [19, 20]. Typical examples include Lagrangean relaxation and decomposition [21], progressive hedging or the Alternating Direction Method of Multipliers [22], Benders decomposition and its variants [23, 24], column generation [25], and cross decomposition [26, 27]. To apply such solution approaches, a decomposition of the optimization problem is necessary. The standard approach is to rely on intuition and/or domain knowledge to obtain a decomposition. This approach is time-consuming and not always possible. Thus, automated methods to identify the structure of a problem without any prior knowledge or insight are of great value in such a solution approach.

Methods from network science have been recently adopted to identify the structure of optimization and control problems automatically. The basic underlying idea is to represent an optimization problem or a dynamical system as a graph and employ structure detection algorithms from graph theory and network science to partition the graph into blocks. Typical examples of such approaches include community detection [28–33] and hypergraph partitioning [34–38]. These methods require some a-priory assumptions about the structure of the graph (e.g. organization into communities) and/or the number of blocks. We have proposed a more generic approach to learning the structure of optimization problems through the application of stochastic block modeling (SBM) and Bayesian inference. This approach has been applied to generic benchmark MINLPs as well as problems that arise in the operation of chemical processes (integration of scheduling and control) revealing the complex structure of these problems [39, 40].

In this chapter, we discuss the application of SBM and Bayesian inference for learning the structure of optimization problems that arise in sustainability-related applications. In Section 2.2, we present graph representations of an optimization problem and introduce SBM and Bayesian inference as a tool for learning the structure of the problem. In Section 2.3, we show how the learned structure can be used as the basis for the implementation of decomposition-based optimization algorithms. Finally, in Section 2.4, we consider the application of network-based decomposition to a capacity expansion problem under uncertainty related to transitioning conventional ammonia supply chains into renewable ones.

2.2 Learning the Structure of Optimization Problems

2.2.1 Optimization Problems as Graphs

We will consider the following optimization problem:

$$\begin{aligned} \underset{x}{\text{minimize}} \quad & f(x) \\ \text{subject to} \quad & g(x) \leq 0 \\ & x \in \mathbb{Z}^{N_d} \times \mathbb{R}^{N_c}, \end{aligned} \tag{2.1}$$

where N_d, N_c is the number of discrete and continuous variables, respectively, $N_c + N_d$ is the total number of variables, and M the number of constraints. Based on the functional form of the objective function and constraints, different classes of optimization problems arise. For example, if $f(x) = c^\top x$ and $g = Ax - b$, the above problem is a Mixed Integer Linear Programming (MILP) problem, whereas if f, g are generic nonconvex functions, then Eq. (2.1) corresponds to a Mixed Integer Nonlinear Programming (MINLP) problem. The interaction pattern between the variables and the constraints determines the structure of an optimization problem.

An approach to capture this interaction pattern is to represent an optimization problem as a network. A network or a graph $\mathcal{G}$ is a discrete mathematical object that represents the interaction pattern among a set of objects called nodes. We define as $V = \{v_i\}_{i=1}^N$ the set of nodes in the graph with v_i corresponding to node i. The connectivity among the nodes is captured via edges $E = \{e_{ij}\}_{i \in V, j \in V}$, where $e_{ij} = 1$ is an edge exists between node i and j. A graph can also be represented by a matrix, called the adjacency matrix, $A \in \mathbb{R}^N \times \mathbb{R}^N$ where $A_{ij} = e_{ij}$, that is $A_{ij} = 1$ if an edge exists between nodes i and j.

Three types of graphs can be used to represent an optimization problem [28]. The first is a variable-constraint bipartite graph $\mathcal{G}_b(V_n, V_c, E)$ which has two sets of nodes, ones representing the constraints V_c and the other the variables V_n. The edges capture the presence of a variable in a constraint. The second type of graph is the variable graph $\mathcal{G}_v(V_n, E_n)$, where the nodes are the variables of the problem and an edge exists between two nodes (variables) if they appear in one or more constraints, i.e. the edges represent constraints that couple to variables. The last type of graph is a constraint graph $\mathcal{G}_c(V_c, E_c)$ where the nodes are the constraints of the optimization problem and the edges capture the variables that couple two constraints. Given the graph representation of an optimization problem, the next task is to find its structure, namely, the partition of the nodes into a set of blocks which will form the basis for the application of decomposition-based optimization algorithms.

2.2.2 Learning the Structure via Stochastic Blockmodeling

An stochastic block model (SBM) is a random graph generative model that can generate random graphs with a specific block structure. A block is defined as a set of nodes that have similar connectivity pattern in the graph. We assume an undirected graph $\mathcal{G}(V, E)$ ($|V| = N_v$, $|E| = N_e$) with adjacency matrix $A \in \mathbb{R}^{N_v \times N_v}$ and a partition of the nodes into B blocks. We define a partition vector $b \in \mathbb{R}^{N_v}$ with $b_i = \{1, \dots, B\}$ denoting the group membership of node i. We assume that the number of edges between all the nodes in block r and all the nodes in block s is a Poisson distributed variable with expected value ω_{rs}. Under these assumptions, the probability of observing a graph with adjacency matrix A is [41]

$$P(A|b, \omega) = \prod_{1 \le i,j \le N_v} \frac{\omega_{b_i b_j}^{A_{ij}}}{A_{ij}!} e^{-\omega_{b_i b_j}} \prod_{i=1}^{N_v} \frac{\left(\frac{1}{2}\omega_{b_i b_i}\right)^{A_{ij}}}{\left(\frac{1}{2}A_{ij}\right)!} e^{-\frac{1}{2}\omega_{b_i b_j}}, \tag{2.2}$$

where the first term represents the probability of an edge between two nodes i and j and the second term is the probability of a self-edge.

Given a graph $\mathcal{G}$ with adjacency matrix A and assuming that a SBM generated it, we are interested in estimating the parameters of the model, namely, the block partition of the nodes and the connectivity pattern between the blocks. First, we focus on the probability of observing a partition b for a given graph with adjacency matrix A, i.e. $P(b|A)$. From Bayes theorem, we have

$$P(b|A) = \frac{P(A|b)P(b)}{P(A)}, \tag{2.3}$$

where $P(A|b) = P(A|\omega, b)P(\omega|b)$ is the marginal likelihood and $P(A) = \sum_b P(A|b)$ $P(b)$ is the evidence which is a constant. The numerator in the above expression can be written as

$$P(A|b)P(b) = P(A|b, \omega)P(\omega|b)P(b) = P(A|b, \omega)P(\omega, b). \tag{2.4}$$

Given this expression, we are interested in finding the partition b that will maximize $P(b|A)$. Since the evidence is constant, the task of maximizing $P(b|A)$ is equivalent to maximizing the logarithm of $P(b|A)$, which, using Eq. (2.4) leads to

$$\underset{b,\omega}{\text{maximize}} \;\; \log P(A|b, \omega) + \log P(b, \omega). \tag{2.5}$$

This objective function has an information theoretical interpretation. Specifically, Eq. (2.4) can be written as follows:

$$P(A|b, \omega)P(\omega, b) = 2^{-\Sigma} \rightarrow \Sigma = -\log_2 P(A|b, \omega) - \log_2 P(\omega, b), \tag{2.6}$$

where Σ is called the description length, and the problem in Eq. (2.5) is equivalent to minimizing the description length

$$\underset{b,\omega}{\text{minimize}} \;\; -\log_2 P(A|b, \omega) - \log_2 P(b, \omega). \tag{2.7}$$

The first term in this objective is equal to the amount of information required to describe the data using the model and the second term is the amount of information required to describe the model itself. An increase in the number of blocks will make the model fit the data better, that is $-\log P(A|b,\omega)$ will be decreased. However, the model will become more complex, that is $-\log P(b,\omega)$ will increase. Therefore, this objective function avoids overfitting.

The estimation of the b, ω parameters using the Bayesian approach poses two challenges. First, the prior distributions must be defined based on our beliefs about the characteristics of the SBM. Since the structure of the graph is not known a-priory, one can follow an uninformative approach regarding the number of blocks and their connection pattern. The connectivity pattern of the blocks, ω, can be substituted by the expected number of edges e_{rs}. Assuming that all the ways of allocating N_e edges among $B(B+1)/2$ blocks are equally probable, $P(e)$ is equal to

$$P(e) = \binom{B(B+1)/2 + m - 1}{m}^{-1}. \tag{2.8}$$

Regarding the partition of the network, we assume that all block sizes are equally probable and that the distribution is agnostic with respect to the size of the blocks. We define as $\mathbf{n} = (\mathbf{n}_1, \dots, \mathbf{n}_B)$ the number of nodes in block r with $\sum_r \mathbf{n}_r = N_v$. Under these assumptions, $P(b)$ can be written as

$$P(b) = P(b|\mathbf{n})P(\mathbf{n}|B)P(B), \tag{2.9}$$

where $P(B) = 1/N_v$ since all number of block (between 1 and N_v) are equally probable and

$$P(\mathbf{n}|B) = \binom{N-1}{B-1}^{-1}, \tag{2.10}$$

since all block sizes are equally probable (forbidding empty groups). Finally, $P(b|n)$ is equal to the reciprocal of the total number of permutations that keep $\mathbf{n}$ unchanged and given by the following equation:

$$P(b|\mathbf{n}) = \frac{\prod_r \mathbf{b}_r!}{N_v!}. \tag{2.11}$$

Once these prior distributions are defined, the next step is the solution of the problem in Eq. (2.7). This is a combinatorial problem since $b_i \in \{1, \dots, B\}$. An approach to solve this problem heuristically is to use a Markov chain Monte Carlo-based algorithm [42]. In this approach, the nodes are initially partitioned randomly b_0, and for each node, a move proposal is made $b_i \rightarrow b_{i'}$. The move proposal is accepted with a certain probability p using a Metropolis-Hastings criterion. This algorithm has quasilinear complexity with respect to the number of nodes $\mathcal{O}(N_v \ln^2 N_v)$. The number of blocks B can either be estimated from the data or fixed and considered a tunable hyperparameter. We refer the reader to [41, 42] for a detailed description of the algorithm and to [43] for access to the Python code of the algorithm.

2.3 Network-Based Decomposition of Optimization Problems

The inference results from the Section (2.2.2) can be used as the basis for the application of decomposition-based optimization algorithms. We show how the inferred structure in the variable and constraint graph can be used as the basis for the application of decomposition-based optimization algorithms.

2.3.1 Benders Decomposition Based on the Variable Graph

In this section, we show how a core periphery structure in the variable graph can be used as the basis for the application of Benders decomposition. We will consider the case, where an optimization problem given by Eq. (2.1) must be solved and the inference results on the variable graph of the problem are presented in Figure 2.1.

This structure is known as core–periphery since the graph has a block of highly connected nodes (variables) that form the core (light gray) and a set of sparsely connected nodes (variables) in the periphery (dark gray). Under this partition of the graph, the variables can be partitioned into two sets x_c (core) and x_p (periphery). Three sets of constraints, i.e. edges, can be identified in the graph: ones that depend only on the variables in the core g_c (edges that connect the light gray nodes with each other), ones that depend only on the variables in the

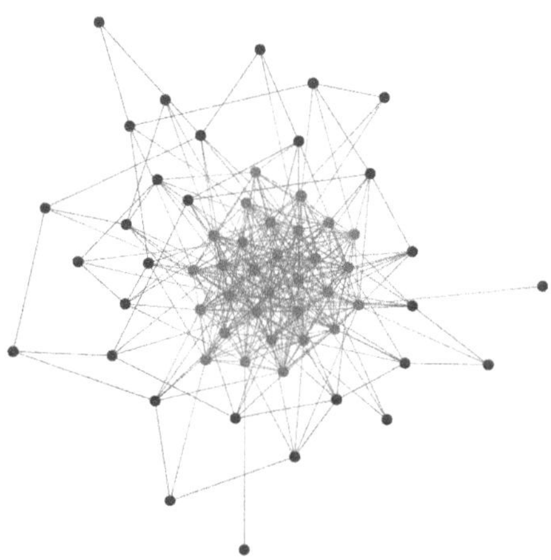

Figure 2.1 An example of core periphery structure.

periphery g_p (edges that connect the dark gray nodes with each other), and ones that depend on both the variables in the core and periphery g_{cp}. The latter set of constraints corresponds to edges in the graph that connect a node belonging to the core (light gray) with a node that belongs in the periphery (dark gray). Given this partition of the variables and constraints, the original problem can be written as

$$\begin{aligned} \underset{x_c, x_p}{\text{minimize}} \quad & f(x_c, x_p) \\ \text{subject to} \quad & g_c(x_c) \leq 0 \\ & g_p(x_p) \leq 0 \\ & g_{cp}(x_c, x_p) \leq 0 \\ & x_c \in \mathbb{Z}^{N_d} \times \mathbb{R}^{N_c^1}, x_p \in \mathbb{R}^{N_c^2}, \end{aligned} \tag{2.12}$$

Assuming that all the integer variables are in the core (master problem), Benders or Generalized Benders Decomposition [23, 24] can be applied to solve the problem with x_c being the complicated variables.

The subproblem is obtained by fixing the variables in the master problem and is equal to

$$\begin{aligned} S(x_c) := \underset{x_p, \overline{x}_c}{\text{minimize}} \quad & f(\overline{x}_c, y) \\ \text{subject to} \quad & g_p(x_p) \leq 0 \\ & g_{cp}(\overline{x}_c, x_p) \leq 0 \\ & \overline{x}_c = x_c \ : \ \lambda \\ & x_p \in \mathbb{R}^{N_c^2}, \overline{x}_c \in \mathbb{R}^{N_d} \times \mathbb{R}^{N_c^1}, \end{aligned} \tag{2.13}$$

where λ is the Lagrangean multiplier for the constraint $\overline{x}_c = x_c$. The master problem is

$$\begin{aligned} \underset{x_c, \eta}{\text{minimize}} \quad & \eta \\ \text{subject to} \quad & g_c(x_c) \leq 0 \\ & \eta \geq S(x_c^l) - \lambda^l(x_c - x_c^l) \quad \forall l = 1, \dots, \mathcal{L} \\ & \eta \in [\underline{\eta}, \overline{\eta}] \\ & x_c \in \mathbb{Z}^{N_d} \times \mathbb{R}^{N_c^1}, \end{aligned} \tag{2.14}$$

where the second constraint is the Benders optimality cut, and l is the iteration number. For cases where the subproblem can become infeasible, Benders feasibility cuts must be added [23, 24]. The solution of the master problem provides a lower bound and the solution of the subproblem an upper bound (when it is feasible). The master problem and subproblem are solved iteratively until the bounds converge. Convergence is guaranteed for cases where the subproblem is convex [23, 24, 44].

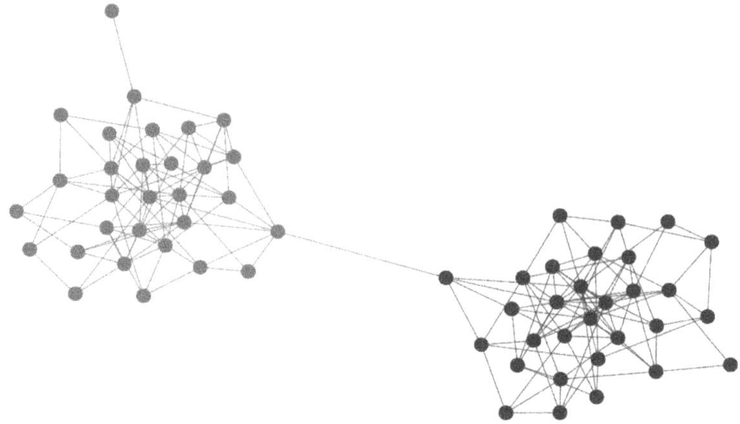

Figure 2.2 An example of a graph with community structure.

2.3.2 Lagrangean Decomposition Based on the Structure of the Constraint Graph

We consider the application of Lagrangean decomposition based on the underlying structure of the constraint graph. Let's assume that the graph presented in Figure 2.2 corresponds to the constraint graph of an optimization problem. The constraint graph is partitioned into two blocks that are loosely connected.

In this partition of the graph, two sets of constraints are identified, g_1 and g_2, which correspond to the red (light gray) and blue (dark gray) nodes respectively. The variables (edges) are assigned into three sets. The first, x_1, contains variables that couple the constraints that correspond to the red nodes (red edges in the graph), the second set x_2 contains the blue edges, and the third set x_3 the edges that connect blue nodes with red nodes. Given this decomposition of the constraint graph, the original optimization problem can be written as

$$\begin{aligned} \underset{x_1,x_2,x_3}{\text{minimize}} \quad & f_1(x_1,x_3) + f_2(x_2,x_3) \\ \text{subject to} \quad & g_1(x_1,x_3) \leq 0 \\ & g_2(x_2,x_3) \leq 0 \\ & x_1 \in \mathcal{X}_1, x_2 \in \mathcal{X}_2, x_3 \in \mathcal{X}_3. \end{aligned} \tag{2.15}$$

This decomposition of the variables and constraints lends itself naturally to the application of Lagrangean decomposition since the problem can be solved independently with respect to x_1 and x_2 if the x_3 variables are fixed. From a graph perspective, this is equivalent to removing the edge that connects the two communities. This can be achieved by creating a copy of the x_3 variable, $\overline{x}_3$, and reformulating the problem as follows:

$$\begin{aligned}
\underset{x_1,x_2,x_3,\overline{x}_3}{\text{minimize}} \quad & f_1(x_1,x_3) + f_2(x_2,\overline{x}_3) \\
\text{subject to} \quad & g_1(x_1,\overline{x}_3) \leq 0 \\
& g_2(x_2,\overline{x}_3) \leq 0 \\
& x_3 = \overline{x}_3 \\
& x_1 \in \mathcal{X}_1, x_2 \in \mathcal{X}_2, x_3 \in \mathcal{X}_3, \overline{x}_3 \in \mathcal{X}_3.
\end{aligned} \tag{2.16}$$

Dualizing the last constraint, we obtain the following problem:

$$\begin{aligned}
\underset{x_1,x_2,x_3,\overline{x}_3}{\text{minimize}} \quad & f_1(x_1,x_3) + f_2(x_2,\overline{x}_3) + \lambda^\top(x_3 - \overline{x}_3) \\
\text{subject to} \quad & g_1(x_1,x_3) \leq 0 \\
& g_2(x_2,\overline{x}_3) \leq 0 \\
& x_1 \in \mathcal{X}_1, x_2 \in \mathcal{X}_2, x_3 \in \mathcal{X}_3, \overline{x}_3 \in \mathcal{X}_3.
\end{aligned} \tag{2.17}$$

This problem can be decomposed into two problems P_1 and P_2 as follows:

$$\begin{aligned}
P_1(\lambda) := \underset{x_1,x_3}{\text{minimize}} \quad & f_1(x_1,x_3) + \lambda^\top x_3 \\
\text{subject to} \quad & g_1(x_1,x_3) \leq 0 \\
& x_1 \in \mathcal{X}_1, x_3 \in \mathcal{X}_3
\end{aligned} \tag{2.18}$$

$$\begin{aligned}
P_2(\lambda) := \underset{x_2\overline{x}_3}{\text{minimize}} \quad & f_2(x_2,\overline{x}_3) - \lambda^\top \overline{x}_3 \\
\text{subject to} \quad & g_2(x_2,\overline{x}_3) \leq 0 \\
& x_2 \in \mathcal{X}_2, \overline{x}_3 \in \mathcal{X}_3.
\end{aligned} \tag{2.19}$$

The solution of problems P_1 and P_2 depends on the values of the Lagrangean multipliers λ. Multiple methods can be used to update the multipliers, such as gradient ascent, bundle methods, and trust region methods [19].

In this section, we presented the application of Benders Decomposition based on the core–periphery structure of the variable graph and the application of Lagrangean decomposition based on the structure of the constraint graph. The solution of an optimization problem using a network-based decomposition has three steps: (i) represent the optimization problem as a graph, (ii) learn (or estimate) the structure of the graph, and (iii) implement a decomposition-based algorithm based on the underlying structure. These steps have been automated in DecODe [40], a Python-based package that uses Pyomo [45] for modeling the optimization problem, NetworkX [46] for representing the problem as a graph, and graph-tool [43] for estimating the structure. Steps 1 and 2 have also been implemented in Matlab as presented in [28]. In general, certain structure in the constraint graph can be used for Benders decomposition, and certain structure in the variable graph can be used for Lagrangean decomposition and relaxation.

We refer to [40] for a detailed discussion on the relation between different decomposition methods and the variable and constraint graph structure.

2.4 Case Study: Transition to Green Ammonia Supply Chain Networks

In this section, we use the network-based representation and decomposition of an optimization problem to understand and solve a two-stage stochastic optimization problem that considers the incorporation of renewable ammonia production facilities in an existing supply chain network while considering price uncertainty. The model is adapted from [47].

2.4.1 Two-Stage Stochastic Programming Problem Formulation

We consider the transition of an existing ammonia supply chain network with C ($\mathcal{C} = \{c_i\}_{i=1}^{C}$) customers, D ($\mathcal{D} = \{D_i\}_{i=1}^{D}$) distribution centers, P ($\mathcal{P} = \{P_i\}_{i=1}^{P}$) conventional producers, R ($\mathcal{R} = \{R_i\}_{i=1}^{R}$) renewable ammonia candidate locations, and S ($\mathcal{S} = \{s_i\}_{i=1}^{S}$) are the set of scenarios. We assume that the planning horizon is N_t years ($\mathcal{K} = \{K_i\}_{i=1}^{N_t}$) and in each time period (duration is one year) the renewable ammonia facilities are available in discrete sizes ($\mathcal{R}_x = \{X_1, \dots, X_N\}$).

We define a binary variable z_{rki} which is equal to one if a plant with capacity i is installed at location r at time period k and zero otherwise. We also define variable x_{rk} which is equal to the capacity installed at location r at time period k. The capacity installed at location r and time period k is equal to

$$\begin{aligned} x_{rk} &= \sum_{i \in \mathcal{R}_x} X_i z_{rki} \ \forall r \in \mathcal{R}, k \in \mathcal{K} \\ &\sum_{i \in \mathcal{R}_x} z_{rki} \leq 1 \ \forall r \in \mathcal{R}, k \in \mathcal{K}. \end{aligned} \tag{2.20}$$

The capacity that is installed in each location is bounded by the available wind resources and electrolysis capacity. This is enforced using the following constraints:

$$\begin{aligned} &\sum_{k'=1}^{k} x_{rk'} \omega_{rk'} \leq \Omega_r \ \forall r \in \mathcal{R}, k \in \mathcal{K} \\ &\sum_{r \in \mathcal{R}} x_{rk} \xi_{rk} \leq \Xi_k \ \forall k \in \mathcal{K}, \end{aligned} \tag{2.21}$$

where Ω_r is the wind capacity at location r, ω_{rk} is a conversion factor that couples ammonia production with wind generation, Ξ_k is the electrolysis capacity at year k, and ξ_{rk} is a conversion factor between mt/y ammonia production and MW of electrolysis.

The ammonia that is purchased from conventional producers is first shipped to distribution centers and then to the final customers whereas the ammonia produced from renewable resources can be directly transported to the customers. We define as $y_{rcks} \geq 0$ the amount of ammonia shipped from candidate location r to customer c in year k and scenario s, $y_{dcks} \geq 0$ the amount of ammonia shipped from distribution center d to customer c in year k and scenario s, and $y_{pdks} \geq 0$ the amount of ammonia shipped from conventional producer p to distribution center d in year k and scenario s. The constraints related to the flow of ammonia in the network are

$$\sum_{p\in\mathcal{P}} y_{rcks} + \sum_{d\in\mathcal{D}} y_{dcks} \geq \delta_{ck} \ \forall s \in \mathcal{S}, c \in \mathcal{C}, k \in \mathcal{K}, \tag{2.22}$$

$$\sum_{p\in\mathcal{P}} y_{pdks} \geq \sum_{c\in\mathcal{C}} y_{dcks} \ \forall s \in \mathcal{S}, d \in \mathcal{D}, k \in \mathcal{K}, \tag{2.23}$$

$$\sum_{d\in\mathcal{D}} y_{pdks} \leq \Lambda_p \ \forall s \in \mathcal{S}, p \in \mathcal{P}, k \in \mathcal{K}, \tag{2.24}$$

$$\sum_{c\in\mathcal{C}} y_{rcks} \leq \sum_{k'=1}^{k} x_{rk'} \ \forall s \in \mathcal{S}, r \in \mathcal{R}, k \in \mathcal{K}. \tag{2.25}$$

$$\sum_{p\in\mathcal{P}} \sum_{d\in\mathcal{D}} y_{pd,N_t,s} = 0 \ \forall s \in \mathcal{S}. \tag{2.26}$$

The first constraint (Eq. (2.22)) guarantees that the demand for each customer, time period, and scenario is satisfied. The second constraint (Eq. (2.23)) enforces that the amount of conventionally produced ammonia shipped from the distribution centers to the customers is greater than the amount of ammonia purchased. The third constraint (Eq. (2.24)) poses an upper bound on the amount of ammonia purchased from conventional producers. Equation (2.25) guarantees that the amount of renewable ammonia shipped to the customer does not exceed the installed capacity and the last constraint (Eq. (2.26)) guarantees that at the end of the horizon, only renewable ammonia is used to satisfy the demand.

The objective function of the optimization problem is the sum of the capital and operating cost for the new facilities, the cost of purchasing ammonia from conventional resources, and the cost of transporting ammonia to the customers. The uncertain parameter is the ammonia price which appears in the ammonia purchasing cost in the objective function. The detailed expression for the objective function can be found in [47].

2.4.2 Structure of the Optimization Problem

We consider the renewable ammonia supply chain network problem for the state of Minnesota, which has 82 counties (customers), 10 conventional ammonia producers, 3 distribution centers, 26 candidate locations, and 3 available capacities for each renewable location. For a planning period of 9 years and 10 scenarios, the optimization problem has 218,034 variables (702 binary) and 11,987 constraints. We will consider different instances of this problem in order to build intuition about the coupling between the variables and the constraints of the two-stage program. Also, we consider small instances in order to keep the graphs simple and readable.

First, we consider the case where the planning horizon is five years, the number of candidate locations is one, the number of counties is two (customers), the number of conventional producers is two, and the number of scenarios is 1, that is this corresponds to the deterministic problem. We apply SBM with six blocks, and the inference results are presented in Figure 2.3.

From these results, we observe that the investment decisions are grouped together and the decisions regarding the transportation of ammonia between renewable facilities, counties, distribution centers, and conventional producers

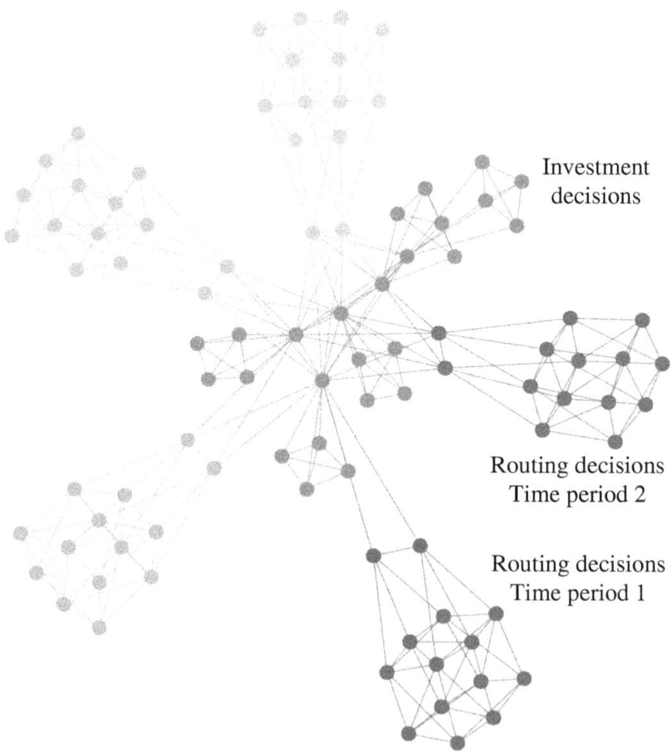

Figure 2.3 Block structure of the deterministic ammonia supply chain transition problem for one candidate location and five time periods.

are grouped by time period. This structure is a core–community one with the investment decisions forming the core and the group (or communities) for each time period being in the periphery. The core and the periphery are connected via edges that correspond to the constraints that limit the amount of ammonia shipped from a candidate location by the installed capacity.

Next, we increase the number of candidate locations to two and decrease the number of periods to three. The inference results for this case are presented in Figure 2.4. We observe that the variable graph is partitioned into five blocks. Specifically, the investment decisions for each location are partitioned into different blocks and the ammonia transportation decisions for each time period are partitioned into different blocks. This is a multicore community structure. The two blocks with the investment decisions form the core (multicore) and are connected with the routing decisions for each time period.

Finally, we consider the two-stage programming problem with three time periods, two candidate locations, and two scenarios. From Figure 2.5, we observe that the graph is more complicated compared to the previous two cases and is partitioned into eight blocks. This partition reveals a multicore community structure. The investment decisions for each location are partitioned into different blocks

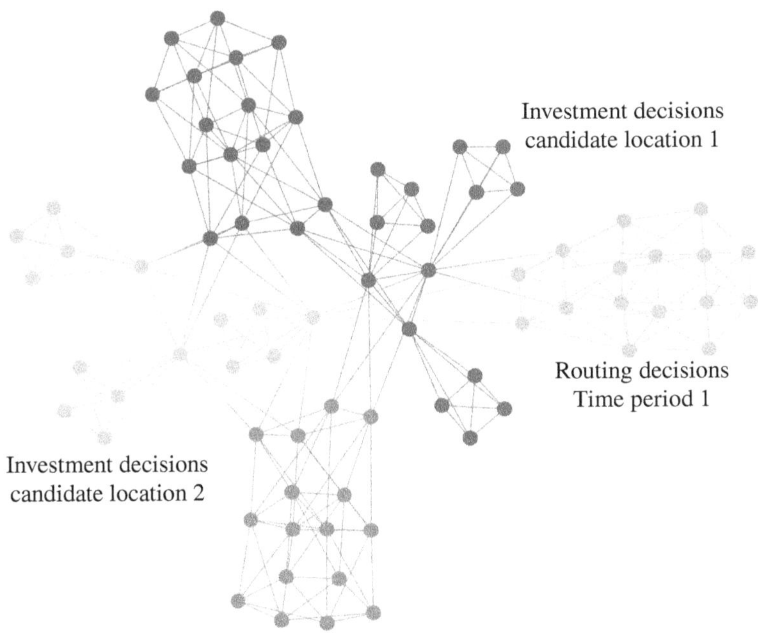

Figure 2.4 Block structure of the deterministic ammonia supply chain transition problem for two candidate locations and three time periods.

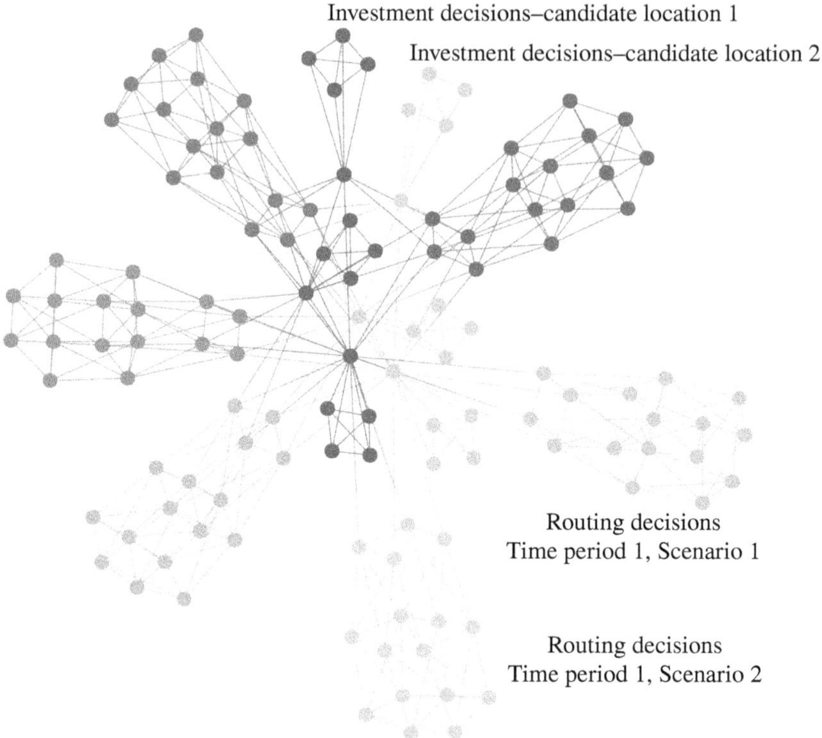

Figure 2.5 Block structure of the two-stage stochastic ammonia supply chain transition problem for two candidate locations and three time periods.

and form the core, whereas the routing decisions for each time period and scenario are partitioned into different blocks.

These results show that SBM and Bayesian inference can provide insights regarding the structure of the problem. The variable graph of this two-stage stochastic capacity expansion problem has a multicore community structure. The investment decisions for each candidate location form the multicore component of the graph since, for each location, the investment decisions are assigned to different blocks. However, these blocks are connected via edges which correspond to the constraints that connect the installed capacity with the available electrolysis capacity. Additionally, each block in the core is connected with the blocks in the periphery which correspond to ammonia routing decisions for each time period and scenario. This coupling arises due to the constraints that guarantee that the amount of ammonia shipped from the renewable facilities is less than or equal to the installed capacity.

Table 2.1 Solution statistics for the monolithic and decomposition-based solution approach.

Solution strategy	Solution time (s)	Objective function	Optimality gap	Number of nodes explored
Monolithic – CPLEX	170	2423	<0.01%	3241
Benders decomposition	15	2423	<0.01%	420

2.4.3 Numerical Results

The partition of the two-stage stochastic problem can be used as the basis for the application of Benders decomposition [23]. Specifically, we assign the investment decisions (nodes that belong to groups that form the multicore) to the master problem and the ammonia routing decisions to the subproblem (the modes that form the blocks in the periphery). The objective function and the values of the parameters of the two-stage problem, as well as the probability of each scenario, are taken from Mitrai et al. [47]. The two-stage stochastic problem is implemented in Pyomo [45]. We compare the computational performance of a monolithic solution approach using CPLEX 22.1.1.0 and Benders decomposition (branch and benders cut) as implemented in CPLEX 22.1.1.0. Both algorithms are executed on an Apple MacBook Pro M1 with 8 physical cores and 16 GB of RAM. We note that the x_{rk} variables are redundant since they are always equal to $\sum_{i\in\mathcal{R}_x} X_i z_{rki}$. Thus, each x_{rk} term is substituted with $\sum_{i\in\mathcal{R}_x} X_i z_{rki}$ in the constraints and the objective function.

The solution time of the different approaches is presented in Table 2.1. From the results, we observe that Benders decomposition leads to a 91% reduction in solution time. Specifically, the solution time of the monolithic approach is 170 seconds, whereas the solution time of Benders Decomposition is 15 seconds. This difference in solution time is also evident in the number of nodes that are explored since the monolithic approach explores 3241 nodes and Benders decomposition approach only 420.

2.5 Conclusions

The operation and design of chemical process systems depend on decisions spanning a wide range of scales. This leads to large optimization problems whose monolithic solution is challenging. In this chapter, we considered the application of SBM and Bayesian inference for analyzing and estimating the structure of an optimization problem. Also, we discussed the application of Benders and

Lagrangean decomposition algorithms based on the underlying structure. Finally, SBM was used to obtain insights and learn the structure of a capacity expansion problem under uncertainty regarding ammonia supply chain networks.

References

1 Cegla, M., Semrau, R., Tamagnini, F., and Engell, S. (2023). Flexible process operation for electrified chemical plants. *Current Opinion in Chemical Engineering* 39: 100898.

2 El Wajeh, M., Mhamdi, A., and Mitsos, A. (2024). Optimal design and flexible operation of a fully electrified biodiesel production process. *Industrial and Engineering Chemistry Research* 63 (3): 1487–1500.

3 Giannikopoulos, I., Skouteris, A., Allen, D.T. et al. (2024). Thermal electrification of chemical processes using renewable energy: economic and decarbonization impacts. *Industrial and Engineering Chemistry Research* 63 (27): 12064–12082.

4 Zhang, Q. and Grossmann, I.E. (2016). Enterprise-wide optimization for industrial demand side management: fundamentals, advances, and perspectives. *Chemical Engineering Research and Design* 116: 114–131.

5 Allen, R.C., Baratsas, S.G., Kakodkar, R. et al. (2023). A multi-period integrated planning and scheduling approach for developing energy systems. *Optimal Control Applications and Methods* 44 (2): 355–372.

6 Lara, C.L., Mallapragada, D.S., Papageorgiou, D.J. et al. (2018). Deterministic electric power infrastructure planning: mixed-integer programming model and nested decomposition algorithm. *European Journal of Operational Research* 271 (3): 1037–1054.

7 Palys, M.J. and Daoutidis, P. (2022). Power-to-x: a review and perspective. *Computers and Chemical Engineering* 165: 107948.

8 Avraamidou, S., Baratsas, S.G., Tian, Y., and Pistikopoulos, E.N. (2020). Circular economy-a challenge and an opportunity for process systems engineering. *Computers and Chemical Engineering* 133: 106629.

9 Garcia, D.J. and You, F. (2015). Supply chain design and optimization: challenges and opportunities. *Computers & Chemical Engineering* 81: 153–170.

10 Luo, Y. and Ierapetritou, M. (2024). Design and operation of modular biorefinery supply chain under uncertainty using generalized benders decomposition. *AIChE Journal* 70 (8): e18458.

11 Wang, S. and Maravelias, C.T. (2024). Optimization methods for plastics management supply chain design. *AICHE Journal* 70: e18464.

12 Biegler, L.T. (2024). Multi-level optimization strategies for large-scale nonlinear process systems. *Computers and Chemical Engineering* 185: 108657.

13 Boukouvala, F., Misener, R., and Floudas, C.A. (2016). Global optimization advances in mixed-integer nonlinear programming, MINLP, and constrained derivative-free optimization, CDFO. *European Journal of Operational Research* 252 (3): 701–727.

14 Tawarmalani, M. and Sahinidis, N.V. (2005). A polyhedral branch-and-cut approach to global optimization. *Mathematical Programming* 103 (2): 225–249.

15 Apap, R.M. and Grossmann, I.E. (2017). Models and computational strategies for multistage stochastic programming under endogenous and exogenous uncertainties. *Computers and Chemical Engineering* 103: 233–274.

16 Grossmann, I.E., Apap, R.M., Calfa, B.A. et al. (2016). Recent advances in mathematical programming techniques for the optimization of process systems under uncertainty. *Computers and Chemical Engineering* 91: 3–14.

17 Gabrel, V., Murat, C., and Thiele, A. (2014). Recent advances in robust optimization: an overview. *European Journal of Operational Research* 235 (3): 471–483.

18 Trespalacios, F. and Grossmann, I.E. (2014). Review of mixed-integer nonlinear and generalized disjunctive programming methods. *Chemie Ingenieur Technik* 86 (7): 991–1012.

19 Conejo, A.J., Castillo, E., Minguez, R., and Garcia-Bertrand, R. (2006). *Decomposition Techniques in Mathematical Programming: Engineering and Science Applications*. Springer Science & Business Media.

20 Flippo, O.E. and Rinnooy Kan, A.H.G. (1993). Decomposition in general mathematical programming. *Mathematical Programming* 60 (1): 361–382.

21 Monique Guignard (2003). Lagrangean relaxation. *Top* 11 (2): 151–200.

22 Boyd, S., Parikh, N., Chu, E. et al. (2011). Distributed optimization and statistical learning via the alternating direction method of multipliers. *Foundations and Trends in Machine Learning* 3 (1): 1–122.

23 Benders, J.F. (1962). Partitioning procedures for solving mixed-variables programming problems. *Numerische Mathematik* 4 (1): 238–252.

24 Geoffrion, A.M. (1972). Generalized benders decomposition. *Journal of Optimization Theory and Applications* 10 (4): 237–260.

25 Desrosiers, J. and Lübbecke, M.E. (2005). A primer in column generation. In: *Column Generation* (ed. J. Desrosiers and M.E. Lübbecke), 1–32. New York: Springer.

26 Ogbe, E. and Li, X. (2017). A new cross decomposition method for stochastic mixed-integer linear programming. *European Journal of Operational Research* 256 (2): 487–499.

27 Van Roy, T.J. (1983). Cross decomposition for mixed integer programming. *Mathematical Programming* 25: 46–63.

28 Allman, A., Tang, W., and Daoutidis, P. (2018). DeCODe: a community-based algorithm for generating high-quality decomposition of optimization problems. *Optimization and Engineering* 20: 1067–1084.

29 Bergner, M., Caprara, A., Ceselli, A. et al. (2015). Automatic Dantzig-Wolfe reformulation of mixed integer programs. *Mathematical Programming* 149: 391–424.

30 del Rio-Chanona, E.A., Fiorelli, F., and Vassiliadis, V.S. (2016). Automated structure detection for distributed process optimization. *Computers and Chemical Engineering* 89: 135–148.

31 Khaniyev, T., Elhedhli, S., and Erenay, F.S. (2018). Structure detection in mixed-integer programs. *INFORMS Journal on Computing* 30 (3): 570–587.

32 Mitrai, I. and Daoutidis, P. (2020). Decomposition of integrated scheduling and dynamic optimization problems using community detection. *Journal of Process Control* 90: 63–74.

33 Tang, W., Allman, A., Pourkargar, D.B., and Daoutidis, P. (2018). Optimal decomposition for distributed optimization in nonlinear model predictive control through community detection. *Computers and Chemical Engineering* 111: 43–54.

34 Aykanat, C., Pinar, A., and Çatalyürek, Ü.V. (2004). Permuting sparse rectangular matrices into block-diagonal form. *SIAM Journal on Scientific Computing* 25 (6): 1860–1879.

35 Ferris, M.C. and Horn, J.D. (1998). Partitioning mathematical programs for parallel solution. *Mathematical Programming* 80 (1): 35–61.

36 Jalving, J., Shin, S., and Zavala, V.M. (2022). A graph-based modeling abstraction for optimization: concepts and implementation in plasmo. jl. *Mathematical Programming Computation* 14 (4): 699–747.

37 Michelena, N.F. and Papalambros, P.Y. (1997). A hypergraph framework for optimal model-based decomposition of design problems. *Computational Optimization and Applications* 8: 173–196.

38 Wang, J. and Ralphs, T. (2013). Computational experience with hypergraph-based methods for automatic decomposition in discrete optimization. In: *Integration of AI and OR Techniques in Constraint Programming for Combinatorial Optimization Problems: 10th International Conference, CPAIOR 2013, Yorktown Heights, NY, USA, May 18–22, 2013. Proceedings 10*, 394–402. Springer.

39 Mitrai, I. and Daoutidis, P. (2021). Efficient solution of enterprise-wide optimization problems using nested stochastic blockmodeling. *Industrial and Engineering Chemistry Research* 60 (40): 14476–14494.

40 Mitrai, I., Tang, W., and Daoutidis, P. (2022). Stochastic blockmodeling for learning the structure of optimization problems. *AIChE Journal* 68 (6): e17415.

41 Peixoto, T.P. (2019). Bayesian stochastic blockmodeling. In: *Advances in Network Clustering and Blockmodeling* (ed. P. Doreian, V. Batagelj, and A. Ferligoj), 289–332. John Wiley & Sons.

42 Peixoto, T.P. (2014a). Efficient Monte Carlo and greedy heuristic for the inference of stochastic block models. *Physical Review E* 89 (1): 012804.

43 Peixoto, T. P. 2014b). The graph-tool python library. *figshare*. http://figshare.com/articles/graph_tool/1164194 (accessed 1 September 2024).

44 Sahinidis, N.V. and Grossmann, I.E. (1991). Convergence properties of generalized Benders decomposition. *Computers and Chemical Engineering* 15 (7): 481–491.

45 Hart, W.E., Laird, C.D., Watson, J.-P. et al. (2017). *Pyomo–optimization modeling in Python*, 2e. Cham: Springer.

46 Hagberg, A. A., Schult, D. A., and Swart, P. J. (2008). Exploring network structure, dynamics, and function using NetworkX. *Proceedings of the 7th Python in Science Conference*, 11–15.

47 Mitrai, I., Palys, M.J., and Daoutidis, P. (2024). A two-stage stochastic programming approach for the design of renewable ammonia supply chain networks. *Processes* 12 (2): 325.

3

Multi-Objective Bayesian Optimization for Networked Black-Box Systems: A Path to Greener Profits and Smarter Designs

Akshay Kudva[1], *Wei-Ting Tang*[1,2], *and Joel A. Paulson*[1,2]

[1] *Department of Chemical and Biomolecular Engineering, The Ohio State University, Columbus, OH, USA*
[2] *Department of Chemical and Biological Engineering, University of Wisconsin-Madison, Madison, WI, USA*

3.1 Introduction

Over the last two centuries, industrial innovation has prioritized efficiency, scale, and profitability, often overlooking the long-term environmental consequences of mass production and resource extraction [1, 2]. Early advances in petrochemical refining and chemical manufacturing focused on maximizing throughput and profit margins, with little consideration for pollutant emissions and hazardous waste. As environmental concerns became more pressing, government policies – such as the Air Pollution Control Act (1955) and the establishment of the Environmental Protection Agency (1970) [3] – led to stricter regulations, prompting process systems engineering (PSE) practitioners to integrate environmental and safety constraints into their design objectives. By the 1970s, nonlinear programming (NLP) methods were being used to optimize capital cost objectives under uncertainty [4, 5], followed by advances in feasibility and flexibility analysis [6–9] and later integrated design and control (IDC) frameworks [10–12]. These developments laid the foundation for more holistic optimization approaches that simultaneously consider economic, operational, and regulatory factors.

In parallel, the increasing availability of computational resources enabled the rise of high-fidelity simulation tools, commonly referred to as *digital twins* [13, 14], which model complex physical and chemical systems with greater accuracy. However, this increase in model fidelity has come at the cost of *computational expense*, making purely equation-oriented (i.e. "white-box") approaches less tractable for large-scale, multi-physics systems. At the same time, the rise of data-driven machine learning and Bayesian optimization (BO) [15, 16] has provided an alternative by enabling the efficient optimization of expensive, black-box processes. BO has proven particularly valuable for optimizing systems where the internal structure is proprietary, analytically intractable, or hidden behind computationally expensive simulations. This has led to successful applications in process design, reaction optimization, and flexibility analysis [17–23].

Despite BO's successes, treating an entire process as a black-box often discards valuable structural information that could significantly improve optimization efficiency, especially as the number of decision variables increases. Traditional equation-oriented methods exploit model structure but require access to explicit equations – a luxury not always available due to proprietary software or the complexity of modern simulators. To bridge this gap, "grey-box" (Bayesian) optimization methods have been developed within the PSE community [24–30],

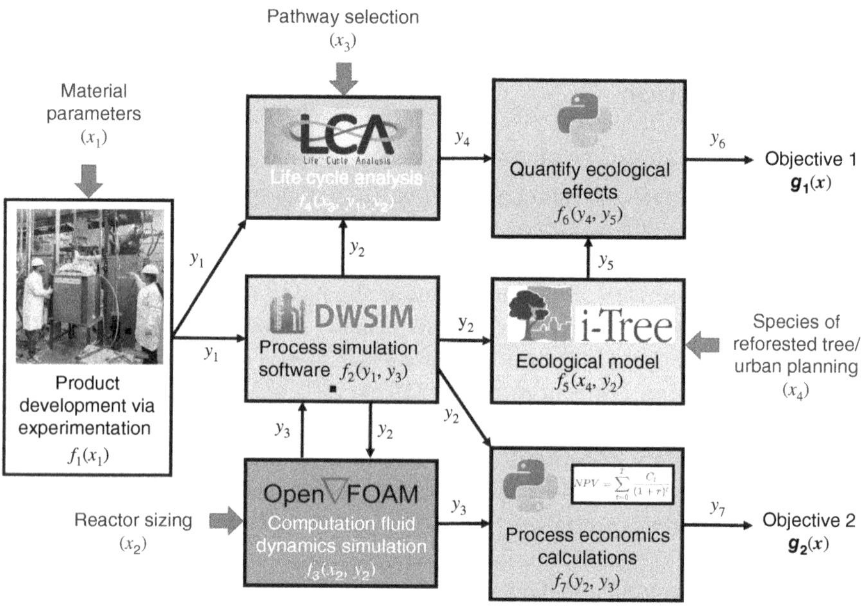

Figure 3.1 Illustration of a complex network system integrating process simulation. *Source:* Medeiros [34] / DWSIM, CFD. *Source:* OpenFOAM Foundation [31], life cycle analysis (LCA), ecological modeling. *Source:* i Tree [33] / i-Tree, and economic evaluation. The design variables, shown with arrows into the various subsystems, impact reactor sizing, material properties, and urban planning, while the models interact dynamically to assess environmental and economic trade-offs. The figure serves as a motivating example, highlighting the challenges inherent in optimizing such systems. We aim to develop algorithms capable of tackling a broad class of these problems, which often involve high-fidelity partial differential equation solvers, embedded optimization modules, and dynamic feedback loops. *Source:* The Ohio State University / https://cbe.osu.edu/news/2025/01/chemical-looping-turns-environmental-waste-fuel / last accessed on September 16, 2025.

which integrate known submodels with data-driven surrogates. While these methods improve efficiency over purely black-box BO, they usually assume a sequential or hierarchical mapping from inputs to outputs, which may not be representative of many real-world systems. Many engineering systems, particularly in PSE, exhibit networked dependencies, where subsystems interact in a complex, often bidirectional manner. Figure 3.1 illustrates such a scenario, where a process system is coupled with environmental and economic subsystems. Here, high-fidelity tools such as computational fluid dynamics (CFD) models [31], life cycle analysis (LCA) [32], ecological models [33], and process simulators [34] interact dynamically, often in a nonhierarchical fashion. A multi-scale simulation – such as one linking a flowsheet model to a detailed heat exchanger CFD simulation – may require back-and-forth data exchange, creating cycles that challenge many existing optimization methods. A naive black-box BO approach would ignore these interconnections, potentially requiring an unmanageable number of function evaluations. Conversely, a strictly equation-oriented approach might be infeasible if key subsystems are black-boxes due to proprietary software restrictions or excessive modeling complexity. This necessitates a more flexible yet structure-aware optimization framework.

Recent work in so-called "composite" BO has sought to extend BO to systems where one function's output feeds into another [26, 29, 35]. However, these methods assume that the function can be expressed as a composite model, $g(\boldsymbol{h}(\boldsymbol{x}))$, where g is known, $\boldsymbol{h}$ is a black-box function, and $\boldsymbol{x}$ are the design variables. A more general representation of the function would be as a network (or graph), which has been explored in recent BO frameworks such as BOFN [36, 37] and BOIS [38, 39]. Graph-based models are known to be powerful representations are seen by their wide applicability in many different fields such as social sciences, biology, chemistry, and engineering [40]. However, these current graph-based BO methods assume that the function network is a directed acyclic graph (DAG) to simplify the uncertainty propagation task. This assumption limits their applicability to process systems with feedback loops, such as recycle streams and interconnected multi-scale simulations.

In this chapter, we extend MOBO to handle complex networked systems, where each node in the network represents a potentially expensive black-box function (e.g. a simulator or data-driven model). We introduce Multi-Objective Bayesian Optimization for Network Systems (MOBONS), which builds upon existing methods while addressing key limitations:

- MOBONS handles multi-objective optimization problems, seeking Pareto-optimal solutions that balance competing objectives such as economic performance and sustainability.
- It explicitly models networked dependencies, including feedback loops and recycle streams, enabling applications in complex industrial systems.

- It naturally accommodates constraints (e.g. safety or regulatory requirements) by treating them as additional outputs of the function network.
- It is readily applicable to batch (parallel) evaluations such that it can accelerate the optimization process by designing multiple candidate evaluation points at each iteration.

To ensure computational efficiency of the algorithm, MOBONS extends the Thompson sampling acquisition function [41], which is significantly cheaper than the improvement-based approach in BOFN [36] and the confidence bound approach in BOIS [38, 39]. Furthermore, we leverage multi-objective genetic algorithms to efficiently solve the subproblems within MOBONS, combining the sample efficiency of BO with the exploration power of genetic search. Note that we have also recently pursued related work in the context of function networks under adversarial uncertainty (i.e., robust optimization) [42], though we do not directly consider that case in this work.

The remainder of this chapter is organized as follows. Section 3.2 formalizes the problem statement. Section 3.3 presents the MOBONS framework, including surrogate modeling and acquisition strategies. Section 3.4 demonstrates the method on case studies. Finally, Section 3.5 provides concluding remarks.

3.2 Problem Formulation

In this chapter, we consider the problem of multi-objective optimization in the presence of expensive function evaluations. Formally, we seek to solve the following optimization problem:

$$\min_{\boldsymbol{x}\in\mathcal{X}} \boldsymbol{G}(\boldsymbol{x}) = [g_1(\boldsymbol{x}), g_2(\boldsymbol{x}), \dots, g_M(\boldsymbol{x})], \tag{3.1}$$

where $\mathcal{X} \subset \mathbb{R}^D$ is a compact design space, $\boldsymbol{x}$ is the decision vector representing all the degrees of freedom in the system, and $\boldsymbol{G} : \mathcal{X} \to \mathbb{R}^M$ is a vector-valued function consisting of M scalar objectives $g_i : \mathcal{X} \to \mathbb{R}$, each of which quantifies a cost or performance metric to be minimized. While the objective functions in (3.1) can represent arbitrary criteria, we are motivated by cases where at least one objective relates to economic considerations (e.g. net present value, return on investment), while others capture environmental sustainability metrics (e.g. pollutant emissions, energy consumption, waste generation).

Multi-objective optimization problems of this form typically do not admit a single globally optimal solution, as improving one objective may degrade another. Instead, the goal is to construct the set of *Pareto-optimal solutions*, forming the Pareto frontier, where no objective can be improved without compromising at least one other. The objective of this work is to develop an algorithm that enables efficient construction of the Pareto frontier by leveraging the underlying structure of $\boldsymbol{G}$, which is often represented as a network of interconnected functions.

The choice of optimization algorithm for solving (3.1) depends on the properties of $\boldsymbol{G}$. The ϵ-constraint method [43] is effective when each single-objective subproblem can be efficiently minimized while imposing constraints on the remaining objectives; sweeping over the constraint parameters then traces out the Pareto frontier. However, this approach requires access to a fully specified model that can be solved using an optimization modeling framework such as Pyomo [44]. When the objective functions are treated as black-box evaluations, evolutionary algorithms like NSGA-II [45] provide an alternative that does not require explicit function derivatives or structural knowledge. While broadly applicable, these methods require a large number of function evaluations and are thus impractical for expensive-to-evaluate functions, such as those arising from high-fidelity process simulations. A more sample-efficient alternative is multi-objective Bayesian optimization (MOBO), which sequentially selects evaluation points based on a surrogate model trained on previously observed data. Popular MOBO strategies include expected hypervolume improvement (EHVI) [46] and Thompson sampling-based methods, such as TSEMO [47] and POTS [48]. While these approaches significantly reduce the number of function evaluations, they still treat $\boldsymbol{G}$ as a fully black-box function, which can limit optimization performance in engineering applications. We argue that this representation is often overly flexible and overlooks valuable structural information that is readily available in many real-world problems.

Building on recent work on function networks [36], we consider a structured representation of $\boldsymbol{G}$ in which the objectives are evaluated through a directed network of interdependent functions $f_1, \dots, f_K$. Each function f_k in the network receives inputs from a subset of external design variables $\boldsymbol{x}$ as well as the outputs of other functions in the network. The evaluation of these functions is assumed to be expensive, meaning that computational or experimental resources are required for function evaluations. To formalize this structure, let $V = \{1, 2, \dots, K\}$ represent the set of nodes in the network, where each node corresponds to a function f_k. The directed edges of the network, denoted by $E = \{(j, k) : f_k \text{ receives input from } f_j\}$, specify the dependencies between functions. Each function f_k can be expressed as an implicit equation

$$y_k = f_k(\boldsymbol{x}_{I(k)}, \{y_j\}_{j \in J(k)}), \quad k = 1, \dots, K, \tag{3.2}$$

where $\boldsymbol{x}_{I(k)}$ represents the subset of decision variables that directly influence f_k, with $I(k) \subseteq \{1, \dots, D\}$ and $J(k) \subseteq \{1, \dots, K\}$ is the set of other functions whose outputs serve as inputs to f_k. Let $\boldsymbol{Y} = (y_1, y_2, \dots, y_K)$ collects all node outputs in the network. For compactness, we define the vector-valued function $\boldsymbol{F}(\boldsymbol{x}, \boldsymbol{Y})$ as the concatenation of the right-hand sides of (3.2), allowing us to rewrite (3.2) as

$$\boldsymbol{Y} = \boldsymbol{F}(\boldsymbol{x}, \boldsymbol{Y}). \tag{3.3}$$

This formulation is more general than prior work that assumes an acyclic network structure. Here, we explicitly allow cycles, enabling the representation of feedback

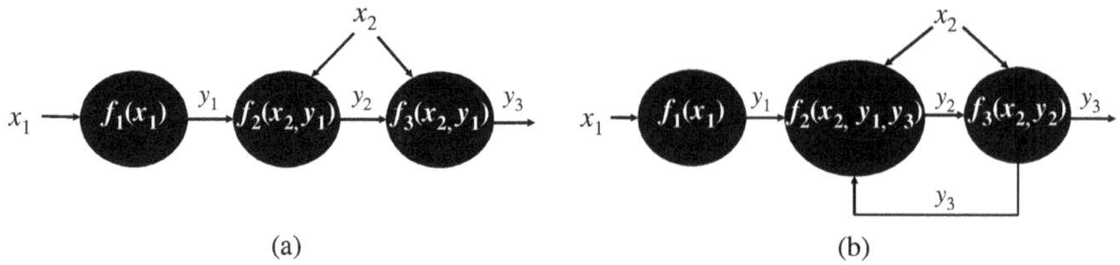

Figure 3.2 Illustrative examples of acyclic (a) and cyclic (b) function networks. The acyclic graph must be capable of being topologically ordered so that future nodes only depend on previous nodes, which simplifies the evaluation. The cyclic graph, on the other hand, may have dependencies that come in the form of equality constraints, which significantly complicates determination of statistical properties of the network when the node functions are modeled as Gaussian processes.

loops and recycle streams, which are common in process systems applications. Figure 3.2 provides illustrative examples of cyclic and acyclic function networks to highlight their differences.

The objective function $\boldsymbol{G}(\boldsymbol{x})$ is then assumed to be a known transformation of the network outputs. Without loss of generality, we assume a linear transformation of the form

$$\boldsymbol{G}(\boldsymbol{x}) = \boldsymbol{C}\boldsymbol{Y}^{\star}(\boldsymbol{x}), \tag{3.4}$$

where $\boldsymbol{C} \in \mathbb{R}^{M \times K}$ is a projection matrix and $\boldsymbol{Y}^{\star}(\boldsymbol{x})$ represents the solution to the implicit system (3.3) for a given $\boldsymbol{x}$. We do not strictly assume $\boldsymbol{Y}^{\star}(\boldsymbol{x})$ is unique for every $\boldsymbol{x}$, though this is expected to improve numerical stability in practice. An interesting direction for future work is to study the impact of properties of the solution map on performance. A key distinction between our approach and conventional MOBO is how the surrogate model is defined. Instead of placing a Gaussian process prior directly on the objective functions $g_1, \dots, g_M$, we propose placing one on the network functions $f_1, \dots, f_K$. While this may appear to be a minor change, it allows us to explicitly exploit the dependency structure in (3.2), which is ignored in standard MOBO methods. By leveraging this network representation, we aim to improve the efficiency of multi-objective optimization in expensive evaluation settings.

In Section 3.3, we describe our proposed approach for incorporating this structured information into a MOBO-like framework.

3.3 Multi-Objective Bayesian Optimization Over Network Systems

In this section, we introduce our approach for incorporating network structure into the multi-objective optimization of expensive functions. We first describe the statistical model used to represent the black-box node functions $f_1, \dots, f_K$, leading to a fundamentally different modeling approach than standard MOBO. We then present the proposed MOBONS algorithm, detailing its key components and workflow. Finally, we discuss extensions that enable parallel evaluations and the handling of inequality constraints.

3.3.1 Statistical Surrogate Model

We assume that each function f_k is a realization from an independent Gaussian process (GP) prior. To simplify notation, we define $\boldsymbol{z}_k = (\boldsymbol{x}_{I(k)}, \boldsymbol{Y}_{J(k)})$ as the vector of inputs to node k, where $\boldsymbol{x} \in \mathbb{R}^D$ represents the external decision variables and $\boldsymbol{Y} \in \mathbb{R}^K$ collects the outputs of all nodes. Each function f_k is then modeled as

$$f_k(\boldsymbol{z}_k) \sim \mathcal{GP}(\mu_{0,k}(\boldsymbol{z}_k), \Sigma_{0,k}(\boldsymbol{z}_k, \boldsymbol{z}_k')), \tag{3.5}$$

where $\mu_{0,k}(\boldsymbol{z}_k) : \mathbb{R}^{n_{z,k}} \to \mathbb{R}$ is the prior mean function and $\Sigma_{0,k}(\boldsymbol{z}_k, \boldsymbol{z}_k') : \mathbb{R}^{n_{z,k}} \times \mathbb{R}^{n_{z,k}} \to \mathbb{R}_+$ is the prior covariance (or kernel) function (note $n_{z,k}$ denotes the dimension of $\boldsymbol{z}_k$ and $\mathbb{R}_+$ denotes the set of positive real numbers). The mean function encodes prior knowledge about the expected behavior of f_k, while the kernel function defines correlations between function values at different inputs, capturing smoothness and variability. A detailed review of kernel choices can be found in [49, Chapter 4]; in this work, we assume an appropriate kernel is selected, potentially learned from data.

We assume that, when the objective functions $\boldsymbol{G}$ are evaluated at a design point $\boldsymbol{x}$, the intermediate node outputs $\boldsymbol{Y}$ are also observed. This assumption holds in many practical settings, such as high-fidelity simulations where intermediate process states can be accessed. Under this assumption, we can infer the value of $f_k(\boldsymbol{z}_k^\star)$ at an arbitrary test input $\boldsymbol{z}_k^\star$ using a GP posterior conditioned on n previously observed function evaluations. Let $\mathcal{D}_{n,k} = \{(\boldsymbol{z}_{i,k}, y_{i,k})\}_{i=1}^n$ denote the dataset of past evaluations for node k. The posterior mean and covariance functions for f_k are given by

$$\mu_{n,k}(\boldsymbol{z}_k^\star) = \mu_{0,k}(\boldsymbol{z}_k^\star) + \Sigma_{0,k}(\boldsymbol{z}_k^\star, \boldsymbol{z}_{1:n,k})\Sigma_{0,k}(\boldsymbol{z}_{1:n,k}, \boldsymbol{z}_{1:n,k})^{-1} \times (y_{1:n,k} - \mu_{0,k}(\boldsymbol{z}_{1:n,k})), \tag{3.6a}$$

$$\Sigma_{n,k}(\boldsymbol{z}_k^\star, \boldsymbol{z}_k^{\star\prime}) = \Sigma_{0,k}(\boldsymbol{z}_k^\star, \boldsymbol{z}_k^{\star\prime}) - \Sigma_{0,k}(\boldsymbol{z}_k^\star, \boldsymbol{z}_{1:n,k})\Sigma_{0,k}(\boldsymbol{z}_{1:n,k}, \boldsymbol{z}_{1:n,k})^{-1} \times \Sigma_{0,k}(\boldsymbol{z}_{1:n,k}, \boldsymbol{z}_k^{\star\prime}). \tag{3.6b}$$

The posterior distributions on $f_1, \dots, f_K$ induce a corresponding posterior on $\boldsymbol{G}$, but this distribution is generally non-Gaussian and difficult to characterize. When the function network is *acyclic*, samples from $\boldsymbol{G}$ can be obtained via a forward propagation procedure, as described in [36, Algorithm 3.1]. However, when the graph contains *cycles*, the outputs $\boldsymbol{Y}$ are defined implicitly by the stochastic fixed-point Eq. (3.3). In such cases, drawing samples requires numerical techniques such as stochastic fixed-point iteration, Markov Chain Monte Carlo

(MCMC), or variational inference, which complicates integration into a sequential optimization framework. We address this challenge by leveraging an efficient heuristic from the multiarmed bandit literature: Thompson sampling (TS) [41].

Remark 3.1 *While we assume that all functions in the network are expensive-to-evaluate black-box functions modeled as Gaussian processes (GPs), in practice, some node functions may be inexpensive and explicitly known (i.e. white-box functions). This scenario can be easily accommodated by directly specifying the known function values in the prior. Specifically, for any white-box function* f_k, *we can set the prior mean function as* $\mu_{0,k}(\boldsymbol{z}_k) = f_k(\boldsymbol{z}_k)$ *and the covariance function as* $\Sigma_k(\boldsymbol{z}_k, \boldsymbol{z}_k') = 0$, *ensuring that the surrogate model exactly recovers the known function values while maintaining uncertainty quantification for the remaining black-box functions.*

3.3.2 Multi-Objective Thompson Sampling for Function Networks

The TS method selects an action (in this case, the next design vector $\boldsymbol{x}_{t+1}$) based on its probability of being optimal. At iteration t, the next query point is sampled from the posterior distribution

$$p_{\mathcal{X}^\star}(\boldsymbol{x}|\mathcal{D}_t) = \int \delta\left(\boldsymbol{x} - \operatorname*{argmin}_{x \in \mathcal{X}} g(\boldsymbol{x})\right) p(g|\mathcal{D}_t) dg, \tag{3.7}$$

where g is the single objective function, $\mathcal{D}_t$ is the history of query-observation pairs up to step t, and $\delta(\cdot)$ is the Dirac delta function, meaning all probability mass is placed on the minimizer of a sampled function realization. Thus, $\boldsymbol{x}_{t+1} = \operatorname{argmin}_{x \in \mathcal{X}} \hat{g}_t(\boldsymbol{x})$ where $\hat{g}_t(\boldsymbol{x}) \sim p(g|\mathcal{D}_t)$ is a function realization from the GP posterior. This idea has been extended to MOBO by sampling from the probability that a design point is Pareto optimal [48]. However, existing methods do not account for the network structure in (3.2).

Our approach, which we refer to as MOBONS, does account for this structure, as shown in Algorithm 3.1. The algorithm proceeds as follows. We start by collecting initial data from the network that is used to train the GP models (Lines 1 and 2). We then enter a loop (Line 3) to sequentially generate the remaining set of samples until our total budget N is exhausted. At any given iteration t, we solve a multi-objective optimization problem inspired by the TS approach (Lines 4–7). In particular, we can interpret (3.10) as generating points according to the probability that they are Pareto optimal, that is

$$\begin{aligned} &p_{X^\star, Y^\star}(\boldsymbol{x}, \boldsymbol{Y}|\{\mathcal{D}_{t,k}\}_{k=1}^K) \\ &= \int \mathbb{I}\left((\boldsymbol{x}, \boldsymbol{Y}) \in \operatorname*{argmin}_{x \in \mathcal{X}, Y} \{\boldsymbol{CY} : \boldsymbol{Y} = \boldsymbol{F}(\boldsymbol{x}, \boldsymbol{Y})\}\right) \\ &\quad \times p(f_1|\mathcal{D}_{t,1}) \cdots p(f_K|\mathcal{D}_{t,K}) df_1 \cdots df_K, \end{aligned} \tag{3.8}$$

where $\mathbb{I}(\boldsymbol{q} \in Q)$ is the indicator function for set Q. Note that $(X^{\star}_{t+1}, Y^{\star}_{t+1})$ from (3.10) are the Pareto sets for the TS version of the network function and thus themselves could contain an infinite set of points in general. As such, we need some selection criteria to pick a single point from these sets to evaluate (Line 8; we discuss the generalization to multi-point selection in Section 3.3.4). We leave the selection function $\Phi(\cdot)$ open since, in principle, any choice from $(X^{\star}_{t+1}, Y^{\star}_{t+1})$ is Pareto optimal for the drawn function samples. A simple strategy is to randomly select a point from this set; however, to promote stronger diversity among the candidate points, it is common to select points that are the furthest away from our past samples, for example [50]. Due to the network structure, we can select points globally (those that maximize distance to the global set of design variables $\boldsymbol{x}$) or locally (those that maximize distance to the input to each local node k). For simplicity, we focus on the global distance in this chapter to serve as a proof of concept – future work should more systematically explore the influence of different choices of selection strategies. To define the global selection approach $\Phi(\cdot)$, let us first define the selected design point from $X^{\star}_{t+1}$ to be the one that has the largest *maximin* distance from all previous selected candidates X_t:

$$\boldsymbol{x}_{t+1} = \underset{\boldsymbol{x} \in X^{\star}_{t+1}}{\operatorname{argmax}} \min_{\boldsymbol{x}_i \in X_t} \|\boldsymbol{x} - \boldsymbol{x}_i\|. \tag{3.9}$$

For this choice of $\boldsymbol{x}_{t+1}$, we can define the corresponding output values $\boldsymbol{Y}_{t+1}$ as those satisfying the TS network equations for this choice of design variables, that is $\boldsymbol{Y}_{t+1} = \widehat{\boldsymbol{F}}_t(\boldsymbol{x}_{t+1}, \boldsymbol{Y}_{t+1})$. The specific input values for every node $k = 1, \ldots, K$ are then defined by $\boldsymbol{z}_{t+1,k} = (\boldsymbol{x}_{t+1,I(k)}, \boldsymbol{Y}_{t+1,J(k)})$ (Line 8). Lastly, the final steps (Lines 9–13) require us to query the nodes of the network function and update the GP posterior with this newly collected data.

3.3.3 Practical Considerations in MOBONS

The conceptual MOBONS approach (Algorithm 3.1) involves several internal steps that require further elaboration for practical implementation.

3.3.3.1 GP Kernel Selection and Tuning

The choice of covariance functions (kernels) in the GP priors is critical in MOBONS, as it governs the properties of the surrogate models used to fit the unknown functions. For example, the radial basis function (RBF) kernel, given by

$$\Sigma_{\text{RBF}}(\boldsymbol{z}, \boldsymbol{z}') = \zeta^2 \exp\left(-\frac{\|\boldsymbol{z} - \boldsymbol{z}'\|^2}{2\ell^2}\right), \tag{3.11}$$

encodes the prior assumption that function values closer in the input space should be more correlated. The lengthscale hyperparameter ℓ controls the degree of smoothness, where smaller values allow more rapid variations in

Algorithm 3.1 MOBONS: Multi-Objective Bayesian Optimization over Network Systems

Required: Design space $\mathcal{X} \subset \mathbb{R}^D$;
Network output space $\mathcal{Y} \subset \mathbb{R}^K$;
GP priors $f_k \sim \mathcal{GP}(\mu_{0,k}, \Sigma_{0,k})$ for all $k = 1, \dots, K$;
Graph representation (V, E) of (3.2);
Matrix $\boldsymbol{C} \in \mathbb{R}^{M \times K}$ that projects network outputs to objectives;
Function $\Phi(\cdot)$ to select candidate points from Pareto set;
Initialization budget n;
Total evaluation budget N.

Initialize data: For every $k = 1, \dots, K$, generate initial set of n sampling points $\boldsymbol{z}_{1:n,k}$ using any space filling design (such as Latin hypercube sampling) inside of the compact set $\mathcal{Z}_k = \mathcal{X}_{I(k)} \times \mathcal{Y}_{J(k)}$. Evaluate the function f_k at these points via (3.2) to construct initial dataset $\mathcal{D}_{n,k} = \{(\boldsymbol{z}_{i,k}, y_{i,k})\}_{i=1}^{n}$.

Train GPs: Train GP surrogate for f_k using $\mathcal{D}_{n,k}$ for all $k = 1, \dots, K$. Let $\mathcal{GP}_{n,k} \leftarrow \mathcal{GP}(\mu_{n,k}, \Sigma_{n,k})$ denote the GP posterior for f_k that can be evaluated via (3.6).

for $t = n, n+1, \dots, N-1$ **do**
  for $k = 1, \dots, K$ **do**
    Generate a Thompson sample $\hat{f}_{t,k} \sim \mathcal{GP}_{t,k}$.
  end for
  Solve the following multi-objective optimization problem:

$$(X^\star_{t+1}, Y^\star_{t+1}) = \underset{(\boldsymbol{x}, \boldsymbol{Y}) \in \mathcal{X} \times \mathbb{R}^K}{\operatorname{argmin}} \; \boldsymbol{C}\boldsymbol{Y} \text{ subject to } \boldsymbol{Y} = \hat{\boldsymbol{F}}_t(\boldsymbol{x}, \boldsymbol{Y}), \tag{3.10}$$

  where $\hat{\boldsymbol{F}}_t$ denotes the compact representation of the network (see 3.3) constructed with the Thompson sample versions of $\hat{f}_{t,k}$ instead of the true (unknown) f_k and $(X^\star_t, Y^\star_t)$ denotes the set of Pareto optimal design and network output values.
  Apply selection rule $\Phi(\cdot)$ to Pareto set $\{\boldsymbol{z}_{t+1,k}\}_{k=1}^{K} = \Phi(X^\star_{t+1}, Y^\star_{t+1})$ to get promising candidates.
  for $k = 1, \dots, K$ **do**
    Evaluate f_k at $\boldsymbol{z}_{t+1,k}$ and update dataset $\mathcal{D}_{t+1,k} \leftarrow \mathcal{D}_{t,k} \cup \{(\boldsymbol{z}_{t+1,k}, y_{t+1,k})\}$.
    Update GP posterior using new data $\mathcal{GP}_{t+1,k} \leftarrow \mathcal{GP}(\mu_{t+1,k}, \Sigma_{t+1,k})$. See (3.6).
  end for
end for

the function. The scale parameter ζ governs the overall magnitude of function variations. More generally, different covariance functions introduce different inductive biases, influencing how the GP extrapolates from observed data. All covariance functions depend on hyperparameters, collectively denoted as $\boldsymbol{\gamma}$, which are typically learned from data by maximizing the marginal log-likelihood (MLL) [49, 51]. The performance of MOBONS relies on well-chosen covariance structures and well-tuned hyperparameters for each function $f_1, \dots, f_K$. For low- to moderate-dimensional inputs (typically 5-10 dimensions), the RBF or Matern class of covariance functions is often sufficient. For significantly higher-dimensional problems, more expressive covariance functions such as deep kernel learning [51] can be beneficial, as they allow the model to capture complex, nonstationary relationships.

3.3.3.2 Thompson Sampling

To formulate the acquisition function (3.10), we require posterior function samples from the GP models. The complexity of this step depends on the cost of evaluating $\widehat{\boldsymbol{F}}_t$. To generate consistent (differentiable) posterior samples, we recommend the efficient procedure in [52], which decomposes a GP posterior sample into two steps: (i) Drawing a realization from the weight-space version of the prior and (ii) Performing a function-space update based on observed data. This requires knowledge of the weight-space interpretation of the GP, where the covariance function is viewed as an inner product in a feature space. For stationary kernels, this feature space can be approximated using random Fourier features (RFF) [53], enabling scalable sampling. A detailed discussion of RFF-based posterior sampling is provided in [47, Section 3].

3.3.3.3 Approximating the Pareto Optimal Set

At each iteration of MOBONS, we solve a multi-objective optimization problem (3.10) based on GP posterior samples (using the TS paradigm). Although the problem is defined in terms of computationally cheap function samples (unlike the original expensive black-box objectives), it remains a nonconvex optimization problem. Consequently, heuristic optimization methods are required for practical implementation. We adopt NSGA-II [45], a widely used evolutionary algorithm that is effective for functions that can be evaluated many times. However, most evolutionary solvers do not natively handle equality or inequality constraints. The most straightforward way to address this is through penalty functions, which adjust the objective values based on constraint violations. A comparison of penalty-based constraint handling techniques is provided in [54].

3.3.3.4 Selection Function

After computing the (approximate) Pareto optimal set $(X^{\star}_{t+1}, Y^{\star}_{t+1})$ in (3.10), we must select specific design points for function evaluations. We use the global selection strategy defined in (3.9), though other strategies are also valid. While (3.9) appears computationally challenging, it is efficient in practice due to the finite discrete representation of the Pareto set obtained from NSGA-II (or another solver). Specifically, the selected solution consists of a discrete point set of cardinality $N_{\star} = |X^{\star}_{t+1}|$, making the maximin distance calculation straightforward. The required computation involves evaluating $N_{\star} \times T$ pairwise distances, where $T = |X_t|$ is the number of previously selected points. The final selection involves: (i) Computing all pairwise distances; (ii) Determining the column-wise minimum distance to previously evaluated points; and (iii) Selecting the candidate that maximizes this minimum distance vector. This approach ensures good exploration of the design space while maintaining diversity among selected candidates.

3.3.4 Handling Parallel Evaluations and Constrained Problems

In many real-world optimization settings, it is desirable to evaluate multiple candidate solutions in parallel. This situation commonly arises in simulation-based optimization, where multiple computing nodes can evaluate different designs simultaneously, and in experimental settings, where multiple physical tests can be conducted concurrently. The MOBONS framework naturally accommodates parallel evaluations due to the structure of its selection process. Since (3.10) already produces a set of Pareto-optimal solutions for a given Thompson sample, multiple points can be selected from this set using a greedy maximin strategy based on (3.9). Specifically, after selecting the first candidate $\boldsymbol{x}_{t+1}$, the next candidate $\boldsymbol{x}_{t+2}$ is chosen as the point that maximizes the worst-case distance to the previously selected points and the first candidate $X_t \cup \{\boldsymbol{x}_{t+1}\}$. This process is repeated until a total of q additional points have been selected. By ensuring that the chosen evaluations are well-distributed, this strategy improves exploration of the objective space while maintaining diversity in the selected points.

If the number of available Pareto-optimal solutions is insufficient to fill all q parallel evaluation slots, additional candidates can be generated by drawing another TS from the function network and re-solving (3.10). This approach has been shown to be effective in single-objective Bayesian optimization [41] and extends naturally to the multi-objective case considered in MOBONS. By leveraging this strategy, the algorithm can efficiently allocate computational resources while maintaining the benefits of its sequential decision-making framework.

Another important consideration in engineering optimization problems is the presence of constraints. Many real-world problems involve inequality constraints that define feasible regions of the design space, such as performance requirements, safety thresholds, or regulatory limits. The function network representation in (3.2) used by MOBONS can be readily extended to accommodate constraints by expanding the definition of the network outputs. Specifically, a separate mapping can be introduced from the outputs $\boldsymbol{Y}$ to a constraint function $\boldsymbol{H}(\boldsymbol{x})$, where the constraints are enforced as $\boldsymbol{H}(\boldsymbol{x}) \leq \boldsymbol{0}$. Similar to how the objective functions are obtained through a projection matrix, we define a constraint projection matrix $\boldsymbol{C}_{\text{cons}}$ that maps the outputs to the constraints via the relation $\boldsymbol{H}(\boldsymbol{x}) = \boldsymbol{C}_{\text{cons}}\boldsymbol{Y}^{\star}(\boldsymbol{x})$ where again $\boldsymbol{Y}^{\star}(\boldsymbol{x})$ is a solution to (3.3) for fixed $\boldsymbol{x}$. By appropriately structuring the function network, constraints can always be represented in this form.

With this formulation, constraints can be incorporated directly into the acquisition function optimization in (3.10), ensuring that candidate solutions satisfy feasibility requirements. In practice, constraints are typically handled using penalty functions, as discussed previously, allowing the optimization process to systematically discourage infeasible solutions. Since constraints are treated in a manner analogous to the objectives, MOBONS can efficiently handle constrained problems without requiring fundamental modifications to the underlying framework.

3.4 Case Studies

To provide a proof-of-concept evaluation of the performance of MOBONS, we apply it to two case studies: a synthetic test problem based on the well-known ZDT4 benchmark function and a sustainable bioethanol production process. The ZDT4 function represents a challenging multi-objective optimization problem with multiple local minima, making it a useful benchmark for assessing the efficiency of MOBONS in a low-data regime. The second case study involves optimizing the design of an ethanol production process to balance economic performance and environmental impact. In both cases, we compare MOBONS against state-of-the-art multi-objective Bayesian optimization baselines and analyze its ability to efficiently identify high-quality Pareto-optimal solutions.

3.4.1 Baseline Methods for Comparison

We compare MOBONS against three baseline methods, representing different approaches to multi-objective optimization:

- **Random:** This method selects each query point uniformly at random, serving as a baseline to assess the relative efficiency of more sophisticated methods.
- **Expected Hypervolume Improvement (qEHVI)** [46]**:** This is a fully black-box Bayesian optimization method that selects points to maximize

expected improvement in hypervolume, guiding the search towards the Pareto front.

- **Pareto Optimal Thompson Sampling (qPOTS)** [48]: This is a fully black-box Thompson sampling-based approach for multi-objective optimization. This method can be thought of as a special case of MOBONS that neglects function network structure. In other words, it treats every objective function as an independent black-box function.

To ensure fair comparisons, all baseline methods and MOBONS are implemented using a consistent framework. We use BoTorch [55] to implement qEHVI including their default choices for hyperparameters. For qPOTS and MOBONS, we train the GP models using GPyTorch [56] using 5/2 Matern covariance function with automatic relevance determination (ARD) [49] (meaning individual lengthscale hyperparameters are included for each dimension). We also employ the pymoo [57] implementation of NSGA-II [45] to numerically optimize the acquisition functions (again using default settings).

3.4.2 Synthetic Test Problem: ZDT4 Benchmark

The first case study is based on the ZDT4 function [58], a widely used synthetic benchmark in multi-objective optimization. This function is defined in a 10-dimensional search space ($D = 10$), where the first variable defines one objective, while the remaining nine variables introduce nonlinear terms resulting in multiple local minima in the objective function surfaces. These characteristics make ZDT4 particularly challenging for optimization algorithms operating under a limited function evaluation budget.

To introduce function network structure, we model the problem as a system of 11 interconnected nodes, where each function output depends on a subset of inputs. The two competing objectives are defined as follows:

$$
\begin{aligned}
y_1 &= f_1(\boldsymbol{x}) = x_1, \\
y_k &= f_k(\boldsymbol{x}) = x_k^2 - 10\cos(4\pi x_k), \quad \forall k \in \{2, \dots, 10\}, \\
y_{11} &= f_{11}(\boldsymbol{x}) = \left(91 + \sum_{k=2}^{10} y_k\right)\left(1 - \sqrt{\frac{y_1}{91 + \sum_{k=2}^{10} y_k}}\right), \\
g_1(\boldsymbol{x}) &= y_1, \\
g_2(\boldsymbol{x}) &= y_{11}.
\end{aligned}
\tag{3.12}
$$

The feasible set for the design variables is constrained within the hyperrectangle: $0 \leq x_1 \leq 1$ and $-10 \leq x_k \leq 10$ for $k = 2, \dots, 10$. We assume that the node functions $f_{1,\dots,10}$ are black-box functions (expensive and unknown), while f_{11} is a white-box function (computationally inexpensive and known).

To evaluate algorithm performance, we use hypervolume as the primary metric, measuring the quality of the approximated Pareto front with respect to a reference point $\boldsymbol{r}$. Following [46], the hypervolume indicator for an approximate Pareto set $\mathcal{P}$ with a finite number of elements is given by the Lebesgue measure λ_M of the space dominated by $\mathcal{P}$ and bounded from below by $\boldsymbol{r}$:

$$\mathrm{HV}(\mathcal{P}, \boldsymbol{r}) = \lambda_M \left(\cup_{i=1}^{|\mathcal{P}|} [\boldsymbol{r}, \boldsymbol{p}_i] \right), \tag{3.13}$$

where $[\boldsymbol{r}, \boldsymbol{p}_i]$ is the hyper-rectangle defined by the reference point $\boldsymbol{r}$ and a Pareto-optimal solution $\boldsymbol{p}_i$. For this case study, we select a reference point of $\boldsymbol{r} = [1, 500]$. Each algorithm is initialized with $n = 21$ randomly chosen samples in $\mathcal{X}$, and the total evaluation budget is $N = 121$ (100 additional sequentially selected points). To account for stochastic variability due to random initialization, we repeat all experiments 30 times, ensuring consistent random seeds across all algorithms.

Figure 3.3 compares the hypervolume progression of each algorithm. As expected, random sampling exhibits the slowest improvement, providing a baseline for comparison. qPOTS performs moderately well, but it only begins to outperform random sampling after approximately 60 iterations. qEHVI initially demonstrates rapid improvement, quickly increasing the hypervolume, but its progress plateaus after 25 iterations, which is an indication of under-exploration.

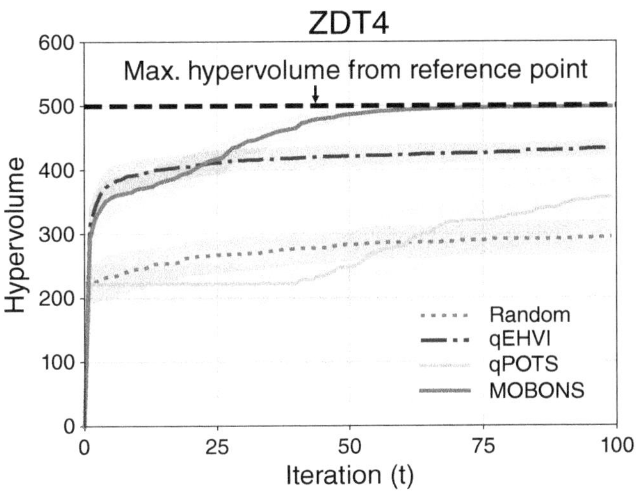

Figure 3.3 Comparison of multi-objective optimization algorithms on the ZDT4 problem, showing hypervolume progression over iterations. The dashed black line indicates the estimated maximum hypervolume from the reference point. MOBONS reaches the maximum hypervolume faster than all baseline methods, demonstrating its effectiveness in leveraging function network structure when selecting candidates.

In contrast, MOBONS maintains steady progress throughout the optimization process, leveraging the function network structure to efficiently navigate the search space. Notably, MOBONS converges to near-optimal hypervolume in approximately 50 iterations, significantly outperforming the baselines. This result highlights the advantage of incorporating network structure in Bayesian optimization, leading to faster convergence and improved Pareto front approximation compared to traditional black-box methods.

To further analyze the distribution of function evaluations across the Pareto front, Figure 3.4 presents scatter plots of the two objective values, $g_1(\boldsymbol{x})$ and $g_2(\boldsymbol{x})$, obtained over the 100 sequential evaluations for each algorithm in the median-performing replicate. MOBONS clearly achieves a better approximation of the Pareto front, as more samples are pushed toward the lower-left corner of the plot, indicating improved trade-offs between the objectives. Interestingly,

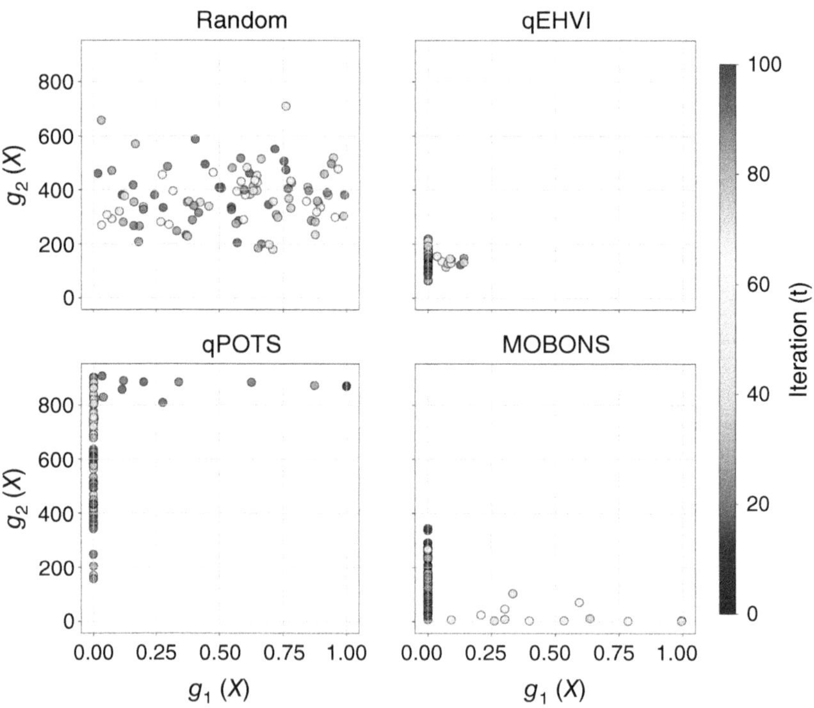

Figure 3.4 Comparison of function evaluations for each algorithm on the ZDT4 problem, showing the distribution of evaluated points over the two objectives. Each subplot presents the median performing replicate across 30 independent runs. MOBONS successfully pushes samples toward the true Pareto front, demonstrating superior efficiency in balancing exploration and exploitation. qEHVI shows limited exploration, leading to a clustered sampling pattern, while qPOTS and Random fail to fully capture the Pareto-optimal region.

qEHVI, the second-best-performing method in terms of hypervolume, tends to cluster evaluations in certain regions of the objective space, exhibiting limited exploration. In contrast, MOBONS maintains a more diverse distribution of evaluated points, likely contributing to its superior performance in reaching the global Pareto front.

3.4.3 Design of Sustainable Bioethanol Process

The transition to renewable energy sources is a pressing global challenge, driven by the environmental consequences of fossil fuel combustion. Among various alternatives, bioethanol has emerged as a promising renewable fuel due to its production from biomass feedstocks via fermentation. Unlike fossil fuels, ethanol combustion can be partially carbon-neutral, as CO_2 emissions are offset by the CO_2 uptake of the biomass feedstocks during growth. However, large-scale ethanol production still presents sustainability challenges, including energy-intensive separation processes and inefficient feedstock utilization. To address these issues, optimization methods must balance economic and environmental trade-offs to develop cost-effective and sustainable ethanol production systems.

3.4.3.1 Process Description and Implementation

In this case study, we apply MOBONS to optimize a steady-state bioethanol production process with competing economic and environmental objectives. This example serves as a proof of concept, demonstrating the potential of MOBONS in process systems engineering. The process consists of a fermentation reactor followed by two distillation columns, as shown in Figure 3.5. The simulation is implemented using BioSTEAM [59], an open-source biochemical process simulation platform that enables the evaluation of complex thermodynamic and economic trade-offs.

3.4.3.2 Problem Formulation and Function Network Representation

We consider eight decision variables, including the feed flowrates of water and glucose (F), the reaction temperature and pressure of the fermentation reactor (T_{rxn}, P_{rxn}), the operating pressures of the two distillation columns (P_1, P_2), the reflux-to-minimum-reflux ratios for both distillation columns (RR_1, RR_2), and the ethanol purity in the distillate of the first distillation column (p). The bounds on these variables are listed below at the top of Table 3.1. The two competing objectives are:

- Economic performance, measured by revenue generated from high-purity ethanol sales. This metric is maximized.
- Environmental impact, quantified using global warming potential (GWP), which accounts for CO_2 emissions. This metric is minimized.

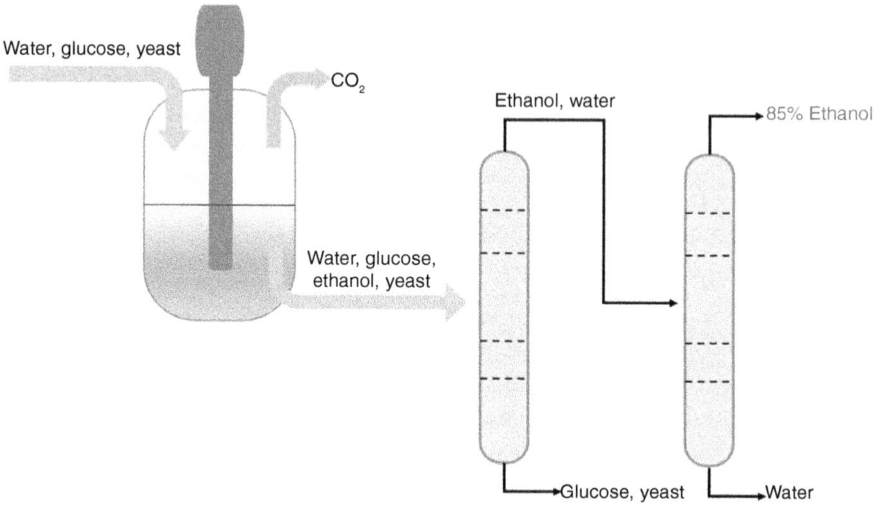

Figure 3.5 Process diagram for the ethanol fermentation system, consisting of a fermentation reactor followed by two distillation columns. The optimization goal is to balance economic and environmental performance by adjusting process design variables.

Table 3.1 Bounds and optimal values at the four Pareto optimal solutions shown in Figure 3.8 for all decision variables in the bioethanol process design problem.

	F (kmol/hr)	T_{rxn} (°C)	P_{rxn} (atm)	P_1 (atm)	RR_1	P_2 (atm)	RR_2	p
Range	[90, 110]	[30, 40]	[1, 5]	[1, 5]	[0.1, 10]	[1, 5]	[2, 10]	[0.5, 0.7]
Point A	90.5	30.0	5.0	4.0	6.9	3.7	2.0	0.7
Point B	95.4	30.6	5.0	4.6	5.4	2.5	2.0	0.7
Point C	98.2	35.1	4.9	5.0	1.8	2.5	3.0	0.7
Point D	101.0	36.7	5.0	2.3	5.1	3.3	3.5	0.7

It is important to note that MOBONS can be applied to any user-defined objective functions; the choices above serve as reasonable illustrative examples.

MOBONS models this ethanol production system as a function network, depicted in Figure 3.6. The network consists of (i) Decision variables ($x_{1,\ldots,8}$) that define the process configuration; (ii) Intermediate process outputs ($y_{1,\ldots,14}$) that include reaction and separation unit responses; and (iii) Objective functions ($g_{1,2}$), where the final calculations of revenue and GWP are explicitly modeled as white-box calculations. Thus, the network consists of a total of $K = 5$ node, with 3 being black-box and 2 being white-box function nodes. To facilitate

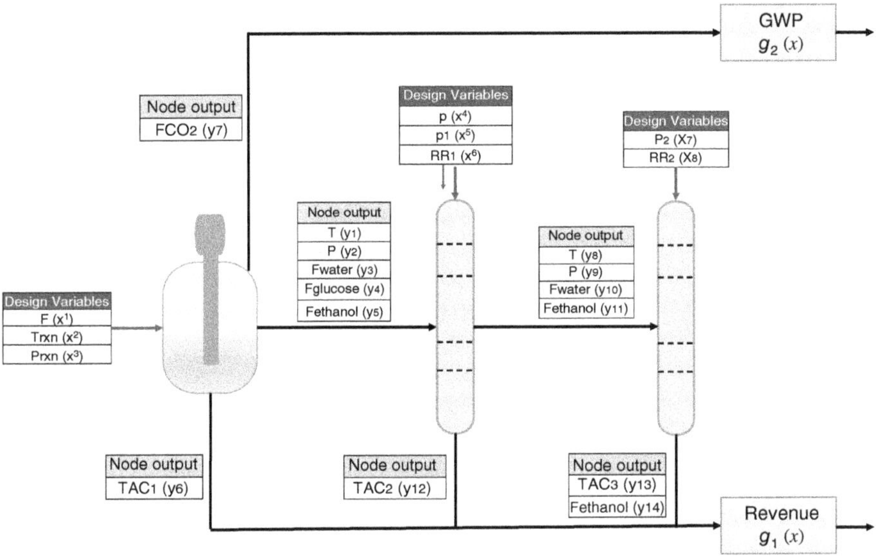

Figure 3.6 Function network representation of the ethanol fermentation process, showing decision variables (x), intermediate outputs (y), and objective functions (g). Revenue and GWP are modeled as explicit white-box calculations, leveraging known process relationships to improve optimization efficiency.

reproducibility, we provide an open-source implementation of this case study on Github at the following link: https://github.com/PaulsonLab/MOBONS. Detailed descriptions of the white-box equations can be found in the provided code.

3.4.3.3 Optimization Performance and Hypervolume Analysis

As in the previous case study, we compare MOBONS against three baseline algorithms. Each algorithm is initialized with 20 random samples and allowed a total budget of 70 function evaluations. The optimization results are evaluated using the hypervolume metric, which quantifies the volume dominated by the obtained Pareto front relative to a predefined reference point $\boldsymbol{r} = [28.3, 13600]$ (where the first element is revenue in units of million USD, and the second element is GWP). Figure 3.7 presents the hypervolume distributions across 20 optimization replicates for each method. MOBONS consistently achieves the highest hypervolume, demonstrating its ability to efficiently identify high-quality Pareto-optimal solutions. Figure 3.8 shows the Pareto front identified by MOBONS in one of the best optimization runs. Notably, the Pareto front exhibits four distinct clusters, reflecting different trade-off regimes between revenue and GWP.

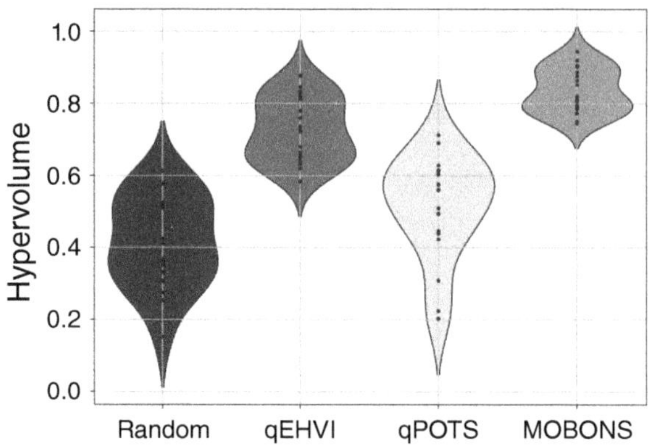

Figure 3.7 Comparison of hypervolume achieved by different multi-objective optimization methods across 20 independent runs. MOBONS consistently outperforms qPOTS and qEHVI, demonstrating its ability to efficiently explore and exploit the search space.

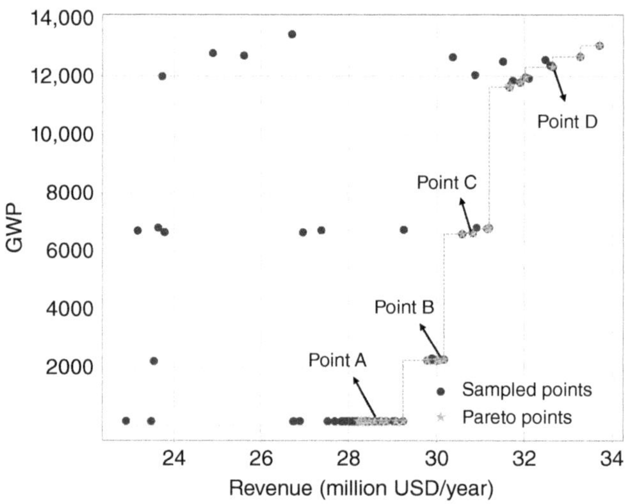

Figure 3.8 Pareto front and design points identified by MOBONS. The front exhibits four distinct regions, highlighting different trade-offs between revenue and environmental impact. The discontinuities arise due to the stepwise penalties applied to GWP as CO_2 emissions increase.

3.4.3.4 Local Sensitivity Analysis

Another advantage of MOBONS is that it produces a surrogate model that incorporates network structure, enabling additional analyses such as sensitivity analysis (SA). Here, we use SA to examine the impact of different decision variables on

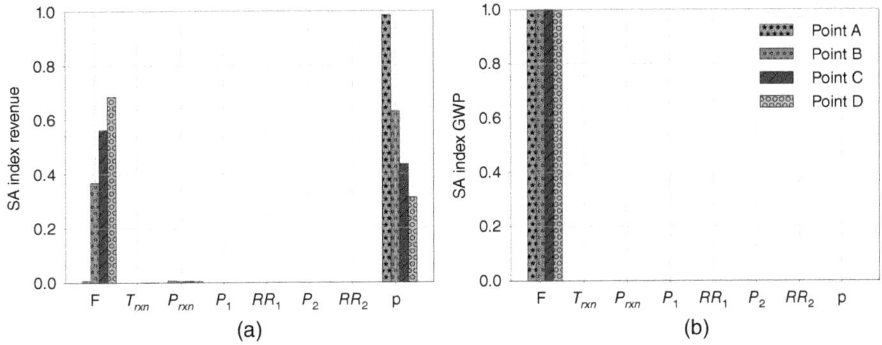

Figure 3.9 Sobol sensitivity analysis results for revenue (a) and GWP (b) across four Pareto-optimal points shown in Figure 3.8. Revenue is primarily influenced by water/glucose feedflow rate (*F*) and ethanol purity in the first distillation column (*p*), while GWP is only sensitive to *F*.

the two objectives (revenue and GWP). Since the function network GP model may not be globally accurate, we focus on a local SA around solutions along the Pareto frontier. We conduct local SA on four selected Pareto solutions (points A, B, C, and D in Figure 3.8) by perturbing each decision variable within a ±5% range. To improve model accuracy in these local regions, we generate an additional 24 function evaluations centered around each Pareto point. First-order Sobol indices [60] are then computed to quantify the relative influence of each decision variable on revenue and GWP. Figure 3.9 presents the results of the local SA. We see that revenue is primarily influenced by water/glucose feed flowrate (F) and ethanol purity in the first distillation column (p). Notably, as revenue increases along the Pareto front (moving from left to right), its sensitivity to F grows, while its sensitivity to p diminishes. On the other hand, we see that GWP is predominantly impacted by F, with negligible (local) sensitivity to other variables. This result aligns with expectations, as CO_2 emissions scale directly with the amount of material fed into the process.

One of the significant advantages of this type of approach is that it enables SA in a way that would be virtually impossible with a purely black-box optimization framework. In traditional black-box optimization, obtaining the Pareto front is already computationally expensive, and conducting local SA at key points on the front would require an impractically large number of additional function evaluations. By leveraging the function network GP model learned by MOBONS, we can efficiently approximate the local response of the system to small perturbations in decision variables, allowing us to gain meaningful insights at a fraction of the computational cost. It is worth emphasizing that this analysis focuses on local sensitivity around selected Pareto-optimal solutions rather than a global sensitivity analysis across the entire design space. While global SA can provide a broader understanding of variable importance, it is often infeasible for high-dimensional, computationally expensive problems. In contrast, local SA, as performed here, is still valuable for decision-making because it highlights which variables have the most influence on system performance near optimal designs. This helps designers (or stakeholders) more quickly identify bottlenecks, assess trade-offs, and estimate design variable importance.

3.5 Conclusion

In this chapter, we presented a flexible and general framework, **M**ulti-**O**bjective **B**ayesian **O**ptimization for **N**etwork **S**ystems (MOBONS), for handling complex, networked models involving expensive black-box function evaluations. Unlike most existing approaches that either rely on fully equation-oriented (white-box) representations or treat the entire problem as a monolithic black-box, our method leverages a network (graph) structure to capture key interactions and information flows between subsystem functions. While MOBONS builds on prior work that uses function networks in Bayesian optimization, it provides two primary advances: to our knowledge, it is the first algorithm to address both multi-objective problems and function networks simultaneously, and it is also the first to accommodate cyclic graphs (e.g. recycle streams). These features are incorporated in a manner that keeps the algorithm both computationally efficient and straightforward to implement.

We provided detailed implementation guidelines and practical considerations, including strategies for incorporating constraints on network outputs and for supporting batch (parallel) evaluations when multiple resources are available. In addition, we supply supporting code to facilitate the application of MOBONS in diverse engineering and scientific problems. Our numerical results, encompassing both a synthetic test system and a green ethanol production case study, illustrate the value of exploiting structural information in networked models. Even in relatively high-dimensional settings, MOBONS identifies Pareto-optimal solutions that effectively balance competing objectives, such as economic performance and environmental impact. We also demonstrate how the resulting function network surrogate model can be used to perform sensitivity analysis around the identified Pareto frontier designs, offering valuable insights for decision makers.

Although these results are promising, the present work serves primarily as a proof of concept, leaving several avenues open for future research. First, a more rigorous theoretical analysis of MOBONS's convergence properties – particularly in the presence of cycles and complex constraints – would help establish broader guarantees. Second, the efficiency of the multi-objective subproblem could be further enhanced by developing specialized solvers capable of handling both equality and inequality constraints more effectively. Third, exploring more sophisticated covariance modeling approaches (such as deep kernel learning or node-specific kernels) may capture each node function's unique characteristics more precisely. Finally, the impact of different selection rules for choosing candidate points from the approximate Pareto front warrants systematic study, given the well-known exploration–exploitation trade-off. We hope these directions will inspire future developments and encourage the broader adoption of network-aware Bayesian optimization methods in both industry and academia.

References

1 Boulding, K.E. (1966). *The Economics of the Coming Spaceship Earth*. In: *Environmental Quality in a Growing Economy* (ed. H. Jarrett), 3–14. Baltimore, MD: Resources for the Future/Johns Hopkins University Press.

2 Bakshi, B.R. (ed.) (2023). *Engineering and Ecosystems: Seeking Synergies Toward a Nature-Positive World*. Springer International Publishing https://doi.org/10.1007/978-3-031-35692-6.

3 U.S. Environmental Protection Agency (2025). U.S. environmental protection agency – scientific and technical information. https://cfpub.epa.gov/si/si_public_record_report.cfm?Lab=NERL&dirEntryId=319430 (accessed 19 February 2025).

4 Takamatsu, T., Hashimoto, I., and Ohno, H. (1970). Optimal design of a large complex system from the viewpoint of sensitivity analysis. *Industrial and Engineering Chemistry Process Design and Development* 9 (3): 368–379. https://doi.org/10.1021/i260035a004.

5 Nishida, N., Ichikawa, A., and Tazaki, E. (1974). Synthesis of optimal process systems with uncertainty. *Industrial and Engineering Chemistry Process Design and Development* 13 (3): 209–214. https://doi.org/10.1021/i260051a003.

6 Grossmann, I.E. and Sargent, R.W.H. (1978). Optimum design of chemical plants with uncertain parameters. *AIChE Journal* 24 (6): 1021–1028. https://doi.org/10.1002/aic.690240612.

7 Halemane, K.P. and Grossmann, I.E. (1983). Optimal process design under uncertainty. *AIChE Journal* 29 (3): 425–433. https://doi.org/10.1002/aic.690290312.

8 Swaney, R.E. and Grossmann, I.E. (1985). An index for operational flexibility in chemical process design. Part I: formulation and theory. *AIChE Journal* 31 (4): 621–630. https://doi.org/10.1002/aic.690310412.

9 Grossmann, I.E. and Floudas, C.A. (1987). Active constraint strategy for flexibility analysis in chemical processes. *Computers & Chemical Engineering* 11 (6): 675–693. https://doi.org/10.1016/0098-1354(87)87011-4.

10 Narraway, L.T., Perkins, J.D., and Barton, G.W. (1991). Interaction between process design and process control: economic analysis of process dynamics. *Journal of Process Control* 1 (5): 243–250. https://doi.org/10.1016/0959-1524(91)85015-b.

11 Luyben, M.L. and Floudas, C.A. (1994). Analyzing the interaction of design and control—1. A multiobjective framework and application to binary distillation synthesis. *Computers & Chemical Engineering* 18 (10): 933–969. https://doi.org/10.1016/0098-1354(94)e0013-d.

12 Burnak, B., Diangelakis, N.A., and Pistikopoulos, E.N. (2019). Towards the grand unification of process design, scheduling, and control—utopia or reality? *Processes* 7 (7): 461. https://doi.org/10.3390/pr7070461.

13 Niederer, S.A., Sacks, M.S., Girolami, M., and Willcox, K. (2021). Scaling digital twins from the artisanal to the industrial. *Nature Computational Science* 1 (5): 313–320. https://doi.org/10.1038/s43588-021-00072-5.

14 Sharma, A., Kosasih, E., Zhang, J. et al. (2022). Digital twins: state of the art theory and practice, challenges, and open research questions. *Journal of Industrial Information Integration* 30: 100383. https://doi.org/10.1016/j.jii.2022.100383.

15 Kushner, H.J. (1964). A new method of locating the maximum point of an arbitrary multipeak curve in the presence of noise. *Journal of Basic Engineering* 86 (1): 97–106. https://doi.org/10.1115/1.3653121.

16 Močkus, J. (1975). *On Bayesian Methods for Seeking the Extremum*, 400–404. Berlin Heidelberg: Springer https://doi.org/10.1007/3-540-07165-2_55.

17 Hickman, R.J., Aldeghi, M., Häse, F., and Aspuru-Guzik, A. (2022). Bayesian optimization with known experimental and design constraints for chemistry applications. *Digital Discovery* 1 (5): 732–744. https://doi.org/10.1039/D2DD00028H.

18 Kudva, A., Sorourifar, F., and Paulson, J.A. (2022). Constrained robust Bayesian optimization of expensive noisy black-box functions with guaranteed regret bounds. *AIChE Journal* 68 (12): e17857. https://doi.org/10.1002/aic.17857.

19 Banerjee, I., Pal, S., and Maiti, S. (2010). Computationally efficient black-box modeling for feasibility analysis. *Computers & Chemical Engineering* 34 (9): 1515–1521.

20 Rogers, A. and Ierapetritou, M. (2015). Feasibility and flexibility analysis of black-box processes part 2: surrogate-based flexibility analysis. *Chemical Engineering Science* 137: 1005–1013.

21 Geremia, M., Bezzo, F., and Ierapetritou, M.G. (2023). A novel framework for the identification of complex feasible space. *Computers & Chemical Engineering* 179: 108427. URL: https://doi.org/10.1016/j.compchemeng.2023.108427.

22 Kudva, A., Tang, W.-T., and Paulson, J.A. (2024). Robust Bayesian optimization for flexibility analysis of expensive simulation-based models with rigorous uncertainty bounds. *Computers & Chemical Engineering* 181: 108515, 10.1016/j.compchemeng.2023.108515.

23 Paulson, J.A. and Tsay, C. (2024). Bayesian optimization as a flexible and efficient design framework for sustainable process systems. *Current Opinion in Green and Sustainable Chemistry* 51: 100983.

24 Boukouvala, F. and Floudas, C.A. (2016). Argonaut: algorithms for global optimization of constrained grey-box computational problems. *Optimization Letters* 11 (5): 895–913. https://doi.org/10.1007/s11590-016-1028-2.

25 Beykal, B., Boukouvala, F., Floudas, C.A., and Pistikopoulos, E.N. (2018). Optimal design of energy systems using constrained grey-box multi-objective optimization. *Computers & Chemical Engineering* 116: 488–502. https://doi.org/10.1016/j.compchemeng.2018.02.017.

26 Paulson, J.A. and Congwen, L. (2022). COBALT: COnstrained Bayesian optimizAtion of computationaLly expensive grey-box models exploiting derivaTive information. *Computers & Chemical Engineering* 160: 107700.

27 Kudva, A., Sorouifar, F., and Paulson, J. A. (2022). Efficient robust global optimization for simulation-based problems using decomposed Gaussian processes: application to MPC calibration. In 2022 American Control Conference (ACC), 2091–2097. https://doi.org/10.23919/ACC53348.2022.9867777.

28 Schweidtmann, A.M., Bongartz, D., Grothe, D. et al. (2021). Deterministic global optimization with Gaussian processes embedded. *Mathematical Programming Computation* 13 (3): 553–581. https://doi.org/10.1007/s12532-021-00204-y.

29 Congwen, L. and Paulson, J.A. (2023). No-regret constrained Bayesian optimization of noisy and expensive hybrid models using differentiable quantile function approximations. *Journal of Process Control* 131: 103085. https://doi.org/10.1016/j.jprocont.2023.103085.

30 Lu, Y.-A., Hu, W.-S., Paulson, J.A., and Zhang, Q. (2025). Bo4io: a Bayesian optimization approach to inverse optimization with uncertainty quantification. *Computers & Chemical Engineering* 192: 108859.

31 OpenFOAM Foundation (2025). OpenFOAM: the open source computational fluid dynamics (CFD) toolbox. https://www.openfoam.org (accessed 3 February 2025).

32 Guinée, J.B. (2002). *Handbook on Life Cycle Assessment: Operational Guide to the ISO Standards*. Dordrecht: Springer https://doi.org/10.1007/BF02978784.

33 i Tree (2023). i-tree eco (version 6.0.32) [computer software]. https://www.itreetools.org/i-tree-tools-download (accessed 3 February 2025).

34 Medeiros, D. (n.d.). Dwsim. http://dwsim.inforside.com.br (accessed 19 February 2025).

35 Astudillo, R. and Frazier, P. (2019). Bayesian optimization of composite functions. In International Conference on Machine Learning, 354–363. PMLR.

36 Astudillo, R. and Frazier, P. (2021). Bayesian optimization of function networks. *Advances in Neural Information Processing Systems* 34: 14463–14475.

37 Buathong, P., Wan, J., Astudillo, R., Daulton, S., Balandat, M., and Frazier, P.I. (2024). Bayesian optimization of function networks with partial evaluations.. In: *Proceedings of the 41st International Conference on Machine Learning*, (ed. R., Salakhutdinov, Z., Kolter, K., Heller, A., Weller, N., Oliver, J., Scarlett, and F., Berkenkamp). (Vol. 235) Proceedings of Machine Learning Research (21–27 Jul), 4752–4784. PMLR. https://proceedings.mlr.press/v235/buathong24a.html.

38 González, L.D. and Zavala, V.M. (2024). BOIS: Bayesian optimization of interconnected systems. *IFAC-PapersOnLine* 58 (14): 446–451.

39 González, L.D. and Zavala, V.M. (2025). Implementation of a Bayesian optimization framework for interconnected systems. *Industrial & Engineering Chemistry Research* 64 (4): 2168–2182.

40 Khoshraftar, S. and An, A. (2024). A survey on graph representation learning methods. *ACM Transactions on Intelligent Systems and Technology* 15 (1): 1–55.

41 Kandasamy, K., Krishnamurthy, A., Schneider, J., and Póczos, B.2018). Parallelised Bayesian optimisation via Thompson sampling. In International Conference on Artificial Intelligence and Statistics, 133–142. PMLR.

42 Kudva, A. and Paulson, J. A. (2025). BONSAI: structure-exploiting robust Bayesian optimization for networked black-box systems under uncertainty. *Computers & Chemical Engineering* 202: 109393.

43 Mavrotas, G. (2009). Effective implementation of the ε-constraint method in multi-objective mathematical programming problems. *Applied Mathematics and Computation* 213 (2): 455–465.

44 Hart, W.E., Watson, J.-P., and Woodruff, D.L. (2011). PYOMO: modeling and solving mathematical programs in python. *Mathematical Programming Computation* 3 (3): 219–260.

45 Deb, K., Pratap, A., Agarwal, S., and Meyarivan, T. (2002). A fast and elitist multiobjective genetic algorithm: NSGA-II. *IEEE Transactions on Evolutionary Computation* 6 (2): 182–197. https://doi.org/10.1109/4235.996017.

46 Daulton, S., Balandat, M., and Bakshy, E. (2020). Differentiable expected hypervolume improvement for parallel multi-objective Bayesian optimization. *Advances in Neural Information Processing Systems* 33: 9851–9864. https://doi.org/10.48550/ARXIV.2006.05078.

47 Bradford, E., Schweidtmann, A.M., and Lapkin, A. (2018). Efficient multiobjective optimization employing Gaussian processes, spectral sampling and a genetic algorithm. *Journal of Global Optimization* 71 (2): 407–438. https://doi.org/10.1007/s10898-018-0609-2.

48 Renganathan, S. A. and Carlson, K. E. (2023). QPOTS: efficient batch multiobjective Bayesian optimization via Pareto optimal Thompson sampling, https://arxiv.org/abs/2310.15788 (accessed 19 February 2025), https://doi.org/10.48550/ARXIV.2310.15788.

49 Williams, C.K.I. and Rasmussen, C.E. (2006). *Gaussian Processes for Machine Learning*, vol. 2. Cambridge, MA: MIT Press.

50 Sun, F., Gramacy, R.B., Haaland, B. et al. (2019). Synthesizing simulation and field data of solar irradiance. *Statistical Analysis and Data Mining* 12 (4): 311–324.

51 Wilson, A. G., Hu, Z., Salakhutdinov, R., and Xing, E. P. (2016). Deep kernel learning. In Artificial Intelligence and Statistics, 370–378. PMLR.

52 Wilson, J., Borovitskiy, V., Terenin, A., et al. (2020). Efficiently sampling functions from Gaussian process posteriors. In International Conference on Machine Learning, 10292–10302. PMLR.

53 Rahimi, A. and Recht, B.2007). Random features for large-scale kernel machines. Advances in Neural Information Processing Systems, 20.

54 Hobbie, J.G., Gandomi, A.H., and Rahimi, I. (2021). A comparison of constraint handling techniques on NSGA-II. *Archives of Computational Methods in Engineering* 28 (5): 3475–3490.

55 Balandat, M., Karrer, B., Jiang, D. R., et al. (2019). Botorch: a framework for efficient Monte-Carlo Bayesian optimization. https://arxiv.org/abs/1910.06403 (accessed 19 February 2025), https://doi.org/10.48550/ARXIV.1910.06403.

56 Gardner, J. R., Pleiss, G., Bindel, D., et al. (2018). Gpytorch: blackbox matrix-matrix Gaussian process inference with GPU acceleration. In Advances in Neural Information Processing Systems.

57 Blank, J. and Deb, K. (2020). pymoo: multi-objective optimization in python. *IEEE Access* 8: 89497–89509.

58 Zitzler, E., Deb, K., and Thiele, L. (2000). Comparison of multiobjective evolutionary algorithms: empirical results. *Evolutionary Computation* 8 (2): 173–195. https://doi.org/10.1162/106365600568202.

59 Cortes-Pea, Y., Kumar, D., Singh, V., and Guest, J.S. (2020). Biosteam: a fast and flexible platform for the design, simulation, and techno-economic analysis of biorefineries under uncertainty. *ACS Sustainable Chemistry & Engineering* 8 (8): 3302–3310. https://doi.org/10.1021/acssuschemeng.9b07040.

60 Sobol, I.M. (2001). Global sensitivity indices for nonlinear mathematical models and their Monte Carlo estimates. *Mathematics and Computers in Simulation* 55 (1–3): 271–280.

4

A Tutorial on Multi-time Scale Optimization Models and Algorithms

Asha Ramanujam and Can Li

Davidson School of Chemical Engineering, Purdue University, West Lafayette, IN, USA

4.1 Introduction

Large process systems often comprise numerous interdependent processes. Managing these systems requires decisions at multiple time scales, ranging from long-term, high-level planning, such as supply chain design, capacity expansion, and production targets – made on yearly, monthly, or weekly bases, to short-term, detailed scheduling and control actions that occur on daily, hourly, or even second-level intervals [1–4].

Optimizing a system requires determining both high-level and low-level decisions. One approach is to sequentially solve models that involve high-level and low-level decisions. In this approach, a high-level problem is first formulated and solved to derive the high-level decisions. These high-level decisions are then fixed in the low-level problems to obtain the low-level decisions. This process creates a one-way, top-down communication within the hierarchical structure.

However, making high-level decisions without any input or feedback from low-level problems often leads to infeasible or suboptimal solutions. This occurs because the decision layers are interconnected and influence one another. Therefore, to ensure a feasible and optimal solution, it is crucial to integrate feedback from low-level problems into the high-level decision-making process. This integration acknowledges the interdependencies between decision layers and enhances overall system performance.

This integration can be achieved by formulating and solving a multi-time scale optimization model, which links decision-making processes occurring across different time horizons [1–4]. Multi-time scale optimization models have been applied in a wide range of domains, including plant production and scheduling [5], electricity market participation [6], and environmental management [7].

A particularly important application lies in the decarbonization and electrification of chemical process systems [8, 9]. Electrification involves using electricity to power chemical processes, ideally from renewable resources. However, renewable energy resources are inherently variable across both space and time, and electricity prices also fluctuate considerably. To capture the full economic potential of electrified chemical plants and their supporting renewable energy infrastructure, it is essential to incorporate these variations into the overall system planning. The most effective strategy is to integrate planning and scheduling decisions for plants and energy resources within a unified multi-time scale optimization model [9].

This tutorial explores multi-time scale optimization models and the algorithms designed to solve them. It is structured as follows: in Section 4.2, we give a representation for a general multi-time scale optimization model. We describe *Value of the Multi-scale Model* (VMM), a metric used to quantify the benefits of using a multi-time scale optimization model in Section 4.3. In Section 4.4, we discuss the algorithms used in literature to solve multi-time scale optimization models. We provide an example of a multi-time scale optimization model in Section 4.5 and discuss the implementation of a few of the algorithms on the example. Finally, in Section 4.6, we draw the conclusions.

4.2 Multi-time Scale Optimization Models

As discussed before, multi-time scale optimization models take decisions at different time scales. We divide the decision variables into two sets: high-level decision variables and low-level decision variables. The high-level decision variables represented by $\boldsymbol{x}$ correspond to the long time scales and low-level decision variables denoted by $\boldsymbol{y}_s$ correspond to all the other shorter time scales defined for subperiods s. Multi-time scale optimization models can be represented in the following way:

$$\text{MM} = \min_{\boldsymbol{x},\boldsymbol{y}_s} \quad f(\boldsymbol{x}) + \sum_{s \in S} w_s q(\boldsymbol{x}, \boldsymbol{y}_s, \theta_s) \tag{4.1a}$$

$$\text{s.t.} \quad g(\boldsymbol{x}) \leq 0 \tag{4.1b}$$

$$h(\boldsymbol{x}, \boldsymbol{y}_s, \theta_s) \leq 0 \quad \forall s \in S \tag{4.1c}$$

$$\tilde{h}(\boldsymbol{x}, \boldsymbol{y}_1, \boldsymbol{y}_2, \dots, \boldsymbol{y}_{|S|}, \theta) \leq 0. \tag{4.1d}$$

The notation used for the formulation above (4.1) is given in Table 4.1.

To understand the formulation, consider an integrated design and scheduling problem. The problem can be formulated as a multi-time scale optimization model. In the multi-time scale model, the design decisions are the high-level decisions $\boldsymbol{x}$ and the scheduling decisions are the low-level decisions $\boldsymbol{y}_s$ for various

Table 4.1 Notation used in Eq. (4.1).

Notation	Meaning
$\boldsymbol{x}$	High-level decisions
s	Subperiod
S	Set of subperiods
w_s	Probability/weight of subperiod s
$\boldsymbol{\theta}_s$	Parameters associated with subperiod s
$\boldsymbol{y}_s$	Low-level decision variables associated with subperiod s
$\boldsymbol{\theta}$	Complete set of parameters
f	Cost function connected to the high-level decisions
q	Cost function connected to the low-level decisions
g	Constraints pertaining only to the high-level decisions
h	Constraints concerning only a single subperiod s
$\tilde{h}$	Constraints that link decisions across different subperiods

subperiods s. The model can have various constraints which can be classified in the following way:

- Equation (4.1b) represents constraints solely on the high-level decisions. Examples include constraints on the bounds of the design decisions, such as the maximum possible capacity for reactor vessels.
- Constraints concerning the high-level decisions and the low-level decisions of a single subperiod s are represented by Eq. (4.1c). Examples include constraints on the bounds of scheduling decisions like maximum possible production.
- Constraints concerning the low-level decisions for multiple single subperiods and possibly the high-level decisions are represented by Eq. (4.1d). Such constraints are often called complicating constraints. Examples include constraints that combine the production of multiple subperiods to satisfy the demand for the entire duration.

This method of writing the constraints is general and will help us further to understand whether different algorithms can be applied to different multi-time scale optimization model in Section 4.4.

Differences with two-stage stochastic programming: two-stage stochastic programming models have a formulation similar to (4.1) with $\boldsymbol{x}$ representing the first-stage variables, and $\boldsymbol{y}_s$ representing the second-stage variables for scenario s. The major difference between these models and multi-time scale models is that these models do not have complicating constraints that connect decisions in different subperiods that is, no $\tilde{h}$. However, multi-time scale models can have complicating constraints.

4.3 Value of the Multi-scale Model (VMM)

While solving a multi-time scale algorithm might be very difficult as compared to a traditional one-way top-down approach, it provides a more economic solution. To assess the benefits of using a multi-time scale optimization model instead of the traditional one-way top-down approach, we can use the metric *VMM* [9]. The VMM is the difference between the objective derived from the one-way top-down approach (Multi-scale Performance of the Single-scale Solution [MPSS]) and the optimal objective from solving the multi-time scale model (MM) as shown in Figure 4.1. The VMM helps evaluate the enhancements in system efficiency and feasibility gained by integrating decision-making across various time scales.

We now define the VMM more formally. Let's start by introducing some notations. The optimal objective obtained from solving the multi-time scale optimization model will be referred to as the performance of the multi-time scale model (MM).

Next, consider a simpler, high-level model that is used in the traditional top-down approach. This model can be formulated as follows:

$$\mathrm{SM} = \min_{\boldsymbol{x},\boldsymbol{z}} \quad f(\boldsymbol{x}) + Q(\boldsymbol{x}, \boldsymbol{z}, \tilde{\theta}) \tag{4.2a}$$

$$\text{s.t.} \quad g(\boldsymbol{x}) \leq 0 \tag{4.2b}$$

$$H(\boldsymbol{x}, \boldsymbol{z}, \tilde{\theta}) \leq 0. \tag{4.2c}$$

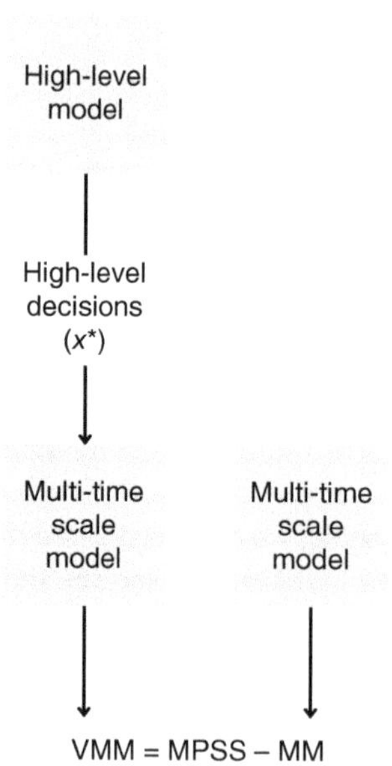

Figure 4.1 Value of the multi-scale model.

In this formulation, $\boldsymbol{z}$ serves as a surrogate for the low-level decisions, and $\tilde{\boldsymbol{\theta}}$ as a surrogate parameter for the original parameters $\boldsymbol{\theta}_s$. Similarly, Q is a surrogate for the function q, while H encompasses surrogates for both h and $\tilde{h}$, that is, the equivalent function and constraints in terms of $\boldsymbol{z}$ and $\tilde{\boldsymbol{\theta}}$.

These high-level decisions are then fixed in the multi-time scale model to obtain the low-level decisions. The model shown below is the multi-time scale model with these decisions fixed:

$$\text{MPSS} = \min_{\boldsymbol{y}_s} \quad f(\boldsymbol{x}^*) + \sum_{s \in S} w_s q(\boldsymbol{x}^*, \boldsymbol{y}_s, \boldsymbol{\theta}_s) \tag{4.3a}$$

$$\text{s.t.} \quad h(\boldsymbol{x}^*, \boldsymbol{y}_s, \boldsymbol{\theta}_s) \leq 0 \quad \forall s \in S \tag{4.3b}$$

$$\tilde{h}(\boldsymbol{x}^*, \boldsymbol{y}_1, \boldsymbol{y}_2, \dots, \boldsymbol{y}_{|S|}, \boldsymbol{\theta}) \leq 0, \tag{4.3c}$$

where $\boldsymbol{x}^*$ signifies the high-level decisions obtained from solving the high-level model in (4.2). The model (4.3) can also be referred to as the low-level model.

The MPSS is defined as the optimal objective obtained from solving (4.3) as shown in Figure 4.1. In case, the (4.3) is infeasible, we set MPSS to be a very large positive value like 10^{10}.

Finally, the *VMM* is calculated as the difference between the MPSS and the performance of the multi-time scale optimization model (MM).

$$\text{VMM} = \text{MPSS} - \text{MM}.$$

This metric quantifies the benefits of using a multi-time scale model over the traditional one-way top-down approach. A higher VMM indicates greater improvements in system efficiency and feasibility, justifying the additional complexity of the multi-time scale optimization.

The VMM metric was proposed based on the concept of the Value of the Stochastic Solution (VSS), originally introduced by the stochastic programming community to measure the advantages of stochastic solutions over the optimal solutions of deterministic models [10]. Computing VSS involves solving an *Expected Value* problem for one nominal scenario corresponding to the expected value of $\boldsymbol{\theta}_s$, that is, $\overline{\boldsymbol{\theta}} = \sum_{s \in S} w_s \boldsymbol{\theta}_s$

$$\text{EV} = \min_{\boldsymbol{x}\overline{\boldsymbol{y}}} \ f(\boldsymbol{x}) + q(\boldsymbol{x}, \overline{\boldsymbol{y}}, \overline{\boldsymbol{\theta}}), \tag{4.4a}$$

$$\text{s.t.} \quad g(\boldsymbol{x}) \leq 0 \tag{4.4b}$$

$$h(\boldsymbol{x}, \overline{\boldsymbol{y}}, \overline{\boldsymbol{\theta}}) \leq 0. \tag{4.4c}$$

In the above formulation, $\overline{\boldsymbol{y}}$ represents the low-level decisions for the nominal scenario. Typically, the size of the expected value model (4.4) is larger than the size of the high-level model used to compute VMM (4.2) since it still involves multiple time scales.

The performance of EV is then evaluated by substituting the solution $\boldsymbol{x}^{\mathrm{EV}}$ obtained from (4.4) in the original full-space problem with the optimal objective being called the *expected result of using the EV solution* (EEV) [10].

$$\mathrm{EEV} = \min_{\boldsymbol{y}_s} \quad f(\boldsymbol{x}^{\mathrm{EV}}) + \sum_{s \in S} w_s q(\boldsymbol{x}^{\mathrm{EV}}, \boldsymbol{y}_s, \boldsymbol{\theta}_s) \tag{4.5a}$$

$$\text{s.t.} \quad h(\boldsymbol{x}^{\mathrm{EV}}, \boldsymbol{y}_s, \boldsymbol{\theta}_s) \leq 0 \quad \forall s \in S. \tag{4.5b}$$

VSS is then defined by

$$\mathrm{VSS} = \mathrm{EEV} - \mathrm{MM}.$$

The difference between VMM and VSS is the way in which the high-level decisions are obtained for the top-down approach. While VMM uses an aggregated model with a lesser number of time scales involved, VSS uses a nominal variable involving the same number of time scales. Thus, VMM shows the value of having variables in multiple time scales, while VSS illustrates the value of having multiple scenarios/subperiods.

4.4 Algorithms to Solve Multi-time Scale Optimization Models

Multi-time scale optimization models can be of different sizes (from thousands of variables and constraints to billions of variables and constraints) and different structures. Different algorithms have been proposed to solve multi-time scale optimization models. We classify these algorithms in the following way:

1) Full-space methods
2) Decomposition algorithms
3) Metaheuristic algorithms
4) Matheuristic algorithms
5) Data-driven methods
6) Parametric Autotuning Multi-time Scale Optimization (PAMSO) algorithm

In Section 4.4, we will be discussing these algorithms. Note that in this chapter, when we refer to solving a model with an algorithm, we refer to the process of generating good feasible solutions for the problem. Apart from full-space methods and conventional decomposition algorithms, the other algorithms used to address multi-time scale optimization models generally do not provide theoretical guarantees of optimality or valid bounds (on their own) for assessing solution quality.

4.4.1 Full-Space Methods

One method to solve multi-time scale optimization models is to solve them directly using exact optimization techniques. In other words, we can solve the proposed mixed integer linear programs (MILP)/mixed integer nonlinear programming (MINLP) integrated models directly using solvers like Gurobi, CPLEX, or BARON. For example, in [11], the authors proposed an MILP to solve production planning and scheduling/batching for multisite batch process industries. A case study involving 20 final products in 2 different production plants was solved using the CPLEX 11.0 solver. The MILP for the case study involved 1444 continuous variables, 5256 binary variables, and 6744 constraints and was solved in 430 seconds.

While this can be done for small-size problems, it becomes very difficult and time-consuming to use these solvers to solve problems with millions of variables and constraints. This is because MILP and MINLP problems are NP-hard problems in general making the problem more intractable as the problem sizes increase.

4.4.2 Decomposition Algorithms

Decomposition algorithms involve breaking the multi-time scale optimization model into smaller subproblems. They can be used on problems moderately large (up to tens of millions of variables [12]) and tackle specific types of problems. The most commonly used decomposition algorithms are the following:

1) Bi-level decomposition
2) Benders decomposition
3) Lagrangian decomposition
4) Branch and price

We discuss the decomposition algorithms below. Furthermore, we discuss in detail the implementation of some of the algorithms to solve a simplified linear multi-time scale optimization model (4.6). We use a simplified problem as it helps illustrate some of the algorithm's classical versions, which are then suitably modified to solve more complex multi-time scale optimization models. The model of the simplified problem is shown below:

$$\min \quad c^T \boldsymbol{x} + \sum_{s \in S} q_s^T \boldsymbol{y}_s \tag{4.6a}$$

$$\text{s.t.} \quad A\boldsymbol{x} = b, \quad \boldsymbol{x} \geq 0 \tag{4.6b}$$

$$T_s \boldsymbol{x} + W_s \boldsymbol{y}_s = h_s, \quad \boldsymbol{y}_s \geq 0, \quad \forall s \in S. \tag{4.6c}$$

In the above representation, $\boldsymbol{x}$ represents the high-level decisions, and $\boldsymbol{y}_s$ represents the low-level decisions for subperiod s. Furthermore, (4.6b) signifies the constraints only on the high-level decisions, and (4.6c) denotes the constraints connecting the high-level and low-level decisions. In the model we consider, we do not have complicating constraints, that is, no $\tilde{h}$. This is because some of the decomposition algorithms do not work on problems with such constraints.

As discussed before, without $\tilde{h}$, the model (4.6) can be treated as a two-stage stochastic linear program, with the subperiods being the scenarios, $\boldsymbol{x}$ being the first-stage variables and $\boldsymbol{y}_s$ being the second-stage variables. In this work, we treat this problem as a multi-time scale optimization problem as opposed to a stochastic program.

4.4.2.1 Bi-level Decomposition

Bi-level decomposition involves decomposing the original model into a high-level model and a low-level model with the performance of the high-level model improved using feedback from the low-level model. The high-level model, typically a relaxation of the optimization model, provides bounds for the model and some of the high-level decisions. The low-level model provides a feasible solution taking into account the high-level decisions obtained from the high-level model as shown in Figure 4.2. The algorithm involves iteratively solving the high-level and low-level models to reduce the difference in the objective values obtained from the models. Problem-specific logic and integer cuts based on the solution from the low-level problem are added to the high-level model to improve the bounds and high-level decisions generated.

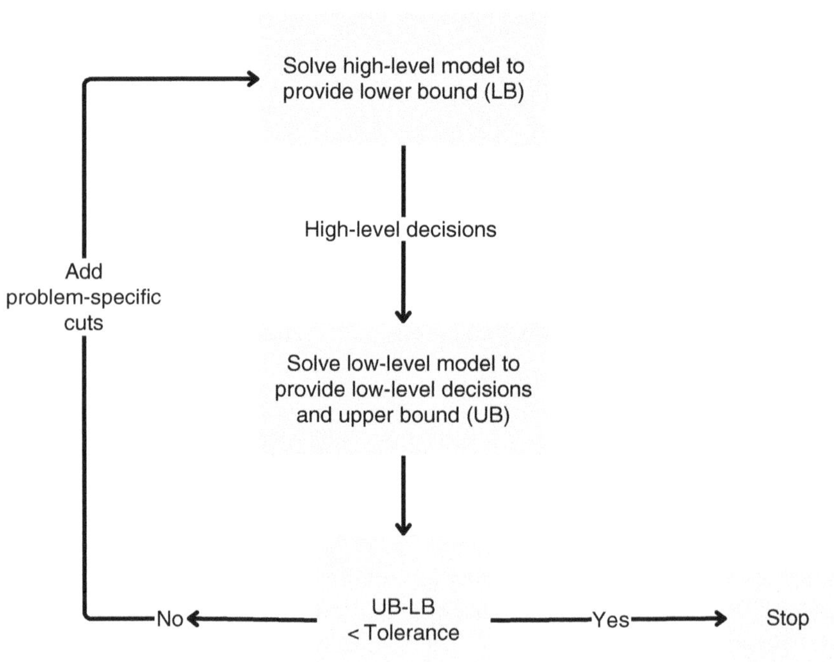

Figure 4.2 Schematic for bi-level decomposition.

The steps involved are shown in Algorithm 4.1.

Algorithm 4.1 Bi-level Decomposition Algorithm

Initialize: Set the Upper Bound (UB) $\leftarrow \infty$ and Lower Bound (LB) $\leftarrow -\infty$.
Decompose the problem into a high-level model and a low-level model.
while True **do**
    Solve the high-level model to obtain high-level decisions and update the Lower Bound (LB).
    Fix the high-level decisions in the low-level model and solve the low-level model to obtain low-level decisions and update the Upper Bound (UB).
    if UB − LB $\leq$ Tolerance **then**
        Stop: Convergence achieved.
    else
        Add problem-specific cuts to the high-level model.
    end if
end while

Bi-level decomposition algorithms have been used in literature to solve multi-time scale optimization models. For example, in [13], the authors presented a multi-period MILP integrating the planning and scheduling of single-stage multi-product continuous plants with parallel units. To solve the model with longer time horizons, a bi-level decomposition algorithm was implemented, decomposing the problem into a high-level and low-level model. The high-level model relaxes the detailed scheduling constraints but incorporates advanced sequencing restrictions derived from the traveling salesman problem, providing precise upper bounds on the profit and some of the product assignments. The low-level model focuses on detailed scheduling by solving subproblems with selected products identified by the upper level. Integer cuts were passed as feedback from the low-level model to the high-level model to exclude supersets of previously obtained feasible configurations. The algorithm was tested on several examples and generally outperformed full-space methods using commercial solvers by achieving faster solutions. The algorithm required only a single iteration to solve these examples. This demonstrates that the high-level model was well-formulated.

The bi-level decomposition algorithm can be extended to the multi-level decomposition algorithm that involves more than two levels. For instance, in [14], the authors presented a multi-level decomposition-based framework to solve the integration of planning and scheduling in a multisite, multi-product plant. Given the complexity inherent in large-scale production networks, their approach involves dividing the overall decision process into distinct levels, with each level abstracting certain lower-level constraints to improve computational efficiency. They illustrated the applicability of this framework specifically within the context of paper production. For smaller problems, a two-tier decomposition was utilized, where the higher level managed the simultaneous assignment and sequencing of orders across various sites, while the lower level focused on reducing trim waste and pattern sequencing. For more extensive problems, a four-tier decomposition was employed, covering order distribution, detailed scheduling, trim loss reduction, and pattern sequencing, respectively.

4.4.2.2 Dual-Based Decomposition Algorithms

In this part, we discuss the decomposition algorithms that involve the use of dual variables namely: Benders decomposition, Lagrangian decomposition, and Branch and price algorithm.

These decomposition methods solve problems with different structures. Benders decomposition solves problems with complicating variables like problems that have a representation as shown in Figure 4.3a. Complicating variables are those which, when fixed, make the remaining problem easier to solve in blocks. On the other hand, Dantzig–Wolfe Decomposition (which serves as a key component of Branch and price algorithm) and Lagrangian Decomposition solve problems with complicating constraints like problems that have a representation as shown in Figure 4.3b. Complicating constraints are constraints that when relaxed make the rest of the problem decomposable by blocks. A specific type of complicating constraint is the nonanticipativity constraint (NAC), which ensures that variables in different blocks or subperiods are equalized.

We describe the algorithms in detail below.

Benders Decomposition Benders decomposition is a decomposition algorithm that can be used to solve multi-time scale optimization models with complicating variables. Benders decomposition involves splitting the problem into two sets of

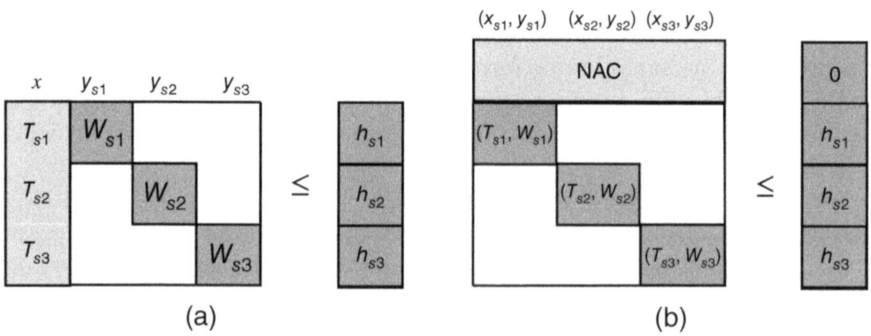

Figure 4.3 Block representation for decomposition algorithms. (a) Benders decomposition and (b) Lagrangian decomposition and Dantzig–Wolfe decomposition.

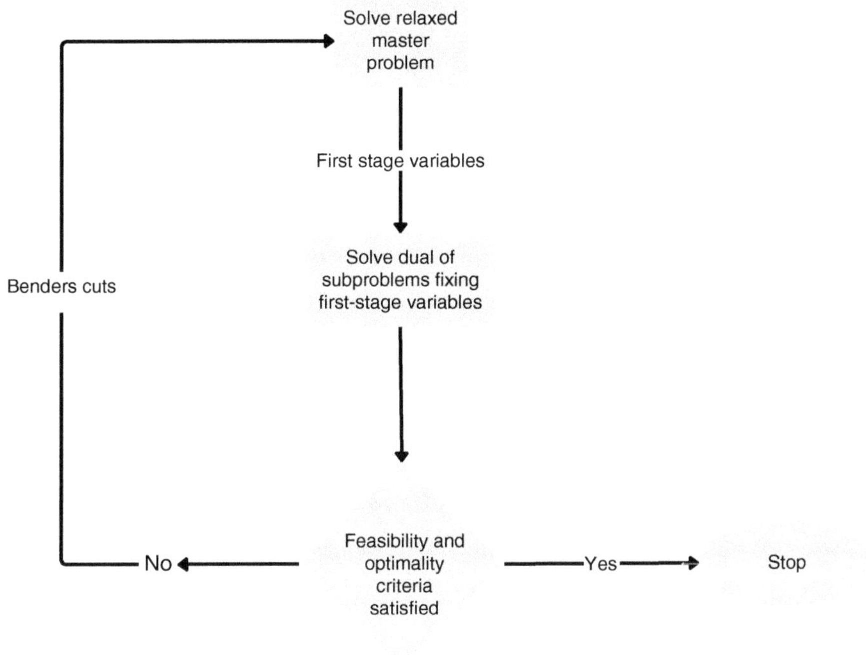

Figure 4.4 Schematic for classical benders decomposition.

problems: a high-level master problem and low-level subproblems. The key idea is to iteratively solve these two problems, using information in the form of cuts derived from the subproblems to refine the master problem. This feedback loop continues until an optimal solution is found as shown in Figure 4.4.

We now explain the classical version of Benders decomposition which requires that the subproblems be linear [15]. Consider solving the problem (4.6) using the Benders Decomposition.

We first decompose the problem into a master problem (4.7) and a set of subproblems (4.8) in the following way:

$$\text{MP}: \quad \min_{\boldsymbol{x},z} c^T\boldsymbol{x} + \sum_{s\in S} z_s \tag{4.7a}$$

$$\text{s.t.} \quad A\boldsymbol{x} = b \tag{4.7b}$$

$$\boldsymbol{x} \geq 0, \tag{4.7c}$$

$$\text{SP}_s: \quad \min_{y} q_s^T\boldsymbol{y} \tag{4.8a}$$

$$\text{s.t.} \quad T_s\boldsymbol{x}^* + W_s\boldsymbol{y} = h_s \tag{4.8b}$$

$$\boldsymbol{y} \geq 0, \tag{4.8c}$$

where $\boldsymbol{x}^*$ represents the high-level decisions obtained from solving the master problem (4.7). If the solutions obtained from the subproblems are feasible and optimal we can stop. If that is not the case, we can improve the performance of the master problem by generating Benders cuts and updating the master problem. To understand the optimality and feasibility of the solution, we investigate the solution obtained from the subproblems. We need the subproblem to be feasible and finite. We can approach this by looking into the dual of the subproblem:

$$\mathrm{SP}_s^{\mathrm{dual}} : \quad \max_{\boldsymbol{p}} \boldsymbol{p}^T(h_s - T_s\boldsymbol{x}^*) \tag{4.9a}$$

$$\text{s.t.} \quad \boldsymbol{p}^T W_s \leq q_s^T \tag{4.9b}$$

with p being the dual variables.

We assume the dual is feasible. This is always true when the primal is bounded.

Three cases can occur for the optimization model corresponding to the dual of the subproblem s:

- **Case 1:** When the dual subproblem (4.9) is unbounded.
  Here, the corresponding primal problem is infeasible. In such a case, there exists an extreme ray $\boldsymbol{w}_s^*$ in the polyhedron $\{\boldsymbol{p}|\boldsymbol{p}^T W_s \leq q_s^T\}$ such that $(\boldsymbol{w}_s^*)^T \left(h_s - T_s\boldsymbol{x}^*\right) > 0$. To handle this case, we generate a cut $(\boldsymbol{w}_s^*)^T \left(h_s - T_s x\right) \leq 0$ which we call the feasibility cut to the master problem (4.7).
- **Case 2:** When the subproblem is feasible and bounded but the solution $\boldsymbol{p}_s^*$ is such that $z_s{}^* < \left(\boldsymbol{p}_s^*\right)^T \left(h_s - T_s\boldsymbol{x}^*\right)$, where $\boldsymbol{p}_s^*$ is an extreme point of the polyhedron $X_s = \{\boldsymbol{p}|\boldsymbol{p}^T W_s \leq q_s^T\}$.
  The optimal dual lower bounds the optimal primal solution. Therefore, we need to have
  $$z_s \geq \max_{\boldsymbol{p} \in X_s} (\boldsymbol{p})^T \left(h_s - T_s x\right).$$
  If this is not ensured, we generate the cut $z_s \geq \left(\boldsymbol{p}_s^*\right)^T \left(h_s - T_s x\right)$ which we call the optimality cut to the master problem.
- **Case 3:** When the dual subproblems are feasible and bounded and $z_s^* \geq \left(\boldsymbol{p}_s^*\right)^T \left(h_s - T_s\boldsymbol{x}^*\right)$ with $\boldsymbol{p}_s^*$ being the solution of the dual subproblem. In such a case, we do not need to add any cut to the master problem.

When all the subproblems are feasible and bounded and provide solutions such that $z_s^* \geq \left(\boldsymbol{p}_s^*\right)^T \left(h_s - T_s\boldsymbol{x}^*\right)$ with $\boldsymbol{p}_s^*$ being the solution of the dual subproblem for subperiod s, the feasibility and optimality criteria are satisfied. That is, the solution derived from the master problem is optimal and we can terminate the algorithm.

The steps to solve a general linear problem with complicating variables are shown in Algorithm 4.2.

Algorithm 4.2 Benders Decomposition Algorithm

1: **Step 1:** Decompose the problem into a master problem (4.7) and a set of subproblems (4.8).
2: **while** true **do**
3: Solve the master problem (4.7) to obtain high-level decisions $\boldsymbol{x}^*$.
4: Using the high-level decisions obtained in Line 3, solve the dual of the subproblems (4.8).
5: **if** The feasibility and optimality criteria are satisfied **then**
6: **Stop:** The current solution is optimal.
7: **else**
8: Generate Benders cuts and update the master problem (4.7).
9: **end if**
10: **end while**

While the master problem gives lower bounds due to its relaxed nature, summing up the subproblems gives upper bounds.

There have been different extensions of Benders decomposition like logic-based Benders decomposition (LBBD) [16], where the subproblems do not have to be linear. These extensions can be used to solve problems like (4.1) with complicating variables. For example, in [17], the authors formulated a mixed integer linear programming model for the integrated process planning and scheduling (IPPS) problem. To address its complexity, they proposed several decomposition methods rooted in LBBD. Their LBBD framework separates the problem into a master problem, which selects process plans and assigns operations to machines, and a subproblem, which handles sequencing and scheduling. To accelerate convergence and strengthen the algorithm, they incorporated relaxations for the makespan objective, introduced a critical-path-based optimality cut, and designed a heuristic to speed up the subproblem solution. The method was tested on 16 well-known benchmark instances, consistently producing either optimal solutions or improvements over the best-known results, thereby demonstrating its advantage over contemporary solution approaches.

Another example is the work [15], which addressed the integration of scheduling and dynamic optimization for batch chemical processes. The authors formulated the integration as a mixed-integer nonlinear dynamic optimization problem and proposed a decomposition strategy based on generalized Benders decomposition (GBD). Within this approach, the master problem handled the scheduling decisions with variable processing times and costs, while the subproblem consisted of separable dynamic optimization problems for individual processing units. Through case studies, they showed that the tailored decomposition scheme dramatically reduced computational effort compared with direct solution methods using general-purpose MINLP solvers, achieving speedups by orders of magnitude.

Lagrangian Decomposition Lagrangian decomposition [18] involves relaxing specific constraints and penalizing their violation. This relaxation simplifies the original problem, allowing it to be decomposed into smaller subproblems, which can be solved independently The relaxation's performance improves through iterative updates based on the solutions of these subproblems as illustrated in Figure 4.5.

Consider solving the optimization model (4.6) by decomposing it by subperiods. The structure of the problem originally looks like Figure 4.3a.

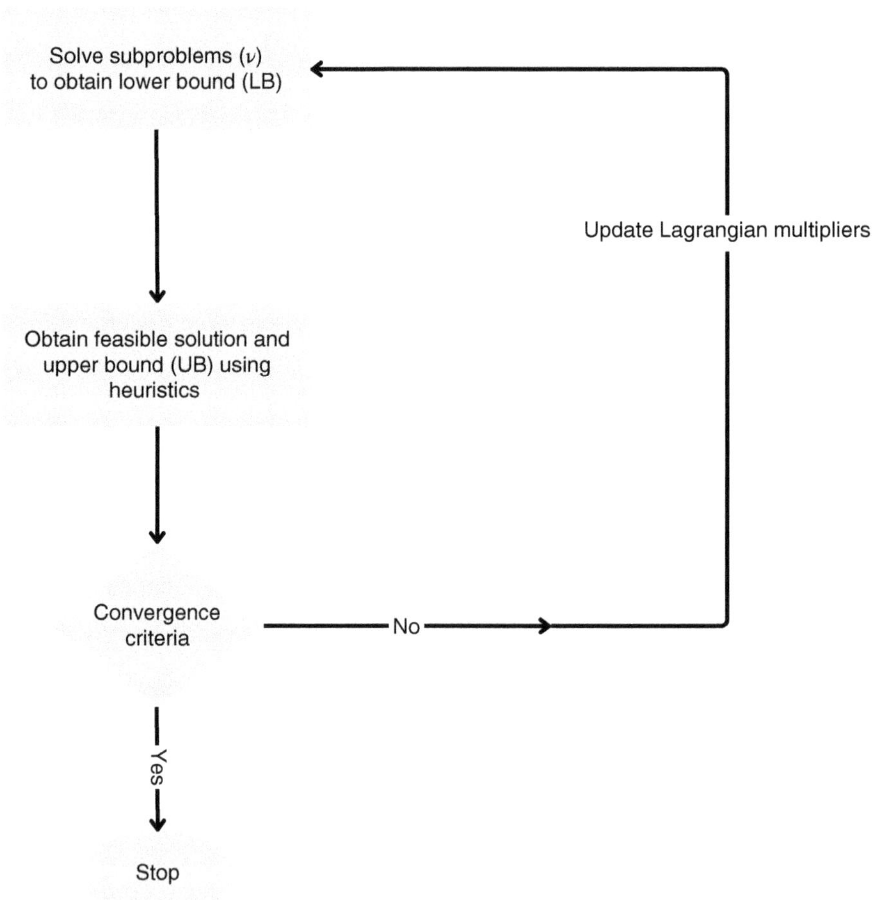

Figure 4.5 Schematic for Lagrangian decomposition.

One way to write this problem is by making copies the variable $\boldsymbol{x}$ for each of the subperiod s, that is, we have a high-level decisions $\boldsymbol{x}_s$ for each of the subperiod s. However, we need to ensure that the high-level decisions are same in all the subperiod s. Therefore, we add the constraints: $\boldsymbol{x}_1 = \boldsymbol{x}_s \quad \forall s \in S, s \neq 1$. These constraints are called Nonanticipativity constraints (NACs). With these modifications, the original problem is written in the following way:

$$\min \quad \sum_{s \in S} \left(\frac{c^T \boldsymbol{x}_s}{|S|} + q_s^T \boldsymbol{y}_s \right) \tag{4.10a}$$

$$\text{s.t.} \quad A\boldsymbol{x}_s = b, \quad \boldsymbol{x}_s \geq 0 \quad \forall s \in S \tag{4.10b}$$

$$T_s \boldsymbol{x}_s + W_s \boldsymbol{y}_s = h_s, \quad \boldsymbol{y}_s \geq 0, \quad \forall s \in S \tag{4.10c}$$

$$\boldsymbol{x}_1 = \boldsymbol{x}_s \quad \forall s \in S, s \neq 1. \tag{4.10d}$$

The structure of the problem now becomes like Figure 4.3b. It is still difficult to decompose the problem (4.10) due to the NACs. To address this, we relax these constraints and introduce a penalty for the deviation in high-level decisions, using weights known as Lagrangian multipliers. The process of relaxing and penalizing the violation of constraints is called "dualizing the constraints" [18]. The relaxed problem is shown below:

$$\min \quad \sum_{s \in S} \left(\frac{c^T \boldsymbol{x}_s}{|S|} + q_s^T \boldsymbol{y}_s \right) + \sum_{s \in S, s \neq 1} \boldsymbol{v}_s^T \left(\boldsymbol{x}_1 - \boldsymbol{x}_s \right) \tag{4.11a}$$

$$\text{s.t.} \quad A\boldsymbol{x}_s = b, \quad \boldsymbol{x}_s \geq 0 \quad \forall s \in S \tag{4.11b}$$

$$T_s \boldsymbol{x}_s + W_s \boldsymbol{y}_s = h_s, \quad \boldsymbol{y}_s \geq 0, \quad \forall s \in S, \tag{4.11c}$$

where $\boldsymbol{v}_s$ are the Lagrangian multipliers.

We now decompose the model (4.11) by subperiod s into the following subproblems:

$$D_1(v) := \min_{\boldsymbol{x}, \boldsymbol{y}} \quad c^T \boldsymbol{x} + q_1^T \boldsymbol{y} + \sum_{s \in S, s \neq 1} \boldsymbol{v}_s^T \boldsymbol{x} \tag{4.12a}$$

$$\text{s.t.} \quad A\boldsymbol{x} = b, \quad \boldsymbol{x} \geq 0 \tag{4.12b}$$

$$T_1 \boldsymbol{x} + W_1 \boldsymbol{y} = h_1, \quad \boldsymbol{y} \geq 0. \tag{4.12c}$$

$$D_{s \in S, s \neq 1}(v) := \min_{\boldsymbol{x}, \boldsymbol{y}} \quad c^T \boldsymbol{x} + q_s^T \boldsymbol{y} - \boldsymbol{v}_s^T \boldsymbol{x} \tag{4.13a}$$

$$\text{s.t.} \quad A\boldsymbol{x} = b, \quad \boldsymbol{x} \geq 0 \tag{4.13b}$$

$$T_s \boldsymbol{x} + W_s \boldsymbol{y} = h_s, \quad \boldsymbol{y} \geq 0. \tag{4.13c}$$

In the above formulation, v is the set of all the Lagrangian multipliers. These subproblems are relaxed as they are independent and do not explicitly take into account the equality of the high-level decisions between different subperiods. Thus, when the objectives of the subproblems are summed up they give a lower bound. We want to get the best lower bound by maximizing over all the possible Lagrangian multipliers. The best lower bound is called the Lagrangian dual which is defined as $v(\text{LD}) = \max_{v} \sum_{s\in S} D_s(v)$. Summing up all the objective functions of the subproblems, the Lagrangian dual can be written as the following way:

$$v(\text{LD}) = \max_{v} \min_{x_s^k, y_s^k} \sum_{s\in S} \left(\frac{c^T \boldsymbol{x}_s^k}{|S|} + q_s^T \boldsymbol{y}_s^k \right) + \sum_{s\in S, s\neq 1} \boldsymbol{v}_s^T \left(\boldsymbol{x}_1^k - \boldsymbol{x}_s^k \right) \quad \forall (\boldsymbol{x}_s^k, \boldsymbol{y}_s^k) \in \mathcal{K}_s, s \in S$$

with $X_s = \{\boldsymbol{x}, \boldsymbol{y} | A\boldsymbol{x} = b, \boldsymbol{x} \geq 0, T_s \boldsymbol{x} + W_s \boldsymbol{y} = h_s, \boldsymbol{y} \geq 0\}$, and K_s = Extreme points (conv(X_s)). We assume that the subproblems are bounded. We use the extreme points of the convex hull because they define the boundaries of the feasible region.

The maximization-minimization problem expressed above can be written in the following way:

$$\begin{aligned} \max_{v,\eta} \quad & \eta \\ \text{s.t.} \quad & \eta \leq \sum_{s\in S} \left(\frac{c^T \boldsymbol{x}_s^k}{|S|} + q_s^T \boldsymbol{y}_s^k \right) + \sum_{s\in S, s\neq 1} \boldsymbol{v}_s^T \left(\boldsymbol{x}_1^k - \boldsymbol{x}_s^k \right) \quad \forall (\boldsymbol{x}_s^k, \boldsymbol{y}_s^k) \in \mathcal{K}_s. \end{aligned} \tag{4.14}$$

The problem (4.14) can also be written in the following way:

$$\begin{aligned} \max_{v,\eta} \quad & \sum_{s\in S} \eta_s \\ \text{s.t.} \quad & \eta_1 \leq \frac{c^T \boldsymbol{x}_1^k}{|S|} + q_1^T \boldsymbol{y}_1^k + \sum_{s\in S, s\neq 1} \boldsymbol{v}_s^T \boldsymbol{x}_1^k \quad \forall (\boldsymbol{x}_1^k, \boldsymbol{y}_1^k) \in \mathcal{K}_1 \\ & \eta_s \leq \frac{c^T \boldsymbol{x}_s^k}{|S|} + q_s^T \boldsymbol{y}_s^k - \boldsymbol{v}_s^T \boldsymbol{x}_s^k \quad \forall (\boldsymbol{x}_s^k, \boldsymbol{y}_s^k) \in \mathcal{K}_s, s \in S, s \neq 1. \end{aligned} \tag{4.15}$$

Equation (4.14) is called the Lagrangian master problem with (4.15) being another way of writing the Lagrangian master problem.

The Lagrangian master problem (4.15) can be solved to get the optimal Lagrangian multipliers. It is very difficult to obtain all the extreme points and there can also be an exponential number of extreme points which makes the Lagrangian master problem intractable. Therefore we can start with a subset of extreme points and iteratively update the master problem by obtaining more extreme points/feasible solutions from the subproblems. That is, we can start with a set of multipliers, solve the subproblems using the set multipliers, and update the master problem by adding the constraints $\eta_1 \leq \frac{c^T \boldsymbol{x}_1^k}{|S|} + q_1^T \boldsymbol{y}_1^k + \sum_{s\in S, s\neq 1} v_s \boldsymbol{x}_1^k, \eta_s \leq \frac{c^T \boldsymbol{x}_s^k}{|S|} + q_s^T \boldsymbol{y}_s^k - v_s \boldsymbol{x}_s^k \ \forall s \in S,\ s \neq 1$ using the solution obtained $\boldsymbol{x}_s^k, \boldsymbol{y}_s^k$ from the subproblems. We can then solve the updated master problem to obtain new multipliers. The constraint added is called the cutting plane, and the method is called the cutting plane method [19].

Another way to find the Lagrangian multipliers is by using subgradients. In the above problem (4.14), $[\boldsymbol{x}_1 - \boldsymbol{x}_s]_{s\in S, s\neq 1}$ represents a direction for the subgradient of the master problem. We can use the subgradients to solve the Lagrangian master problem.

This method involves starting with an initial set of Lagrangian multipliers, solving the subproblems using the set multipliers, and updating the multipliers using the solutions obtained in the following way:

$$\boldsymbol{\nu}_{k+1,s} = \boldsymbol{\nu}_{k,s} + \lambda_k(\boldsymbol{x}_1^k - \boldsymbol{x}_s^k) \quad \forall s \in S, s \neq 1,$$

where $\boldsymbol{x}_1^k, \boldsymbol{x}_s^k$ are the solutions from solving the subproblems D_1, D_s respectively, $\nu_{k,s}$ is the old Lagrangian multiplier, $\nu_{k+1,s}$ is the new Lagrangian multiplier and λ_k is the step size. This method is called the subgradient method.

In practice, a step size that has been proven to be effective is the following:

$$\lambda_k = \frac{v^* - v_k(\mathrm{LD})}{\sum_{s\in S, s\neq 1} ||\boldsymbol{x}_1^k - \boldsymbol{x}_s^k||^2},$$

where $v_k(\mathrm{LD})$ is the Lagrangian dual obtained, and v^* is the upper bound obtained from heuristics.

We can use the solutions obtained from the subproblems to obtain feasible solutions and correspondingly update upper bounds on the optimal objective. This iterative process is continued till the gap between the upper bound and lower bound is within a set a limit of tolerance or we obtain the same Lagrangian multipliers in two consecutive iterations.

The above can be extended to solve (4.1) using the steps in Algorithm 4.3.

Algorithm 4.3 Lagrangian Decomposition

1: **Initialize:** Set the Upper Bound (UB) $\leftarrow \infty$, Lower Bound (LB) $\leftarrow -\infty$, k = 1.
2: Decompose the problem into independent subproblems by making copies of the high-level variables x and low-level decision variables y connected through the complicating constraints.
3: Start with an initial point for the Lagrangian multipliers ν_k.
4: **while** true **do**
5: Solve the subproblems using the Lagrangian multipliers ν_k and update the Lower Bound (LB).
6: Use heuristics to obtain upper bounds on the optimal objective.
7: **if** UB − LB $\leq$ Tolerance or $k \geq 2, \nu_k = \nu_{k-1}$ **then**
8: **Stop:** Convergence achieved.
9: **else**
10: Update $k \rightarrow k+1$ and the Lagrangian multipliers ν_k.
11: Go back to step 3.
12: **end if**
13: **end while**

Line 10 of the algorithm can be done using the subgradient method or the cutting plane method.

Different extensions of Lagrangian decomposition/relaxation methods have been used to solve multi-time scale optimization models. For example, in [20], the authors addressed an integrated planning and scheduling problem for multisite batch facilities operating in global markets by formulating it as an MILP. To solve this model, they employed an augmented Lagrangian decomposition strategy, where the coupling constraints linking planning and scheduling variables were dualized. Unlike standard Lagrangian methods that rely only on linear terms, the augmented version also introduces a quadratic penalty term, which enhances convergence and improves numerical stability. Since this quadratic term leads to non-separable cross-product expressions in the augmented Lagrangian function, a diagonal approximation was applied to simplify computations. Tests on three case studies demonstrated that the proposed decomposition method achieved faster solution times relative to directly solving the complete model.

Progressive hedging [21] is another extension that relaxes nonanticipativity conditions, allowing the scenario subproblems to be solved independently while penalizing deviations from averaged decision values. For instance, in [22], the authors introduced a customized progressive hedging-based framework for integrated planning and scheduling under demand uncertainty. They incorporated several enhancements aimed at speeding up the algorithm and ensuring convergence. The method was validated on variations of a standard state–task network benchmark and further applied to an industrial ethylene plant case. Computational experiments demonstrated that the proposed approach achieved superior performance compared with directly applying commercial solvers.

Branch and Price Column Generation methods are techniques used to solve large-scale linear programming (LP) problems that involve an overwhelming number of variables. Instead of solving the entire problem with all variables, these methods start with a restricted subset of variables, and iteratively improve this subset by solving subproblems. The subproblems identify new variables (columns) that can potentially improve the solution. This iterative approach continues until no further improvement can be made, effectively solving the full LP with a manageable number of variables.

The Dantzig–Wolfe Decomposition is a classical algorithm that employs column generation to solve linear programming problems with complicating constraints. Complicating constraints are those that connect variables across different subproblems, often creating dependencies that make solving the problem as a whole difficult.

The Dantzig–Wolfe algorithm reformulates the original problem by decomposing it into a master problem and a set of subproblems. In the Dantzig–Wolfe framework, the master problem initially contains only a subset of the variables, and new variables (columns) are iteratively generated by solving the subproblems. This process continues until an optimal solution is found which is implied by non-negative reduced costs, as shown in Figure 4.6.

We now discuss Dantzig–Wolfe decomposition in detail. Consider solving the problem (4.6). Thc problem can be modified into (4.10). The NACs are

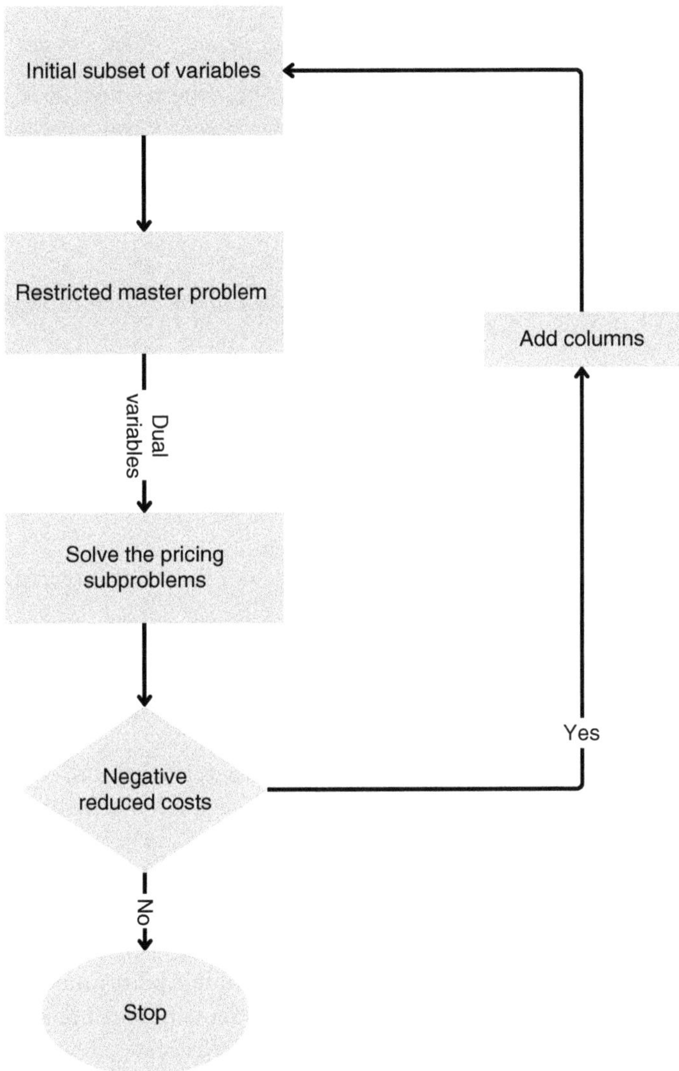

Figure 4.6 Schematic for Dantzig–Wolfe decomposition.

complicating constraints. The structure problem now becomes like Figure 4.3b. The problem can be further written as follows:

$$\min \quad \sum_{s\in S}\left(\frac{c^T\boldsymbol{x}_s}{|S|}+z_s\right) \tag{4.16a}$$

$$\text{s.t.} \quad \boldsymbol{x}_1=\boldsymbol{x}_s \quad \forall s\in S, s\neq 1 \tag{4.16b}$$

$$\boldsymbol{x}_s, z_s \in X_s \quad \forall s \in S, \tag{4.16c}$$

where $X_s = \{\boldsymbol{x}, z|A\boldsymbol{x}=b, \boldsymbol{x}\geq 0, T_s\boldsymbol{x}+W_s\boldsymbol{y}=h_s, \boldsymbol{y}\geq 0, z=q_s^T\boldsymbol{y}\}$.

Let $(\tilde{\boldsymbol{x}}^j_{x,s}, \tilde{\boldsymbol{x}}^j_{z,s})$ be the extreme points and $(\tilde{\boldsymbol{w}}^k_{x,s}, \tilde{\boldsymbol{w}}^k_{z,s})$ be extreme rays of X_s. The points in the polyhedrons X_s can be represented by a convex combination of the extreme points and a conic combination of the extreme rays of the polyhedron [23]. Taking this into account, we can reformulate the LP in the following way:

$$\min \quad \sum_{s\in S}\sum_{j\in J_s}\lambda^j_s\left(\frac{c^T\tilde{\boldsymbol{x}}^j_{x,s}}{|S|}+\tilde{x}^j_{z,s}\right)+\sum_{s\in S}\sum_{k\in K_s}\theta^k_s\left(\frac{c^T\tilde{\boldsymbol{w}}^k_{x,s}}{|S|}+\tilde{w}^k_{z,s}\right) \tag{4.17a}$$

$$\text{s.t.} \quad \sum_{j\in J_1}\lambda^j_1\tilde{\boldsymbol{x}}^j_{x,1}+\sum_{k\in K_1}\theta^k_1\tilde{\boldsymbol{w}}^k_{x,1}=\sum_{j\in J_s}\lambda^j_s\tilde{\boldsymbol{x}}^j_{x,s}+\sum_{k\in K_s}\theta^k_s\tilde{\boldsymbol{w}}^k_{x,s} \quad \forall s\in S, s\neq 1 \tag{4.17b}$$

$$\sum_{j\in J_s}\lambda^j_s=1 \quad \forall s\in S \tag{4.17c}$$

$$\lambda^j_s\geq 0, \quad \forall s\in S,\ j\in J_s, \quad \theta^k_s\geq 0 \quad s\in S,\ k\in K_s. \tag{4.17d}$$

We call this reformulated problem the master problem. The master problem is the linear programming dual of the Lagrangian master problem [24].

Let $\boldsymbol{v}_s$ be the set of dual variables for the constraint (4.17b) corresponding to s and r_s be the dual of the constraint (4.17c) corresponding to s.

To check if the solution obtained is optimal, we need to check the reduced costs of the variables and examine whether any one of them is negative. The reduced cost of λ^j_1 is $\frac{c^T\tilde{\boldsymbol{x}}^j_{x,1}}{|S|}+\tilde{x}^j_{z,1}-\sum_{s\in S,s\neq 1}\boldsymbol{v}_s^T\tilde{\boldsymbol{x}}^j_{x,1}-r_1$ and the reduced cost of θ^k_1 is $\frac{c^T\tilde{\boldsymbol{w}}^k_{x,1}}{|S|}+\tilde{w}^k_{z,1}-\sum_{s\in S,s\neq 1}\boldsymbol{v}_s^T\tilde{\boldsymbol{w}}^k_{x,1}$. Similarly, for $s\in S, s\neq 1$, the reduced cost of λ^j_s is $\frac{c^T\tilde{\boldsymbol{x}}^j_{x,s}}{|S|}+\tilde{x}^j_{z,s}+\boldsymbol{v}_s^T\tilde{\boldsymbol{x}}^j_{x,s}-r_s$, and the reduced cost of θ^k_s is $\frac{c^T\tilde{\boldsymbol{w}}^k_{x,s}}{|S|}+\tilde{w}^k_{z,s}+\boldsymbol{v}_s^T\tilde{\boldsymbol{w}}^k_{x,s}$.

While solving the complete master problem with all the extreme points and rays gives us the solution, it is very difficult to obtain all the extreme rays and points. Therefore, we start with a restricted master problem with a few extreme points and extreme rays and iteratively add them to the restricted master problem.

The goal is to identify if any reduced costs are negative.

The extreme points and extreme rays are columns that can be generated by solving the following pricing sub problems:

$$\mathrm{SP}_1 : \min \frac{c^T \boldsymbol{x}}{|S|} + z - \sum_{s \in S, s \neq 1} \boldsymbol{v}_s^T \boldsymbol{x}$$
$$\text{s.t.} \quad \boldsymbol{x}, z \in X_1.$$

$$\mathrm{SP}_{s \in S, s \neq 1} : \min \frac{c^T \boldsymbol{x}}{|S|} + z + \boldsymbol{v}_s^T \boldsymbol{x}$$
$$\text{s.t.} \quad \boldsymbol{x}, z \in X_s.$$

If the pricing problem gives an objective of $-\infty$, the solution is an extreme ray which can then be added to the master problem. If the pricing problem gives a finite objective which is less than the r_s, the solution is an extreme point that can then be added to the master problem. If the optimal objective from all the pricing subproblems SP_s are finite and greater than r_s, the reduced cost of all the variables λ_s^j, θ_s^k are non-negative, implying that the solution obtained from the master problem is optimal.

The steps for solving a linear programming problem using Dantzig–Wolfe decomposition are shown in Algorithm 4.4.

Algorithm 4.4 Dantzig–Wolfe Decomposition Algorithm

Initialize: Start with a basic feasible solution to the master problem.
while True **do**
    Solve the master problem and obtain the dual variables for the constraints
    Solve the pricing subproblems SP_s.
    if all subproblems yield finite optimal objectives that are greater than corresponding dual r_s **then**
        Stop: Current solution is optimal.
    else
        For subproblems returning an optimal objective of $-\infty$ add the solution as an extreme ray to the master problem.
        For subproblems returning a finite optimal objective less than r_s add the solution as an extreme point to the master problem.
    end if
    Repeat.
end while

While the master problem gives upper bounds because of its restricted nature, summing up the pricing subproblems gives lower bounds due to its relaxed nature. Note that unlike Benders decomposition and Lagrangian decomposition, where the solution of the master problem converges from a smaller value to a larger one, the master problem in Dantzig–Wolfe decomposition moves in the opposite direction, from a larger value to a smaller one. This difference arises because in Dantzig–Wolfe decomposition, the master problem initially starts with a restricted set of variables and, with each iteration, adds more columns (subsets of variables), gradually expanding the feasible region. As more variables are introduced, the solution improves, leading to a lower objective value. In contrast, Benders and Lagrangian decompositions iteratively introduce new constraints or penalties in the form of cuts or relaxations, progressively tightening the feasible region of the master problem. As a result, their master problems start with a broader feasible region and converge upwards as they are constrained more heavily with each iteration.

The steps explained in Algorithm 4.4 can be used directly to solve linear programming problems with complicating constraints. However, they cannot be directly used to solve MILPs with complicating constraints. In such cases, we can embed column generation methods like the Dantzig–Wolfe decomposition algorithm, into the branch and bound framework. The branch and bound method is a method commonly used to solve MILPs [25]. The branch and bound method begins by solving the linear programming (LP) relaxation of the MILP, which ignores the integer constraints and provides an initial solution and bound on the objective value. If the solution to this relaxation includes fractional values for integer variables, the problem is split into child problems by branching on these fractional variables – creating new child problems with additional constraints that force the variable to be either rounded up or down. Each child problem is then solved, and bounds on the objective function are calculated. Problems that are infeasible or have bounds worse than the current best solution are pruned, eliminating them from further consideration. The algorithm selects the next new problem to explore based on strategies like depth-first or best-bound-first search and continues branching, bounding, and pruning until all child problems are either solved or discarded.

To accelerate the branch and bound process, which involves solving multiple LPs, column generation methods can be integrated to solve the LPs more efficiently. This hybrid combination is known as branch and Price. Branch and price can also address some MINLP problems by constructing a Dantzig–Wolfe reformulation of the problem via a discretization approach [26]. Application of branch and price to solve multi-time scale optimization models include [27]. In this work, the authors developed an MILP formulation for an integrated production scheduling and distribution problem with routing considerations in a multisite supply chain. To address the problem, a branch-and-price algorithm was designed, where initial columns were generated using a Bees algorithm. The computational study on a variety of test cases showed that the proposed framework outperformed commercial solvers and produced tighter bounds compared with a conventional branch-and-price approach without the Bees algorithm.

4.4.2.3 Limitations of Decomposition Algorithms

An important limitation encountered while using decomposition algorithms is that these algorithms are not scalable to very large problems. While applying these algorithms to very large problems with say tens of millions of variables, a large number of iterations are needed to converge and that could require a large amount of time. Another important aspect of these algorithms is that these decomposition algorithms are restricted to problems of a particular structure.

4.4.3 Metaheuristic Algorithms

Metaheuristic algorithms are problem-independent methods often inspired by natural processes and designed to find approximate solutions to optimization problems by exploring a search space. They provide a framework that guides the process of searching for a solution toward optimal or near-optimal solutions by combining various heuristic methods and incorporating randomness and iterative improvements. This is done by balancing between exploring new regions of the search space (exploration) and intensifying the search around the best solutions found so far (exploitation).

An example of a metaheuristic algorithm is the genetic algorithm (GA) [28] inspired by the process of natural selection. It involves the steps shown in Algorithm 4.5.

Algorithm 4.5 Genetic Algorithm

1: **Initialize:** Generate an initial population of solutions (chromosomes).
2: **while** stopping criterion not met **do**
3: Evaluate the fitness/objective of each solution in the population.
4: Select some of the solutions based on their fitness using a suitable heuristic
5: Apply crossover and mutation operations to generate new solutions.
6: Replace some or all of the old population with the newly generated solutions.
7: **end while**

For line 4, there have been a lot of heuristics described in literature like proportionate reproduction, ranking selection, and tournament selection [29].

An application of the genetic algorithm to solve multi-time scale optimization problems includes the work in [30]. The authors of this work focused on IPPS, a complex combinatorial problem involving multiple alternatives and constraints. To deal with these challenges, they proposed a new chromosome structure that encodes all IPPS operations and flexibilities within a single representation, enabling an efficient adaptation of the genetic algorithm. Their approach was evaluated on benchmark instances, and the findings indicated that it achieved shorter makespans and faster computation compared with other metaheuristics.

Other metaheuristic algorithms used to solve multi-time scale optimization models include simulated annealing [31] inspired by the annealing process in metallurgy, ant colony optimization [32] inspired by the actions of an ant colony, and variable neighborhood search [33] that involves systematically exploring the neighborhoods of solutions.

Despite their versatility and effectiveness, metaheuristic methods face scalability issues, particularly when dealing with problems involving more than a thousand variables [34].

4.4.4 Matheuristic Algorithms

Another method to solve multi-time scale optimization models involves algorithms that combine mathematical programming and heuristics which are called matheuristics [35]. An example of this is local branching [36], a technique used in mixed-integer programming (MIP) to improve solutions by iteratively exploring a sequence of solution neighborhoods in an MILP model. The steps involved in local branching are shown in Algorithm 4.6.

Algorithm 4.6 Local Branching Algorithm

Use a heuristic method to generate an initial feasible solution x^{init}.
Set initial neighborhood size k, and maximum iterations max_iter.
Initialize iter $\leftarrow 0$ and $x^{\text{best}} \leftarrow x^{\text{init}}$.
while iter < max_iter **do**
    Add local branching constraints to the MILP to limit the neighborhood to at most k changes from x^{best}.
    Solve the modified MILP to obtain a solution x^{new}.
    if an improved solution is found, i.e. $f(x^{\text{new}}) < f(\text{best\_solution})$ **then**
        Update $x^{\text{best}} \leftarrow x^{\text{new}}$.
        Optionally reset or reduce k to intensify the search.
    else
        Increase k to explore a broader region.
    end if
    Increment iteration counter: iter $\leftarrow$ iter $+ 1$.
end while
Return best_solution.

More complex and problem-specific matheuristics have been employed to solve multi-time scale optimization models. For instance, in [9], the authors proposed a MILP model that integrates the planning and scheduling of modular electrified plants, renewable-based generating units, and power lines within a microgrid. The decisions included single-time investment decisions on the location of the plants, renewable resources as well as power lines, monthly transportation and inventory decisions as well as hourly operating decisions. To solve the model, which included variables across three time scales, a K-means clustering-based aggregation-disaggregation matheuristic was introduced. This matheuristic involved two major steps: first, aggregating variables based on their locations into clusters using the K-means clustering algorithm on their coordinates, and second, disaggregating the investment decisions made into these clusters through a heuristic. To further obtain the power lines connecting locations in different clusters, a matching ILP was solved. Finally, to obtain the monthly and hourly decisions, the investment decisions were fixed in the full-space model. The matheuristic effectively reduced the problem's complexity by solving multiple small-sized MILPs instead of one large-sized MILP, making it particularly effective given the NP-hard nature of MILPs. The matheuristic was tested on a 20-location problem and was able to outperform commercial solvers.

Other examples include the relax-and-fix heuristic [37], proposed to solve an integrated multiproduct, multiperiod, and multistage capacitated lot sizing with a hybrid flow shop, as well as the spatial aggregation and decomposition algorithm proposed in [38] to solve large multi-time scale energy problems.

While matheuristics are efficient for problems with small to moderate size [9], they are not scalable to larger problems and are problem-specific.

4.4.5 Data-Driven Methods

Data-driven methods leverage data to solve optimization models. There have been different data-driven methods used such as surrogate models and reinforcement-learning-based algorithms. The most common method used is the surrogate model, which is a simpler model used to represent a complex model [2, 39–44]. An example of a surrogate-based method used to solve multi-time scale optimization models involves using surrogate models to represent the feasible region or carry out feasibility analysis [45], particularly for the low-level variables. The surrogate model for the feasible low-level variables can then be included in a high-level model to obtain the high-level variables. The steps involved are shown in Algorithm 4.7.

Algorithm 4.7 Surrogate Model Approach for High-Level Optimization

1: **Step 1:** Formulate a high-level optimization model with unknown constraints for the low-level decision variables/black box model.
2: **Step 2:** Obtain samples for the high-level decision variables and assess their output/feasibility in the low-level model.
3: **Step 3:** Determine algebraic approximations for the unknown constraints, through feasibility analysis and machine learning methods like regression and classification. These equations form a surrogate model for the detailed low-level model.
4: **Step 4:** Integrate the obtained constraints into the high-level model. Solve the high-level model to obtain good solutions.

An application of this method can be found in the work [42], where the authors utilized this approach to develop a framework for integrating planning, scheduling, and control. Their approach first considered scheduling and control integration, where the control system was treated as a black box with unknown constraints. To approximate the system's behavior, surrogate models were developed to estimate state variables, performance measures, and average flows as functions of control parameters. In the next stage, this integrated scheduling-control problem was itself treated as a black box within the planning–scheduling integration. Here, surrogate models were again employed, ensuring that aggregate production targets were feasible while also predicting energy requirements with respect to the integrated scheduling-control problem. To mitigate computational challenges arising from high dimensionality, the authors incorporated feature selection in constructing the surrogate models, which improved tractability and efficiency. The framework was demonstrated through a case study involving enterprise-wide optimization of air separation plants, highlighting its practical applicability.

Reinforcement learning-based approaches are another subset of data-driven methods used to solve multi-time scale optimization problems with multiple time periods [3, 46]. Reinforcement learning (RL) algorithms learn optimal actions by interacting with an environment to maximize cumulative rewards. To implement RL algorithms for solving multi-time scale optimization models, the optimization model – or a component of it, such as the high-level version – is modeled as a Markov Decision Process (MDP) [47]. In an MDP, the system occupies a state s at each discrete time step. The decision-maker selects an action a that is available. As a consequence, the system transitions to a subsequent state s' and produces a reward, which is denoted by $R_a(s, s')$. A representation of an MDP is shown in Figure 4.7. The goal is to maximize the cumulative rewards. The MDP can then be solved using RL algorithms.

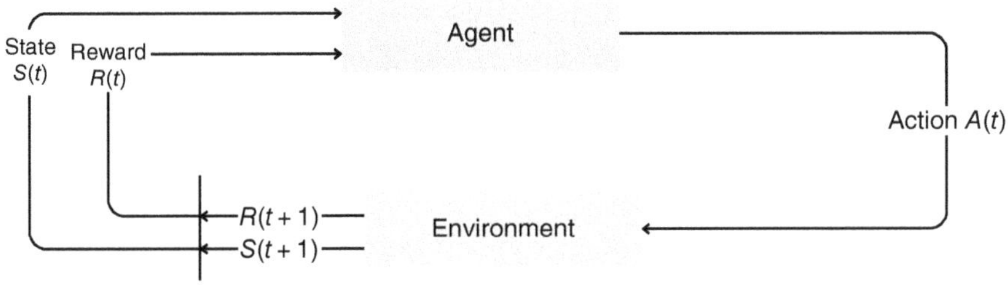

Figure 4.7 Markov decision process.

An example in the literature where this method was used is [3], where the authors proposed a multi-timescale decision-making framework that integrates MDP with mathematical programming (MP) in a complementary manner. To make the resulting MP-embedded MDP problem computationally feasible, they developed a solution strategy based on reinforcement learning (RL). Another example is the work [46], where the authors proposed a novel multi-agent deep reinforcement learning framework for efficient multi-time scale bidding for hybrid power plants.

While these methods are effective for problems of moderate size, they struggle with scalability for larger problems due to the need for extensive parameter dimensions and a large number of samples.

4.4.6 PAMSO

Parametric autotuning multi-time scale optimization (PAMSO) [48] algorithm is a scalable and transferable algorithm used to solve multi-time scale optimization models. It draws inspiration from parametric cost function approximations (CFAs) [49] and from tuning controllers using derivative-free optimization techniques [50, 51]. PAMSO involves tuning selected parameters in a low-fidelity optimization model to achieve a good performance in a high-fidelity optimization model as shown in Figure 4.8.

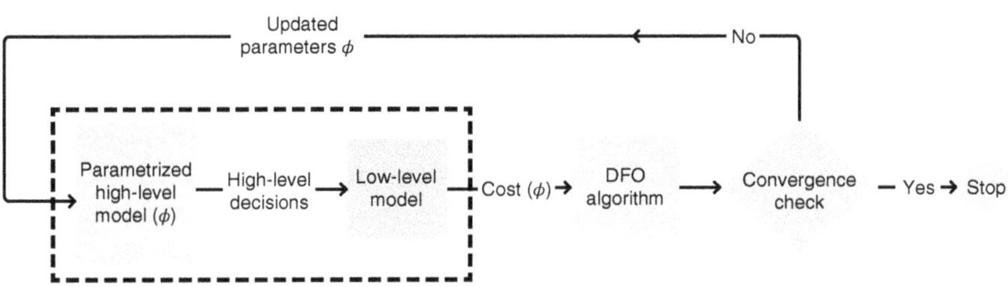

Figure 4.8 Schematic for PAMSO.

PAMSO involves splitting the multi-time scale optimization model into a high-level model and low-level model. The high-level model provides high-level decisions which can then be fixed in the low-level model to obtain the low-level decisions as well as the objective of the multi-time scale optimization model. This is similar to the one-way communication method with (4.2) being the high-level model and (4.3) being the low-level model.

We improve the performance of the high-level model by adding tunable parameters to the high-level model, to help better reflect the mismatch in variations between the high-level and low-level models. The parameterized high-level model is as follows:

$$\min_{\boldsymbol{x},\boldsymbol{z}} \quad \tilde{f}(\boldsymbol{x}, \boldsymbol{\rho}) + \tilde{Q}(\boldsymbol{x}, \boldsymbol{z}, \tilde{\theta}, \boldsymbol{\rho}) \tag{4.18a}$$

$$\text{s.t.} \quad \tilde{g}(\boldsymbol{x}, \boldsymbol{\rho}) \leq 0 \tag{4.18b}$$

$$\tilde{H}(\boldsymbol{x}, \boldsymbol{z}, \tilde{\theta}, \boldsymbol{\rho}) \leq 0. \tag{4.18c}$$

Here, $\boldsymbol{\rho}$ represents the set of tunable parameters we introduce. The functions f, Q, g, and H are modified to $\tilde{f}$, $\tilde{Q}$, $\tilde{g}$, and $\tilde{H}$, respectively.

We now treat the system of high-level and low-level models as a black box that takes in the tunable parameters as input and gives the corresponding objective of the multi-time scale optimization model as an output. This system is called the Multi-time scale black box function (MBBF). We can optimize the MBBF by tuning the parameters using derivative free optimizers (DFOs). The steps used in PAMSO are shown in Algorithm 4.8.

Algorithm 4.8 Optimization of Multi-time Scale Model with Tunable Parameters

1: **Step 1:** Split the multi-time scale optimization model into a high-level model (4.2) and a low-level model (4.3).
2: **Step 2:** Add tunable parameters to the high-level model.
3: **Step 3:** Treat the system comprising the parameterized high-level and low-level models as a black box function. Optimize the black box function by tuning the parameters using Derivative-Free Optimization (DFO) techniques.

An important advantage of PAMSO is its ability to incorporate transfer learning by learning parameters from problems of smaller size to solve problems of large size. This can significantly reduce the time to solve large-sized problems.

While PAMSO has been explored to solve case studies involving high-level decisions that are neither temporally interdependent nor variable over time, there is a need to check its effectiveness in cases where the high-level decisions are temporarily interdependent.

4.5 Illustrative Example

In this section, we provide an example of a multi-time scale optimization problem and attempt to solve it with a few of the algorithms described in Section 4.4.

4.5.1 Problem Statement

An electric utility is installing two generators (indexed by $j = 1, 2$) with different fixed and operating costs, in order to meet the demand within its service region for **3 representative days** s. Each day is divided into three parts of 8 hours each, indexed by $i = 1, 2, 3$. These correspond to parts of the day during which demand takes a base, medium, or peak value, respectively. The fixed cost per unit capacity of generator j is amortized over its lifetime and amounts to c_j per day. The operating cost of generator j during the ith part of the day is $f_{i,j}$. If the demand during the ith part of the day cannot be served due to lack of capacity, additional capacity must be purchased at a cost of g_s on the sth day per unit capacity purchased.

The demand for the ith part of the day on the sth day is $d_{i,s}$. Furthermore, the availability of generator j on the ith part of the day for s^{th} day is $a_{s,i,j}$. To optimize the cost, we need to optimize the installed capacity of the generator as well as the operating levels of the generators and the amount of power purchased to meet the unmet demand at different times of the day in different representative days. For this, we formulate a multi-time scale model with decisions in three-time scales: single-time decisions and decisions for the different parts of different days (corresponding to 2 scales). The integrated model is formally defined in Section 4.5.2.

Note: The problem we use is inspired by Example 6.5 in [23] and has been adapted to fit the multi-time scale setting.

4.5.2 Integrated Model

4.5.2.1 Indices and Sets

$j \in \mathcal{J}$ set of generators
$i \in \mathcal{I}$ set of parts of the day for 8 hours each
$s \in \mathcal{S}$ set of days

4.5.2.2 Variables

x_j installed capacity of generator j (kW)
$y_{s,i,j}$ operating levels of generator j during the ith part of the day on the sth day (kW)
$\tilde{y}_{s,i}$ power purchased in the ith part of the day on the sth day (kW)

4.5.2.3 Parameters

$a_{s,i,j}$ availability of generator j on i^{th} part of the day for s^{th} day
c_j fixed cost per unit capacity of generator j per day amortized over its lifetime (\$/kW)
$d_{s,i}$ demand of power during the ith part of the day on the sth day (kW)
$f_{i,j}$ operating cost of generator j during the ith part of the day (\$/kW)
g_s cost of additional capacity purchased per power purchased on the sth day (\$/kW)

4.5.2.4 Constraints

We add a constraint to ensure that the capacity of each generator j is non-negative.

$$x_j \geq 0 \quad \forall j \in \mathcal{J}. \tag{4.19}$$

The operating variables are non-negative.

$$y_{s,i,j} \geq 0 \quad \forall i \in \mathcal{I}, s \in \mathcal{S}, j \in \mathcal{J}. \tag{4.20}$$

$$\tilde{y}_{s,i} \geq 0 \quad \forall i \in \mathcal{I}, s \in \mathcal{S}. \tag{4.21}$$

The operating level of generators are constrained by $a_{s,i,j}x_j$.

$$y_{s,i,j} \leq a_{s,i,j}x_j \quad \forall i \in \mathcal{I}, j \in \mathcal{J}, s \in \mathcal{S}. \tag{4.22}$$

The demand for power needs to be satisfied and is ensured by the following constraint:

$$\sum_{j \in \mathcal{J}} y_{s,i,j} + \tilde{y}_{s,i} \geq d_{s,i} \quad \forall i \in \mathcal{I}, k \in \mathcal{K}. \tag{4.23}$$

4.5.2.5 Objective

The objective of the model is to minimize the net cost consisting of the fixed cost of setting up the generators $\sum_{s \in \mathcal{S}} \sum_{j \in \mathcal{J}} c_j x_j$, as well as the total operating cost $\sum_{s \in \mathcal{S}} \left(\sum_{i \in \mathcal{I}} \sum_{j \in \mathcal{J}} \left(f_{i,j} y_{s,i,j} \right) + g_s \tilde{y}_{s,i} \right)$. The objective is the sum of these components.

4.5.2.6 Optimization Model

The entire model is given below (4.24):

$$\min \quad \phi = \sum_{s \in \mathcal{S}} \sum_{j \in \mathcal{J}} c_j x_j + \sum_{s \in \mathcal{S}} \left(\sum_{i \in \mathcal{I}} \sum_{j \in \mathcal{J}} \left(f_{i,j} y_{s,i,j} \right) + g_s \tilde{y}_{s,i} \right) \tag{4.24a}$$

$$\text{s.t.} \quad x_j \geq 0 \quad \forall j \in \mathcal{J} \tag{4.24b}$$

$$y_{s,i,j} \leq a_{s,i,j} x_j \quad \forall i \in \mathcal{I}, j \in \mathcal{J}, s \in \mathcal{S} \tag{4.24c}$$

$$\sum_{j \in \mathcal{J}} y_{s,i,j} + \tilde{y}_{s,i} \geq d_{s,i} \quad \forall\, i \in \mathcal{I}, s \in \mathcal{S} \tag{4.24d}$$

$$y_{s,i,j} \geq 0 \quad \forall\, i \in \mathcal{I}, s \in \mathcal{S}, j \in \mathcal{J} \tag{4.24e}$$

$$\tilde{y}_{s,i} \geq 0 \quad \forall\, i \in \mathcal{I}, s \in \mathcal{S}. \tag{4.24f}$$

4.5.3 Solving the Problem

We now solve the problem using a few of the algorithms described in Section 4.4. The algorithms are implemented in Python using pyomo models, and the models are solved using Gurobi. The code for the implementation is given in https://github.com/li-group/MultiScaleOpt-Tutorial.git.

4.5.3.1 Using Full-Space Method

We solve the multi-time scale optimization model and we obtain a cost of US$357,408.98 with $x_1 = 2515.15$ kW, $x_2 = 909.09$ kW.

4.5.3.2 Using Benders Decomposition

To implement the Benders decomposition, we formulate a master problem as follows:

$$\min \quad \phi = \sum_{s \in \mathcal{S}} \sum_{j \in \mathcal{J}} c_j x_j + \sum_{s \in \mathcal{S}} z_s \tag{4.25a}$$

$$\text{s.t.} \quad x_j \geq 0 \quad \forall\, j \in \mathcal{J} \tag{4.25b}$$

$$z_s \geq -M \quad \forall\, s \in \mathcal{S}, \tag{4.25c}$$

where M is a large positive number. Constraint (4.25c) is added to keep the master problem bounded. We formulate subproblems as follows:

$$SP_s = \min \quad z_s = \sum_{j \in \mathcal{J}} c_j x_j + \sum_{i \in \mathcal{I}} \left(\sum_{j \in \mathcal{J}} (f_{i,j} y_{i,j}) + g_s \tilde{y}_i \right) \tag{4.26a}$$

$$\text{s.t.} \quad y_{i,j} \leq a_{s,i,j} x_j^* \quad \forall\, i \in \mathcal{I}, j \in \mathcal{J} \tag{4.26b}$$

$$\sum_{j \in \mathcal{J}} y_{i,j} + \tilde{y}_i \geq d_{s,i} \quad \forall\, i \in \mathcal{I} \tag{4.26c}$$

$$y_{i,j} \geq 0 \quad \forall\, i \in \mathcal{I}, j \in \mathcal{J} \tag{4.26d}$$

$$\tilde{y}_i \geq 0 \quad \forall\, i \in \mathcal{I}, \tag{4.26e}$$

where x^* is the solution obtained from the master problem. We then use the benders decomposition algorithm as shown in Section "Benders Decomposition" and get a solution of $x_1 = 2515.15$ kW, $x_2 = 909.09$ kW with a cost of US$357,408.97 which is the global minimum. The algorithm runs for nine iterations with the upper and lower bounds converging to the optimal solution as shown in Figure 4.9.

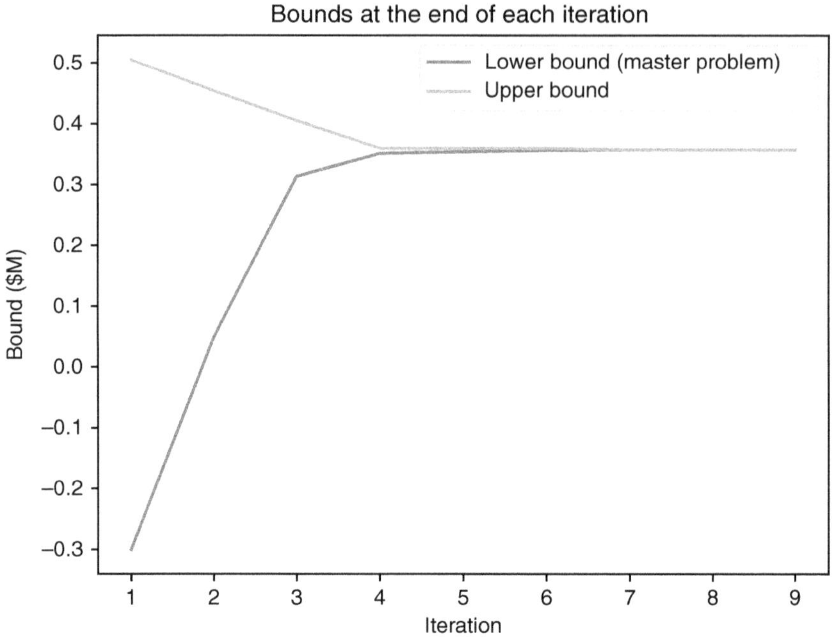

Figure 4.9 Optimal objective of master problem obtained during Benders decomposition.

4.5.3.3 Using Lagrangian Decomposition

To implement the Lagrangian decomposition, we decompose the problem into a set of subproblems as follows:

$$D_1(\nu) := \min \quad \sum_{j\in\mathcal{J}} c_j x_j + \sum_{i\in\mathcal{I}} \left(\sum_{j\in\mathcal{J}} (f_{i,j} y_{i,j}) + g_1 \tilde{y}_i \right) + \sum_{s\in\mathcal{S}, s\neq 1} \sum_j \nu_{s,j} x_j \tag{4.27a}$$

$$\text{s.t.} \quad x_j \geq 0 \quad \forall j \tag{4.27b}$$

$$y_{i,j} \leq a_{1,i,j} x_j \quad \forall\, i \in \mathcal{I}, j \in \mathcal{J}, k \in \mathcal{K} \tag{4.27c}$$

$$\sum_{j\in\mathcal{J}} y_{i,j} + \tilde{y}_i \geq d_{1,i} \quad \forall\, i \in \mathcal{I} \tag{4.27d}$$

$$y_{i,j} \geq 0 \quad \forall\, i \in \mathcal{I}, j \in \mathcal{J} \tag{4.27e}$$

$$\tilde{y}_i \geq 0 \quad \forall\, i \in \mathcal{I}, \tag{4.27f}$$

$$D_{s\in\mathcal{S}, s\neq 1}(\nu) := \min \quad \sum_{j\in\mathcal{J}} c_j x_j + \sum_{i\in\mathcal{I}} \left(\sum_{j\in\mathcal{J}} (f_{i,j} y_{i,j}) + g_s \tilde{y}_i \right) - \sum_j \nu_{s,j} x_j \tag{4.28a}$$

$$\text{s.t.} \quad x_j \geq 0 \quad \forall j \tag{4.28b}$$

$$y_{i,j} \leq a_{s,i,j} x_j \quad \forall i \in \mathcal{I}, j \in \mathcal{J}, k \in \mathcal{K} \tag{4.28c}$$

$$\sum_{j \in \mathcal{J}} y_{i,j} + \tilde{y}_i \geq d_{s,i} \quad \forall i \in \mathcal{I} \tag{4.28d}$$

$$y_{i,j} \geq 0 \quad \forall i \in \mathcal{I}, j \in \mathcal{J} \tag{4.28e}$$

$$\tilde{y}_i \geq 0 \quad \forall i \in \mathcal{I}. \tag{4.28f}$$

We use the cutting plane method and the subgradient method to solve the problem using Lagrangian decomposition. The heuristic we use to update the upper bound and obtain feasible solutions is to fix all the investment decisions obtained from solving the subproblems at various iterations. The cutting plane method gives a lower bound of US\$357, 408.98, which is the global minimum and an upper bound of US\$359, 290.31, which is around 0.5% from the global optimum with $x_1 = 1876.21$ kW, $x_2 = 1619.03$ kW. The decomposition algorithm runs for 12 iterations till we obtain the same Lagrangian multipliers. The upper and lower bounds at the end of each iteration are shown in Figure 4.10.

We run the subgradient method for 60 iterations. We get a lower bound of US\$357, 169.21, which is around approximately the global optimum and an upper

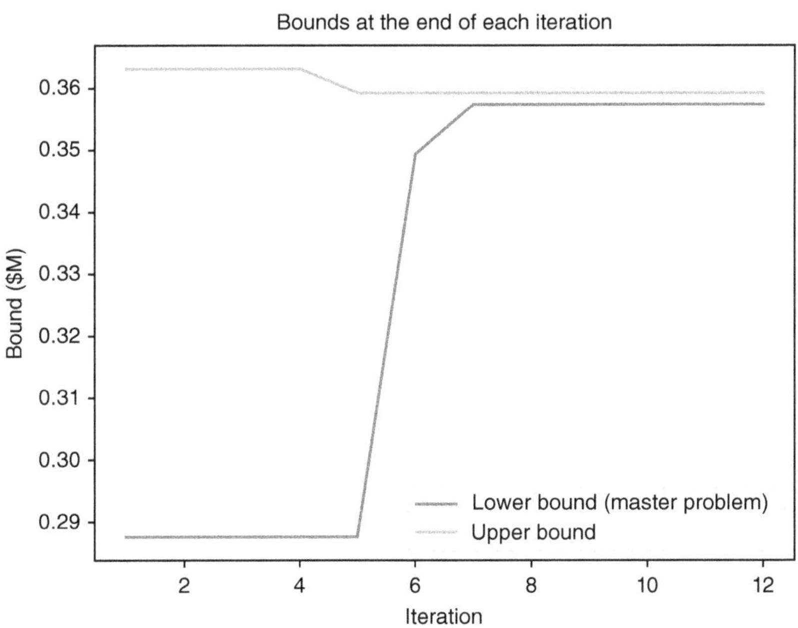

Figure 4.10 Lower bound obtained during Lagrangian decomposition with cutting plane method.

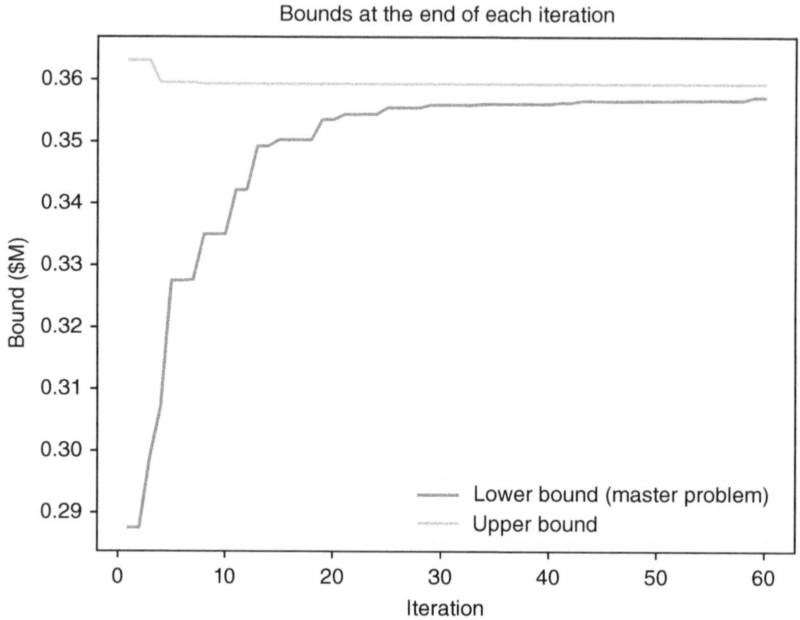

Figure 4.11 Lower bound obtained during Lagrangian decomposition with subgradient method.

bound of US$359, 290.31, which is around 0.5% from the global optimum with $x_1 = 1876.21$ kW, $x_2 = 1619.03$ kW. The upper and lower bounds obtained at the end of each iteration are shown in Figure 4.11.

4.5.3.4 Using Dantzig–Wolfe Decomposition

To implement Dantzig–Wolfe decomposition to the problem, we formulate a master problem as follows:

$$\min \quad \sum_{s\in S}\sum_{k\in K_{\lambda,s}} \lambda_s^k \left(\sum_{j\in\mathcal{J}} c_j \tilde{x}_{x,s,j}^k + \tilde{x}_{z,s}^k\right) + \sum_{s\in S}\sum_{k\in K_{\theta,s}} \theta_s^k \left(\sum_{j\in\mathcal{J}} c_j \tilde{w}_{x,s,j}^k + \tilde{w}_{z,s}^k\right) \tag{4.29a}$$

$$\text{s.t.} \quad \sum_{k\in K_{\lambda,1}} \lambda_1^k \tilde{x}_{x,1,j}^k + \sum_{k\in K_{\theta,1}} \theta_1^k \tilde{w}_{x,1,j} = \sum_{k\in K_{\lambda,a}} \lambda_s^k \tilde{x}_{x,s,j} + \sum_{k\in K_{\theta,s}} \theta_s^k \tilde{w}_{x,1,j} \quad \forall s \in S, s \neq 1 \tag{4.29b}$$

$$\sum_{k\in K_{\lambda,s}} \lambda_s^k = 1 \quad \forall s \in S \tag{4.29c}$$

$$\lambda_s^k \geq 0, \quad \forall k \in S_{\lambda,s},\ s \in S, \quad \theta_s^k \geq 0 \quad k \in S_{\theta,s},\ s \in S. \tag{4.29d}$$

We then formulate the following subproblems as follows:

$$
\begin{aligned}
\mathrm{SP}_1 : \min & \sum_{j\in\mathcal{J}} c_j x_j + z - \sum_{j\in\mathcal{J}} \sum_{s\in\mathcal{S}, s\neq 1} v_{s,j} x_j \\
\text{s.t.} \quad & z = \sum_{j\in\mathcal{J}} c_j x_j + \sum_{i\in\mathcal{I}} \left(\sum_{j\in\mathcal{J}} (f_{i,j} y_{i,j}) + g_1 \tilde{y}_i \right) \\
& y_{i,j} \leq a_{1,i,j} x_j \quad \forall\, i \in \mathcal{I}, j \in \mathcal{J} \\
& \sum_{j\in\mathcal{J}} y_{i,j} + \tilde{y}_i \geq d_{1,i} \quad \forall\, i \in \mathcal{I} \\
& y_{i,j} \geq 0 \quad \forall\, i \in \mathcal{I}, j \in \mathcal{J} \\
& \tilde{y}_i \geq 0 \quad \forall\, i \in \mathcal{I} \\
& x_j \geq 0 \quad \forall\, j \in \mathcal{J} \\
& x_j \leq M \quad \forall\, j \in \mathcal{J},
\end{aligned}
$$

$$
\begin{aligned}
\mathrm{SP}_{s\in\mathcal{S}, s\neq 1} : \min & \sum_{j\in\mathcal{J}} c_j x_j + z + \sum_{j\in\mathcal{J}} v_{s,j} x \\
\text{s.t.} \quad & z = \sum_{j\in\mathcal{J}} c_j x_j + \sum_{i\in\mathcal{I}} \left(\sum_{j\in\mathcal{J}} (f_{i,j} y_{i,j}) + g_s \tilde{y}_i \right) \\
& y_{i,j} \leq a_{s,i,j} x_j \quad \forall\, i \in \mathcal{I}, j \in \mathcal{J} \\
& \sum_{j\in\mathcal{J}} y_{i,j} + \tilde{y}_i \geq d_{s,i} \quad \forall\, i \in \mathcal{I} \\
& y_{i,j} \geq 0 \quad \forall\, i \in \mathcal{I}, j \in \mathcal{J} \\
& \tilde{y}_i \geq 0 \quad \forall\, i \in \mathcal{I} \\
& x_j \geq 0 \quad \forall\, j \in \mathcal{J} \\
& x_j \leq M \quad \forall\, j \in \mathcal{J},
\end{aligned}
$$

where $v_{s,j}$ is the dual for the constraint (4.29b) corresponding to s for the jth generator.

We add constraints upper bounding the investment decisions to keep the subproblem bounded. We apply the Dantzig–Wolfe decomposition algorithm with the above formulations. The algorithm runs for seven iterations before getting all non-negative reduced costs and we get a solution of $\boldsymbol{x}_1 = 2515.15, x_2 = 909.09$ with a cost of US\$357,408.97 which is the global minimum. The algorithm runs for seven iterations with the upper and lower bounds converging to the optimal solution as shown in Figure 4.12.

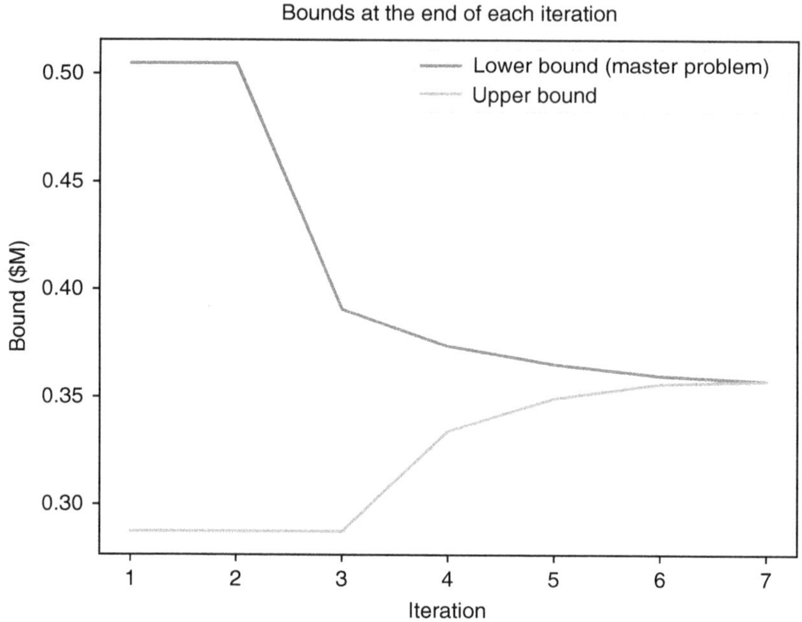

Figure 4.12 Optimal objective of master problem obtained during Dantzig–Wolfe decomposition.

4.5.3.5 Using PAMSO

To implement PAMSO, we split the optimization model into a high-level model taking decisions regarding the capacity of the two generators, and a low-level model deciding the operating levels of generators as well as the power purchased.

The high-level model is adapted from the full space model with the operating decisions for the entire duration aggregated. The high-level model takes decisions in only one time scale, that is, decisions valid for the entire time duration as a whole. With a slight abuse in notation, the operating variables of the aggregated model are summarized below (Table 4.2):

Table 4.2 Variables in high-level model.

Variable in full-space	Physical meaning	Variable in high-level model	Physical meaning
$y_{s,i,j}$	Operating levels of generator j during the ith part of the sth day	y_j	Net operating levels of generator generator j
$\tilde{y}_{s,i}$	Power purchased during the ith part of the sth day	$\tilde{y}$	Net power purchased

From the high-level model, we obtain the installed capacity of the generators x_j. The optimization model is given below:

$$\min \quad \phi = \sum_{s\in\mathcal{S}}\sum_{j\in\mathcal{J}} c_j x_j + \sum_{s\in\mathcal{S}}\left(\sum_{j\in\mathcal{J}}\sum_{i\in\mathcal{I}} (f_{i,j})\, y_j + g_s\tilde{y}\right) \tag{4.30a}$$

$$\text{s.t.} \quad x_j \geq 0 \quad \forall j \in \mathcal{J} \tag{4.30b}$$

$$y_j \leq \sum_{s\in\mathcal{S}}\sum_{i\in\mathcal{I}} a_{s,i,j} x_j \quad \forall ,j \in \mathcal{J} \tag{4.30c}$$

$$\sum_{j\in\mathcal{J}} y_j + \tilde{y} \geq \sum_{s\in\mathcal{S}}\sum_{i\in\mathcal{I}} d_{s,i} \tag{4.30d}$$

$$y_j \geq 0 \quad \forall j \in \mathcal{J} \tag{4.30e}$$

$$\tilde{y} \geq 0. \tag{4.30f}$$

The low-level model is the full-space model with the installed capacity x_j of the generators fixed from the high-level model. The low-level model takes decisions for the parts of the day on the 3 days.

Parameterized high-level model: Based on the data, we can either underestimate or overestimate the effect of the availability in the aggregated model. Therefore we add a prefactor to the availability of each of the generators to compensate for the mismatch. Furthermore, to better capture the differences in the operation of the generators, we add a third parameter which signifies the minimum capacity of each generator, that is, the capacity of each generator is greater than this parameter. This can help in cases when an optimum configuration consists of having both generators installed with a reasonable capacity. Without this parameter, there is a high chance of installing only one generator for cases where there are some intricate differences between the generators. Therefore, the parameters we add to the high-level model are

1) ρ_1 which represents the prefactor to the availability of generator 1,
2) ρ_2 which represents the prefactor to the availability of generator 2,
3) ρ_3 which represents the minimum capacity installed for each of the generators.

$$\min \quad \phi = \sum_{s\in\mathcal{S}}\sum_{j\in\mathcal{J}} c_j x_j + \sum_{s\in\mathcal{S}}\left(\sum_{j\in\mathcal{J}}\sum_{i\in\mathcal{I}} (f_{i,j})\, y_j + g_s\tilde{y}\right) \tag{4.31a}$$

$$\text{s.t.} \quad x_j \geq 0 \quad \forall j \in \mathcal{J} \tag{4.31b}$$

$$y_j \leq \rho_j \sum_{s\in\mathcal{S}}\sum_{i\in\mathcal{I}} a_{i,k,j} x_j \quad \forall j \in \mathcal{J} \tag{4.31c}$$

$$\sum_{j\in\mathcal{J}} y_j + \tilde{y} \geq \sum_{k\in\mathcal{K}}\sum_{i\in\mathcal{I}} d_{i,k} \tag{4.31d}$$

$$x_j \geq \rho_3 \quad \forall j \in \mathcal{J} \tag{4.31e}$$

$$y_j \geq 0 \quad \forall j \in \mathcal{J} \tag{4.31f}$$

$$\tilde{y} \geq 0. \tag{4.31g}$$

We then form the MBBF using the parameterized high-level model and the low-level model and tune the MBBF using DFO solvers.

For our implementation of PAMSO, we use the Bayesian Adaptive Direct Search (BADS) [52] algorithm as a DFO solver via the PyBADS software [53] to solve the MBBF. The range for the parameters is $\rho_1 \in [0, 1.5]$, $\rho_2 \in [0, 1.5], \rho_3 \in [0, 2100]$. After applying the BADS algorithm, we get the optimum parameters as $\rho_1 = 0.609, \rho_2 = 0.279, \rho_3 = 909.092$, which gives a solution of $x_1 = 2515.15$ kW, $x_2 = 909.09$ kW and a cost of US$357,408.98, which is the global minimum. The algorithm takes 478 function evaluations to reach the solution. A convergence plot, showing the evolution of the best cost with function evaluations, is shown in Figure 4.13.

4.5.4 VMM

We finally calculate the VMM of the problem with the high-level model described in Eq. (4.30). The optimal objective from solving the multi-time scale model (MM) turns out to be US$357,408.98. The MPSS is US$407, 520.75. Thus, the VMM is US$50,111.77.

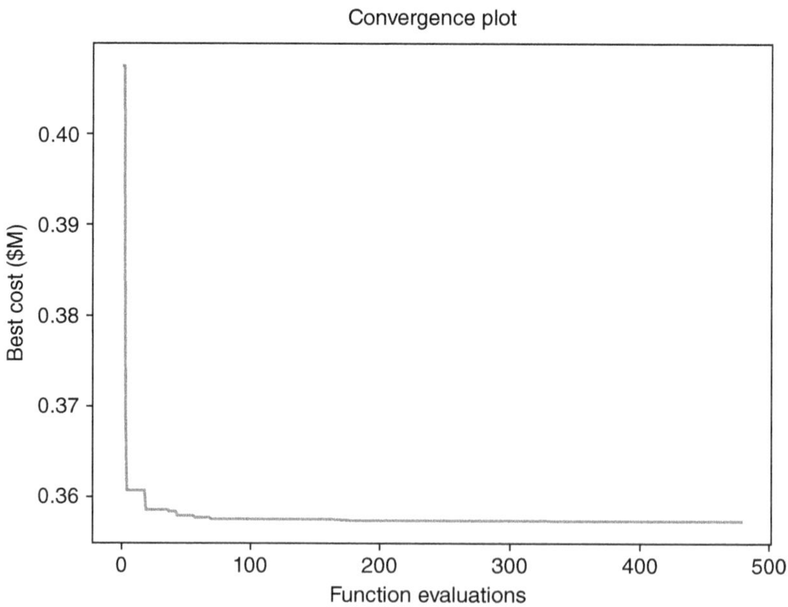

Figure 4.13 Best cost obtained in different function evaluations of PAMSO.

4.6 Conclusion

In this chapter, we have provided an overview of multi-time scale optimization models and the algorithms that have been used in the literature to solve these models. Multi-time scale optimization models involve decision variables in multiple time scales like long-time/high-level planning decisions and short-time/low-level scheduling and control decisions. The applications of these models are widespread, including in economics and environmental management. The *VMM* is a metric introduced to quantify the benefit of using a multi-time scale optimization model as opposed to using separate independent models for the different decision layers.

Multi-time scale optimization models can be of different sizes and different structures. To solve such models, different methods have been used in the literature. We have provided the different methods, their applications, and their limitations. We have also applied some of the algorithms to solve an example optimization problem at https://github.com/li-group/MultiScaleOpt-Tutorial.git.

References

1 Allen, R.C., Baratsas, S.G., Kakodkar, R. et al. (2023). A multi-period integrated planning and scheduling approach for developing energy systems. *Optimal Control Applications and Methods* 44 (2): 355–372. https://doi.org/10.1002/oca.2866.

2 Maravelias, C.T. and Sung, C. (2009). Integration of production planning and scheduling: overview, challenges and opportunities. *Computers & Chemical Engineering* 33 (12): 1919–1930. https://doi.org/10.1016/j.compchemeng.2009.06.007. https://www.sciencedirect.com/science/article/pii/S0098135409001501.

3 Shin, J. and Lee, J.H. (2019). Multi-timescale, multi-period decision-making model development by combining reinforcement learning and mathematical programming. *Computers & Chemical Engineering* 121: 556–573. https://doi.org/10.1016/j.compchemeng.2018.11.020. https://www.sciencedirect.com/science/article/pii/S0098135418308913.

4 Subramanian, K., Rawlings, J.B., Maravelias, C.T. et al. (2013). Integration of control theory and scheduling methods for supply chain management. *Computers & Chemical Engineering* 51: 4–20. https://doi.org/10.1016/j.compchemeng.2012.06.012. https://www.sciencedirect.com/science/article/pii/S0098135412001883.

5 Biondi, M., Sand, G., and Harjunkoski, I. (2017). Optimization of multipurpose process plant operations: a multi-time-scale maintenance and production scheduling approach. *Computers & Chemical Engineering* 99: 325–339.

https://doi.org/10.1016/j.compchemeng.2017.01.007. https://www.sciencedirect.com/science/article/pii/S009813541730008X.

6 Dowling, A.W., Kumar, R., and Zavala, V.M. (2017). A multi-scale optimization framework for electricity market participation. *Applied Energy* 190: 147–164. https://doi.org/10.1016/j.apenergy.2016.12.081. https://www.sciencedirect.com/science/article/pii/S0306261916318487.

7 Tabrizi, B.H. (2018). Integrated planning of project scheduling and material procurement considering the environmental impacts. *Computers & Industrial Engineering* 120: 103–115. https://doi.org/10.1016/j.cie.2018.04.031. https://www.sciencedirect.com/science/article/pii/S0360835218301670.

8 Kim, S., Choi, Y., Park, J. et al. (2024). Multi-period, multi-timescale stochastic optimization model for simultaneous capacity investment and energy management decisions for hybrid micro-grids with green hydrogen production under uncertainty. *Renewable and Sustainable Energy Reviews* 190: 114049. https://doi.org/10.1016/j.rser.2023.114049. https://www.sciencedirect.com/science/article/pii/S1364032123009073.

9 Ramanujam, A., Constante-Flores, G.E., and Li, C. (2023). Distributed manufacturing for electrified chemical processes in a microgrid. *AIChE Journal* 69: e18265. https://doi.org/10.1002/aic.18265.

10 Birge, J.R. and Louveaux, F. (2011). *Introduction to Stochastic Programming*. New York, NY: Springer New York. ISBN: 978-1-4614-0236-7 https://doi.org/10.1007/978-1-4614-0237-4.

11 Kopanos, G.M. and Puigjaner, L. (2009). Multi-site scheduling/batching and production planning for batch process industries. In: *Computer Aided Chemical Engineering*, vol. 27 (ed. R.M. de Brito Alves, C.A.O. do Nascimento, and E.C. Biscaia), 2109–2114. Elsevier. ISBN: 1570-7946 https://doi.org/10.1016/S1570-7946(09)70742-4. https://www.sciencedirect.com/science/article/pii/S1570794609707424.

12 Li, C., Conejo, A.J., Liu, P. et al. (2022). Mixed-integer linear programming models and algorithms for generation and transmission expansion planning of power systems. *European Journal of Operational Research* 297 (3): 1071–1082. https://doi.org/10.1016/j.ejor.2021.06.024. https://www.sciencedirect.com/science/article/pii/S0377221721005397.

13 Erdirik-Dogan, M. and Grossmann, I.E. (2008). Simultaneous planning and scheduling of single-stage multi-product continuous plants with parallel lines. *Computers & Chemical Engineering* 32 (11): 2664–2683. https://doi.org/10.1016/j.compchemeng.2007.07.010. https://www.sciencedirect.com/science/article/pii/S0098135407001998.

14 Munawar, S.A., Kapadi, M.D., Patwardhan, S.C. et al. (2005). Integration of planning and scheduling in multi-site plants: application to paper manufacturing. In: *Computer Aided Chemical Engineering*, vol. 20 (ed. L. Puigjaner

and A. Espua), 1621–1626. Elsevier. ISBN: 1570-7946 https://doi.org/10.1016/S1570-7946(05)80112-9. https://www.sciencedirect.com/science/article/pii/S1570794605801129.

15 Chu, Y. and You, F. (2013). Integrated scheduling and dynamic optimization of complex batch processes with general network structure using a generalized Benders decomposition approach. *Industrial & Engineering Chemistry Research* 52 (23): 7867–7885. https://doi.org/10.1021/ie400475s.

16 Hooker, J.N. and Ottosson, G. (2003). Logic-based Benders decomposition. *Mathematical Programming* 96 (1): 33–60.

17 Barzanji, R., Naderi, B., and Begen, M.A. (2020). Decomposition algorithms for the integrated process planning and scheduling problem. *Omega (Westport)* 93: 102025. https://doi.org/10.1016/j.omega.2019.01.003. https://www.sciencedirect.com/science/article/pii/S0305048318306698.

18 Guignard, M. and Kim, S. (1987). Lagrangean decomposition: a model yielding stronger lagrangean bounds. *Mathematical Programming* 39 (2): 215–228. https://doi.org/10.1007/BF02592954.

19 Bertsekas, D.P. (1997). Nonlinear programming. *Journal of the Operational Research Society* 48: 334.

20 Shah, N.K. and Ierapetritou, M.G. (2012). Integrated production planning and scheduling optimization of multisite, multiproduct process industry. *Computers & Chemical Engineering* 37: 214–226. https://doi.org/10.1016/j.compchemeng.2011.08.007. https://www.sciencedirect.com/science/article/pii/S0098135411002626.

21 Rockafellar, R.T. and Wets, R.J.-B. (1991). Scenarios and policy aggregation in optimization under uncertainty. *Mathematics of Operations Research* 16 (1): 119–147. https://doi.org/10.1287/moor.16.1.119.

22 Peng, Z., Zhang, Y., Feng, Y. et al. (2019). A progressive hedging-based solution approach for integrated planning and scheduling problems under demand uncertainty. *Industrial & Engineering Chemistry Research* 58 (32): 14880–14896. https://doi.org/10.1021/acs.iecr.9b02620.

23 Bertsimas, D. and Tsitsiklis, J. (1997). *Introduction to Linear Optimization*, 1e. Athena Scientific. ISBN: 1886529191.

24 Conforti, M., Cornuéjols, G., and Zambelli, G. (2014). Reformulations and relaxations. In: *Integer Programming* (ed. M. Conforti, G. Cornuéjols, and G. Zambelli), 321–350. Cham: Springer International Publishing. ISBN: 978-3-319-11008-0 https://doi.org/10.1007/978-3-319-11008-0_8.

25 Conforti, M., Cornuéjols, G., and Zambelli, G. (2014). Getting started. In: *Integer Programming* (ed. M. Conforti, G. Cornuéjols, and G. Zambelli), 1–44. Cham: Springer International Publishing. ISBN: 978-3-319-11008-0 https://doi.org/10.1007/978-3-319-11008-0_1.

26 Allman, A. and Zhang, Q. (2021). Branch-and-price for a class of nonconvex mixed-integer nonlinear programs. *Journal of Global Optimization* 81 (4): 861–880. https://doi.org/10.1007/s10898-021-01027-w.

27 Gharaei, A. and Jolai, F. (2019). A branch and price approach to the two-agent integrated production and distribution scheduling. *Computers & Industrial Engineering* 136: 504–515. https://doi.org/10.1016/j.cie.2019.07.050. https://www.sciencedirect.com/science/article/pii/S0360835219304474.

28 Mitchell, M. (1996). *An Introduction to Genetic Algorithms. Complex adaptive systems*. Cambridge, MA, US: The MIT Press. ISBN: 0-262-13316-4 (Hardcover).

29 Goldberg, D.E. and Deb, K. (1991). A comparative analysis of selection schemes used in genetic algorithms. In: *Foundations of Genetic Algorithms*, vol. 1 (ed. G.J.E. Rawlins), 69–93. Elsevier. ISBN: 1081-6593 https://doi.org/10.1016/B978-0-08-050684-5.50008-2. https://www.sciencedirect.com/science/article/pii/B9780080506845500082.

30 Lee, H.C. and Ha, C. (2019). Sustainable integrated process planning and scheduling optimization using a genetic algorithm with an integrated chromosome representation. *Sustainability* 11 (2): https://doi.org/10.3390/su11020502. https://www.mdpi.com/2071-1050/11/2/502.

31 Li, W.D. and McMahon, C.A. (2007). A simulated annealing-based optimization approach for integrated process planning and scheduling. *International Journal of Computer Integrated Manufacturing* 20 (1): 80–95. https://doi.org/10.1080/09511920600667366.

32 Liu, X., Ni, Z., and Qiu, X. (2016). Application of ant colony optimization algorithm in integrated process planning and scheduling. *The International Journal of Advanced Manufacturing Technology* 84 (1): 393–404. https://doi.org/10.1007/s00170-015-8145-4.

33 Leite, M.M.S., Pinto, T.M.P., and Alves, C.M.M. (2023). Solving the integrated planning and scheduling problem using variable neighborhood search based algorithms. *Expert Systems with Applications* 228: 120191. https://doi.org/10.1016/J.ESWA.2023.120191.

34 Hussain, K., Salleh, M.N.M., Cheng, S., and Shi, Y. (2019). Metaheuristic research: a comprehensive survey. *Artificial Intelligence Review* 52 (4): 2191–2233. https://doi.org/10.1007/s10462-017-9605-z.

35 Ball, M.O. (2011). Heuristics based on mathematical programming. *Surveys in Operations Research and Management Science* 16 (1): 21–38. https://doi.org/10.1016/j.sorms.2010.07.001. https://www.sciencedirect.com/science/article/pii/S1876735410000036.

36 Fischetti, M. and Lodi, A. (2003). Local branching. *Mathematical Programming* 98 (1): 23–47. https://doi.org/10.1007/s10107-003-0395-5.

37 Silva, D.M. and Mateus, G.R. (2023). A mixed-integer programming formulation and heuristics for an integrated production planning and scheduling problem. In: *Metaheuristics* (ed. L. Di Gaspero, P. Festa, A. Nakib, and M. Pavone), 290–305. Cham: Springer International Publishing. ISBN: 978-3-031-26504-4.

38 Reinert, C., Nilges, B., Baumgärtner, N., and Bardow, A. (2023). This is SpArta: rigorous optimization of regionally resolved energy systems by spatial aggregation and decomposition. *arXiv e-prints*, arXiv:2302.05222. https://doi.org/10.48550/arXiv.2302.05222.

39 Badejo, O. and Ierapetritou, M. (2022). Integrating tactical planning, operational planning and scheduling using data-driven feasibility analysis. *Computers & Chemical Engineering* 161: 107759. https://doi.org/10.1016/j.compchemeng.2022.107759. https://www.sciencedirect.com/science/article/pii/S0098135422001004.

40 Beykal, B., Avraamidou, S., and Pistikopoulos, E.N. (2022). Data-driven optimization of mixed-integer bi-level multi-follower integrated planning and scheduling problems under demand uncertainty. *Computers & Chemical Engineering* 156: 107551. https://doi.org/10.1016/j.compchemeng.2021.107551. https://www.sciencedirect.com/science/article/pii/S009813542100329X.

41 Chu, Y. and You, F. (2014). Integrated planning, scheduling, and dynamic optimization for batch processes: MINLP model formulation and efficient solution methods via surrogate modeling. *Industrial & Engineering Chemistry Research* 53 (34): 13391–13411. https://doi.org/10.1021/ie501986d.

42 Dias, L.S. and Ierapetritou, M.G. (2020). Integration of planning, scheduling and control problems using data-driven feasibility analysis and surrogate models. *Computers & Chemical Engineering* 134: 106714. https://doi.org/10.1016/j.compchemeng.2019.106714. https://www.sciencedirect.com/science/article/pii/S0098135419306982.

43 Yang, S., Oliver Gao, H., and You, F. (2023). Integrated optimization in operations control and systems design for carbon emission reduction in building electrification with distributed energy resources. *Advances in Applied Energy* 100144. https://doi.org/10.1016/j.adapen.2023.100144. https://www.sciencedirect.com/science/article/pii/S2666792423000239.

44 Ye, W. and You, F. (2015). A simulation-based optimization method for solving the integrated supply chain network design and inventory control problem under uncertainty. In: *Chemical Engineering Transactions*, vol. 45, 499–504. Italian Association of Chemical Engineering - AIDIC https://doi.org/10.3303/CET1545084. http://www.scopus.com/inward/record.url?scp=84946116663partnerID=8YFLogxK. http://www.scopus.com/inward/citedby.url?scp=84946116663partnerID=8YFLogxK.

45 Dias, L.S. and Ierapetritou, M.G. (2019). Data-driven feasibility analysis for the integration of planning and scheduling problems. *Optimization and Engineering* 20 (4): 1029–1066. https://doi.org/10.1007/s11081-019-09459-w.

46 Ochoa, T., Gil, E., Angulo, A., and Valle, C. (2022). Multi-agent deep reinforcement learning for efficient multi-timescale bidding of a hybrid power plant in day-ahead and real-time markets. *Applied Energy* 317: 119067. https://doi.org/10.1016/j.apenergy.2022.119067. https://www.sciencedirect.com/science/article/pii/S0306261922004603.

47 Sutton, R.S. and Barto, A.G. (2018). *Reinforcement Learning: An Introduction*. MIT Press.

48 Ramanujam, A. and Li, C. (2025). Pamso: parametric autotuning multi-time scale optimization algorithm. *Computers & Chemical Engineering* 200: 109160. https://doi.org/10.1016/j.compchemeng.2025.109160. https://www.sciencedirect.com/science/article/pii/S0098135425001644.

49 Perkins, R.T., III, and Powell, W.B. (2017). Stochastic optimization with parametric cost function approximations. *arXiv preprint arXiv:1703.04644*.

50 Coutinho, J.P.L., Santos, L.O., and Reis, M.S. (2023). Bayesian optimization for automatic tuning of digital multi-loop PID controllers. *Computers & Chemical Engineering* 173: 108211. https://doi.org/10.1016/j.compchemeng.2023.108211. https://www.sciencedirect.com/science/article/pii/S0098135423000807.

51 Paulson, J.A., Sorourifar, F., and Mesbah, A. (2023). A tutorial on derivative-free policy learning methods for interpretable controller representations. In: *2023 American Control Conference (ACC)*, 1295–1306. IEEE. ISBN: 979-8-3503-2806-6 https://doi.org/10.23919/ACC55779.2023.10156412.

52 Acerbi, L. and Ma, W.J. (2017). Practical Bayesian optimization for model fitting with Bayesian adaptive direct search. In: *Advances in Neural Information Processing Systems*, vol. 30 (ed. I. Guyon, U. Von Luxburg, S. Bengio, et al.). Curran Associates, Inc. https://proceedings.neurips.cc/paper_files/paper/2017/file/df0aab058ce179e4f7ab135ed4e641a9-Paper.pdf.

53 Singh, G.S. and Acerbi, L. (2024). PyBADS: fast and robust black-box optimization in Python. *Journal of Open Source Software* 9 (94): 5694. https://doi.org/10.21105/joss.05694.

5

Many Objective Optimization Tools for Sustainable Decision-Making

Andrew Allman and Hongxuan Wang

Department of Chemical Engineering, University of Michigan, Ann Arbor, MI, USA

5.1 Introduction

What does it mean to "optimize" for a "sustainable" system? Perhaps the first thing that might come to mind for many is a goal of minimizing greenhouse gas emissions in order to meet global climate pledges and ensure a viable planet for future generations. Limiting other undesirable pollutants, such as nitrates, microplastics, or particulates, is similarly important to maintaining quality of life for both current and future denizens of the world. Of course, resource consumption plays a big role in sustainability, so decisions made should take into account limiting the need for scarce materials such as precious metals, fossil resources, or even water. Recent global events have also sharply emphasized the need of a social perspective in sustainable decision-making; sustainable systems should be safe, promote the well-being of the people using them, and not disproportionately benefit or disadvantage particular communities. Lastly, one cannot forget the importance of cost and economics, as society cannot benefit from processes it cannot afford.

In short, there are a plethora of objectives one may consider when optimizing for a sustainable system. All have some degree of importance, and most will point to different decision-making outcomes. Combining each of these objectives into a single meaningful metric is a nontrivial task, and understanding the landscape of trade-offs between these objectives is essential for determining an appropriate compromise point that can be considered a sustainable decision. Fortunately, the process systems engineering community is well equipped with tools that allow for the analysis of optimization problems with different objective functions, often referred to as "multi-objective optimization problems" (MOPs). This chapter will provide a discussion of these tools and their application to sustainable decision-making, paying particular attention to problems of four or more objectives, referred to as "many-objective optimization problems" (MaOPs), which can be particularly challenging to both solve and interpret. The remainder of the chapter is structured as follows: first, an overview of objectives commonly used when formulating optimization problems for sustainable systems is provided. Next, several methods for the solution of MOPs are presented, as well as a new approach for systematically reducing the dimensionality of MaOPs. Two case studies applying some of these methods are presented: one which looks at comparing cost and emissions driven demand response operation, and another which considers planetary boundaries as competing objectives in a large-scale energy supply chain. Finally, we conclude with perspectives for future research in this area.

5.2 Sustainability Objectives

When considering the multiple objective functions, one might include when optimizing for a sustainable system, a commonly used starting point is to consider the classic "three pillars of sustainability." This paradigm states that a sustainable decision should be high performing in all aspects of economic, environmental, and social performance [1]. A large body of work exists of researchers analyzing the trade-offs between an economic objective (typically, minimizing some measure of cost) and an environmental objective (most commonly greenhouse gas emissions, global warming potential, or a life cycle assessment score). Representative studies applying this approach to sustainable optimization include works in distributed energy systems [2–4], chemical plant design [5–7], supply chain management [8, 9], and chemical process operations [10, 11]. Less commonly, a three objective problem is considered, adding a social objective (most commonly related to the jobs provided by a system or process safety) to the environmental and social objectives [12–14]. A challenge with the three pillar approach comes when the objective corresponding to one of the pillars consists of multiple disjoint goals. Consider operating a process where hydrogen can be obtained either from wind-powered electrolysis of water, or steam reforming of methane. Intuitively, the electrolytic approach will be better from a greenhouse gas emissions perspective; however, in a water-scarce region, minimizing water usage may also be a major environmental objective promoting the use of the steam reformer. However, by combining these aspects into a single environmental objective function (say, through the use of a life cycle assessment tool), valuable information about the sustainability trade-offs is lost, making it more challenging to make an informed sustainable decision. Alternatively, if these aspects are kept as separate objectives, the problem becomes a more challenging MaOP.

Beyond the three pillars of sustainability, there are a number of other multi-objective frameworks for sustainable decision-making. One useful approach is to analyze a set of metrics referred to as "planetary boundaries" [15]. There are nine metrics for which if the total global contribution toward these metrics crosses some threshold, abrupt, catastrophic change to planetary-scale systems is expected which may threaten the continued existence of humanity. These metrics include climate change, ocean acidification, stratospheric ozone, global phosphorous and nitrogen cycles, atmospheric aerosol loading, freshwater use, land use change, biodiversity loss, and chemical pollution. Common approaches for implementing planetary boundary considerations into sustainable decision-making include attributing a small fraction of the planetary boundary threshold to the specific process of interest, and either including planetary boundary thresholds as constraints or lumping threshold violations into a single objective to be minimized [16]. However, one could also consider each of these metrics as its own objective to be minimized, as it is desirable for a sustainable process to make as little contribution toward potentially exceeding global thresholds as possible. Similarly, one can consider the United Nations' Sustainable Development Goals [17] as a set of objectives for sustainable decision-making. This provides a set of 17 sustainability goals broadly agreed upon by global governments: no poverty, zero hunger, good health and well-being, quality education, gender equality, clean water and sanitation, affordable and clean energy, decent work and economic growth, industry, innovation, and infrastructure, reduced inequalities, sustainable cities and communities, responsible consumption and production, climate action, life below water, life on land, peace, justice and strong institutions, and partnerships for the goals. Although not all of these objectives are relevant for decision-making in energy or chemical systems, they in general are aligned with the ideas previously discussed in this section for sustainable decision-making and provide a list of goals which are publicly desirable. While both the planetary boundaries and sustainable development goals provide a set of broadly accepted sustainability objectives, a key challenge in their use is that it can be difficult to quantify how a decision made for a process of interest affects different objectives, particularly given the disparity between decisions made at a (relatively) small scale versus the global scale considered for many of these objectives.

5.3 MOP Solution Methods

This section discusses popular methods for solving general multi-objective optimization problems, like those encountered when optimizing for a sustainable system, which are of the form:

$$\min_{x} \quad \{f_1(x), f_2(x), \dots, f_n(x)\} \tag{5.1a}$$

$$\text{s.t.} \quad g(x) \geq 0 \tag{5.1b}$$

$$h(x) = 0. \tag{5.1c}$$

In general, the n objectives cannot be all be optimized simultaneously; most will point to their own unique solution. Thus, the solution to a MOP is not a single decision, but a manifold of points known as the Pareto frontier. Mathematically, the Pareto frontier is the set of all feasible solutions to (5.1) whose objective values are not dominated by another feasible point; that is to say, the following statement should hold for all solutions on the Pareto frontier with objective values denoted $\hat{f}$:

$$\{f : f_i \leq \hat{f}_i \forall i = \{1, \dots, n\}, \quad f \in \mathcal{F} \backslash \{\hat{f}\}\} = \emptyset, \tag{5.2}$$

where $\mathcal{F}$ is the set of objective values corresponding to all points in the feasible region of (5.1). Stated differently, the Pareto frontier is the set of all points whereby one cannot improve upon one objective from this point without sacrificing performance in at least one other objective. One can also define a weak Pareto frontier as the set of all feasible solutions to (5.1) where the set generated by Eq. (5.2) is defined by strict rather than weak inequalities (often referred to as "strict dominance"). The weak Pareto frontier consists of the Pareto frontier but potentially additional points.

One can classify solution methods for MOPs into two different categories. The first are scalarization approaches, which systematically generate a set of single objective optimization problems that can be solved to generate different points along the Pareto frontier. When combined with state-of-the-art global optimization solvers, scalarization approaches are able to provide guarantees that the points generated are indeed Pareto optimal for the MOP of interest, which is extremely important when optimizing decision-making for energy and chemical systems as even cost improvements in the 1–2% range can make a massive difference in the economic viability of a project. However, as will be shown later, such approaches tend not to scale well for solving MaOPs. In contrast, heuristic evolutionary algorithms [18] for identifying the Pareto frontier do not provide any guarantees of optimality, but are much easier to scale up to higher objective dimensionality and in many practical cases tend to work well. Generally speaking, these approaches do not attempt to solve any individual optimization problems directly, but instead generate a set of points in objective space that are iteratively driven toward the Pareto frontier using, for example, natural selection principles such as mutation and recombination.

Because rigorous methods for provable optimal solutions are highly desirable in sustainable chemical and energy systems, it is useful to go through some of the more popular scalarization methods for solving MOPs. One such approach is the weighted sum approach, which provides the following single-objective formulation for identifying Pareto-optimal points:

$$\min_{x} \quad \sum_{i=1}^{n} \omega_i f_i(x) \tag{5.3a}$$

$$\text{s.t.} \quad g(x) \leq 0 \tag{5.3b}$$

$$h(x) = 0. \tag{5.3c}$$

As the name suggests, the method optimizes a sum of all objectives scaled by weights ω_i. Pareto optimal points for the MOP can be obtained by systematically varying the weights and solving the single objective problem. A common approach is to vary each weight between 0 and 1 such that $\sum_{i=1}^{n} \omega_i = 1$. A graphical interpretation of the weighted sum approach is shown in Figure 5.1a. The method can be thought of as pulling a hyperplane in objective space until it is only just touching the feasible region of objective values. Changing weights adjusts the orientation of the hyperplane and allows for the generation of a different solution. A critical downside of the weighted sum approach, as demonstrated graphically in Figure 5.1b, is that if the feasible objective region is nonconvex, the weighted sum method is unable to generate Pareto optimal points within the nonconvex region. Moreover, a challenge common to all scalarization methods is that the number of single-objective problems needed to generate the full Pareto frontier scales exponentially with the number of objectives.

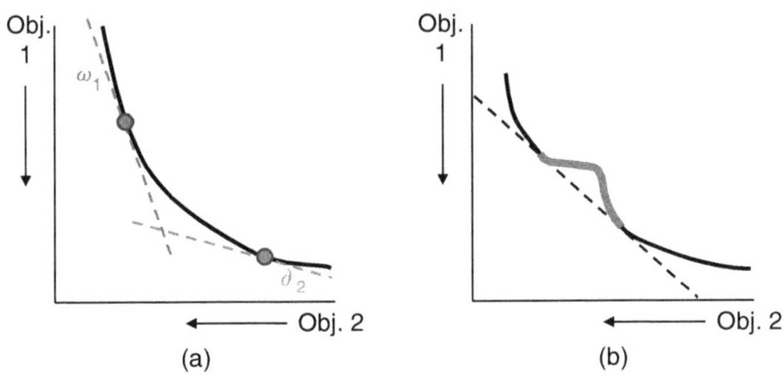

Figure 5.1 (a) Finding Pareto-optimal points using the weighted sum method with two different weights, ω_1 and ω_2. (b) The weighted sum method is unable to generate points in nonconvex regions of a Pareto frontier.

A second popular scalarization approach is the epsilon constraint method, which provides the following single-objective formulation for identifying Pareto-optimal points:

$$\min_{x} \quad f_j(x) \tag{5.4a}$$

$$\text{s.t.} \quad g(x) \leq 0 \tag{5.4b}$$

$$h(x) = 0 \tag{5.4c}$$

$$f_i(x) \leq \epsilon_i \quad \forall i \in \{1, \ldots, n\} \setminus \{j\}. \tag{5.4d}$$

In this approach, one of the n objectives is retained in the objective function while the remaining are converted to new constraints. By varying the ϵ values, weakly Pareto optimal points for the MOP can be obtained. A graphical interpretation of the epsilon constraint approach is shown in Figure 5.2a. By cutting off some of the feasible region of objective values, different solutions on the Pareto frontier can be obtained. Unlike the weighted sum method, the epsilon constraint method is able to identify Pareto-optimal points within nonconvex regions of the feasible objective space. However, it can also generate solutions that are only weakly Pareto optimal, and thus less useful, as shown in Figure 5.2b. An additional challenge, particularly once three or more objectives are being considered, is that choosing bad ϵ values can lead to trying to solve an infeasible problem; however, it is impossible to know what ϵ values will do this without knowing information about the full Pareto frontier.

The Pareto frontier represents the furthest we can go, mathematically, toward decision-making for a MOP, since each solution on the Pareto frontier is mathematically equivalent to every other solution. It provides stakeholders with important information about objective trade-offs that they can use to help inform their values. However, ultimately, decision-makers need to arrive at a single

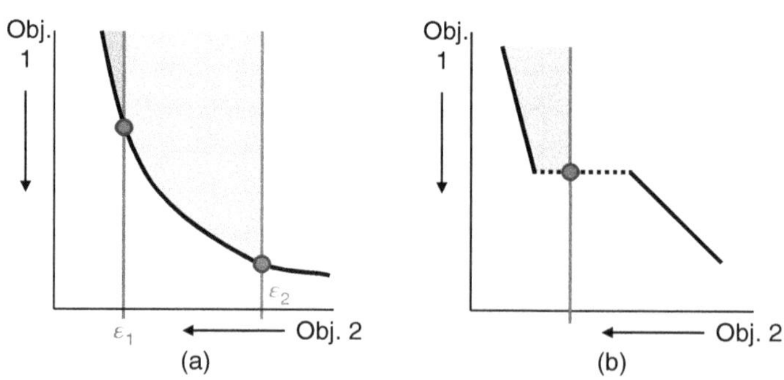

Figure 5.2 (a) Finding Pareto-optimal points using the epsilon constraint method with two different constraint values, ϵ_1 and ϵ_2. (b) The epsilon constraint method can sometimes generate points that are only weakly Pareto optimal (i.e. on the dashed line).

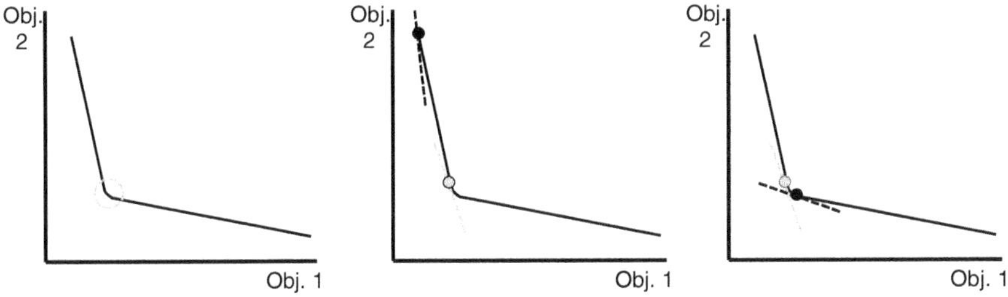

Figure 5.3 An example of a knee region of a Pareto frontier circled. Far from the knee, small changes in objective weights cause large changes in the solution, while near the knee, the solution is more robust to larger changes in objective weights.

action, rather than a manifold of possibilities. While the mathematics cannot take us to this single decision, heuristic approaches are commonly used once a Pareto frontier is obtained. First is to find a "knee" point on the Pareto frontier, as demonstrated on the left of in Figure 5.3, whereby any perturbation from this point causes a disproportionately large degradation in one or more objectives, in comparison to the benefits seen by improving others. Equivalently, as the middle and right parts of Figure 5.3 demonstrate, a solution at a knee point is not very sensitive to changes in objective preferences or weights. Second is to minimize some metric of distance from a "utopia" point, that is, the point corresponding to the best possible value for each objective, individually. It is possible to weight this distance based on a stakeholder's preference toward specific objectives [19]. Ultimately, decision-makers need to carefully consider how the values of their organization are in alignment with the trade-offs presented by the multi-objective solution to determine a best course of action.

5.4 Objective Dimensionality Reduction for MaOPs

In Section 5.3, several rigorous scalarization methods for solving MOPs were presented. A unifying feature of these methods was an exponential scaling of computational effort required to generate a Pareto frontier with number of objectives. Moreover, the ability of human stakeholders to visualize and interpret trade-offs in a Pareto frontier to guide their values and decision-making is severely degraded in problems of four or more dimensions. As such, when encountering a MaOP, it is desirable to be able to systematically reduce the number of objectives to three or fewer whenever possible. Moreover, this should not be done by, that is, grouping environmentally or socially relevant objectives into the same basket, but by identifying objectives that, when grouped together, will result in as little loss of trade-off information as possible.

Recent research efforts [20] have developed a novel method for achieving the aforementioned goal of reducing objective dimensionality for (mixed integer) linear programs, that is, MaOPs of the following form:

$$\min_{x} \quad \{c_1^T x, c_2^T x, \dots, c_n^T x\} \tag{5.5a}$$

$$\text{s.t.} \quad Ax \leq b, \tag{5.5b}$$

where variables x can be continuous or integer. The premise of the method is depicted in Figure 5.4 and stated as follows: in an MaOP, some objectives will point toward similar solutions ("correlating" objectives), while others will point toward very different solutions ("competing objectives). Since in a linear optimization problem, all solutions lie at a vertex of the feasible region, assessing the projections of objective cost vectors along constraint surfaces can help us identify when objectives are likely to be correlating or competing. This information can be embedded into a graph, on which network theoretic tools can be used to partition objectives into groups such that objectives are correlated with others in the same group, and competing with others in different groups.

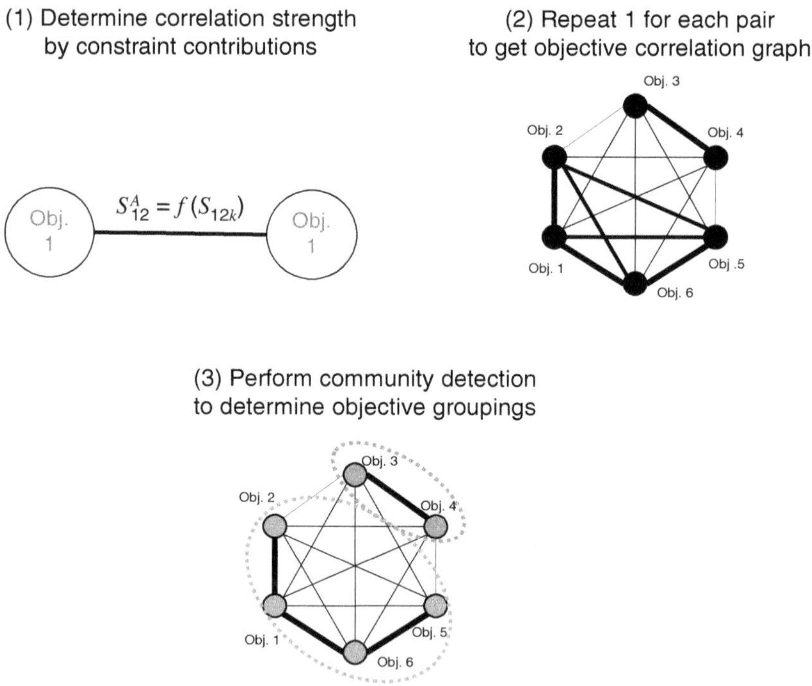

Figure 5.4 Algorithm for systematically grouping objectives of a MaOP.

In what follows, the objective reduction method described generally above is presented in detail. First, an interaction strength between a pair of objectives is calculated for each constraint surface. Achieving this requires projecting cost vectors onto the surface of individual constraints defined by normal vector a_k, a row of the constraint matrix A:

$$c_{ik}^N = \frac{-c_i^T a_k}{\| a_k \|_2^2} a_k, \tag{5.6}$$

$$\hat{c}_{ik}^P = \frac{-c_i - c_{ik}^N}{\| -c_i - c_{ik}^N \|_2}, \tag{5.7}$$

where c_{ik}^N represents the component of the cost vector c_i normal to constraint surface k, and $\hat{c}_{ik}^P$ is the component of the cost vector c_i along the surface of constraint k, normalized to length 1. Then, for two objectives i and j interacting on constraint k, we define a strength of interaction S_{ijk}:

$$S_{ijk} = (\hat{c}_{ik}^P)^T \, \hat{c}_{jk}^P. \tag{5.8}$$

Since projection vectors are normalized, the strength of objective interaction along a constraint can vary between 1, when objectives are perfectly correlated and pointing in the exact same direction, and −1, when objectives are pointing in the exact opposite direction. In some cases, the cost vector c can be large in dimensionality, which can lead to numerical instabilities due to elements of c being small in magnitude after normalization. This can be mitigated by normalizing the ∞, rather than 2 norm, to be one, or (if some normalized cost vector are sparse), setting all elements that are zero in one cost vector to be zero in the other, and renormalizing. To determine the total objective correlation strength, S_{ij}^A, we consider a weighted sum of the constraint interaction strengths. Weights W_{ijk} are zero when both normal vectors c_{ik}^N and c_{jk}^N point into the feasible region, as this implies constraint k is unlikely to be active in optimizing either objective, and are otherwise determined using the following logistic function:

$$W_{ijk} = 1 - \alpha \left(\frac{1}{1 + \exp(-\beta \times S_{ijk})} \right). \tag{5.9}$$

This equation captures the fact that competition on any constraint can lead to an objective trade-off, while correlation needs to occur on all constraints for objectives to be correlated. As such, it makes sense to weight negative values of S_{ijk} more than positive ones. The hyperparameters α and β alter the shape of the logistic curve, such that α, which should lie between zero and one, determines how much correlating constraints are discounted and β, which should be positive, determines the slope of the change from minimum to maximum weight, with larger values giving a more step-like change. In general, the more constraints a problem has, the larger one should consider making α and β such that competition is not obfuscated by a large number of correlating constraints. The total objective interaction strength is determined as follows:

$$S_{ij}^{A} = 0.5\left(1 + \frac{\sum_{k=1}^{M+N} W_{ijk} S_{ijk}}{\sum_{k=1}^{M+N} W_{ijk}}\right). \tag{5.10}$$

The objective correlation strength is scaled to be between 0 and 1, with low values corresponding to highly competing objectives and high values corresponding to highly correlating objectives. The strengths are used to define edge weights in an objective correlation graph, where nodes correspond to each individual objective. Community detection, using, that is, spectral partitioning [21], fast unfolding [22], or the Leiden algorithm [23], can then be applied to the objective correlation graph to identify groups of objectives which are highly correlated to others in the same group, but competing with those in different groups.

5.5 Case Study: Cost Versus Emissions-Driven Demand Response

Demand response is a newer paradigm for chemical systems operation, whereby energy usage is scaled up and down in response to an intermittent driving force such as the cost of power from the grid or availability of on-site renewable power. This idea runs in contrast with traditional paradigms for continuous chemical manufacturing, which suggest to run at a maximum capacity steady state as often as possible. While demand response operation can often have sustainability benefits by supporting the integration of intermittent renewable power generation, in most demand response studies the primary driver is still cost, as savings are usually possible for processes flexible enough to operate at a high capacity when energy is cheap/abundant, and a low capacity when it is not. However, one could posit making demand response decisions directly considering sustainability objectives. In this case, one would want to consider time-varying grid *emissions*, rather than or alongside cost, as an intermittent signal providing a driving force for time-varying demand response operation.

The importance of considering time-varying cost and emissions as separate demand response driving forces is demonstrated in the following case study, adapted from a previous research work [24]. In this study, the goal is to optimize the operation of a small scale ammonia production facility. The hydrogen required for ammonia production can come from either on-site water electrolysis (supplied with power purchased from the grid) or a nearby steam reforming facility. A pressure swing absorption unit for nitrogen generation and small scale Haber-Bosch reactor are also supplied with grid-purchased power. The optimization determines hourly production rates over a 48-hour forecast period, and is repeated hourly in a moving horizon implementation. Four sustainability relevant objectives are considered: cost, carbon emissions, water usage, and process safety. The first two objectives depend on time-varying signals for the cost and emissions associated with purchasing power from the grid, while the latter two objectives are time-independent, always strongly correlated (since both promote using the steam reformer), and act as a control for determining when the cost and emissions objectives are correlating or competing. More details about the model objective functions can be found in [24], while the full formulation of constraints and lists of parameter values can be obtained in [25].

As formulated, the optimal ammonia production problem being considered is a four-objective problem, putting it in the class of many objective problems that are difficult to solve and interpret. However, the previously described objective reduction tool allows us to systematically reduce the dimensionality to a two-objective problem. In particular, using objective reduction to identify when cost and emissions are correlated (thus, optimizing for cost does a good job of keeping emissions low) versus competing (such that it is desirable to find a compromise operating point) is an informative task. In Figure 5.5, the results of such an analysis are shown using real cost and emissions data obtained from CAISO in the month of August 2022. At each hourly time point, the objective correlation strength between cost and emissions objectives, as well as the overall objective groupings determined by the objective reduction tool, are determined for the 48-hour optimal operation problem beginning at that time. In this example, it is apparent that most of the time, cost and emissions are strongly correlating objectives which are grouped together all but a roughly 12 hour period during the month. This should not come as much of a surprise: California has a large amount of solar power installed, and low cost times are often when there is an excess of supply due to solar generation. As such, these low cost times are inherently lower in grid

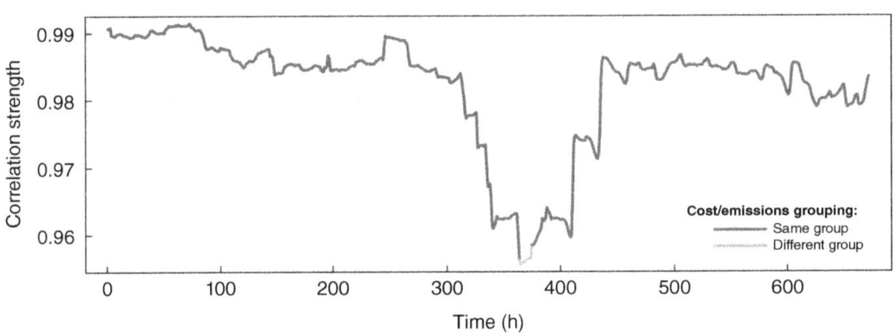

Figure 5.5 Correlation strengths and objective groupings using CAISO power grid pricing and emissions data for August 2022.

emissions. However, this is not always the case. For example, on a really hot day, higher demand from air conditioning can make energy costs increase despite the lower emissions intensity solar energy being generated. Of course, these trends are also largely location dependent: a similar analysis using UK power price and emissions data generally shows lower correlation strengths between cost and emissions objectives.

The objective reduction tool allows for the identification of times where generating a full Pareto frontier is helpful in identifying a sustainable compromise operating point versus times where optimizing for cost is adequate for generating solutions that perform well in other sustainability metrics. Consider two time points from the period analyzed in Figure 5.5. The first, starting at 3:00 PM August 3, has the highest objective correlation strength, while the second, starting at 6:00 AM August 16, has the smallest. Results from the cost-optimal and emissions-optimal solution for each time point are shown in Table 5.1. In the first, we see that the optimal results from the two different objectives are very similar, with costs and emissions varying by no more than 2.1% from the two cases. However, when the correlation strength is weaker, the trade-off becomes greater: in the second time point, we see an emissions increase of 29.4% between the emissions- and cost-optimal solutions. As this difference is significant, it is worthwhile to build and analyze the Pareto frontier (Figure 5.6) to examine the range of trade-offs for a compromise solution point. Here, we can see that multiple potential knee points exist that provide a trade-off which enables reduction of emissions from the cost-optimal solution at a relatively lower added cost. At the two points selected, it is apparent that for an extra 0.8% in cost, we can reduce the emissions increase from the cost-optimal case from 29.4% to 4.6%, while for another 1.1% of cost on top of that, we can further reduce emissions to within 1.1% of their optimal value.

Table 5.1 Cost and emissions values for single-objective optimal solutions at 2:00 PM 8/3 (correlating) and 6:00 AM 8/16 (competing), plus two knee points on Pareto for competing time point.

		Cost ($)	**Emissions (kg CO_2)**
2:00 PM August 3	Cost-optimal	40,974	130,247 (2.1%)
	Emissions-optimal	41,500 (1.3%)	127,506
	Cost-optimal	52,734	234,136 (29.4%)
	Emissions-optimal	57,654 (9.3%)	180,919
6:00 AM August 16	Trade-off Point 1	53,159 (0.8%)	189,285 (4.6%)
	Trade-off Point 2	53,762 (1.9%)	182,895 (1.1%)

Values in parentheses correspond to increase from respective optimal case.

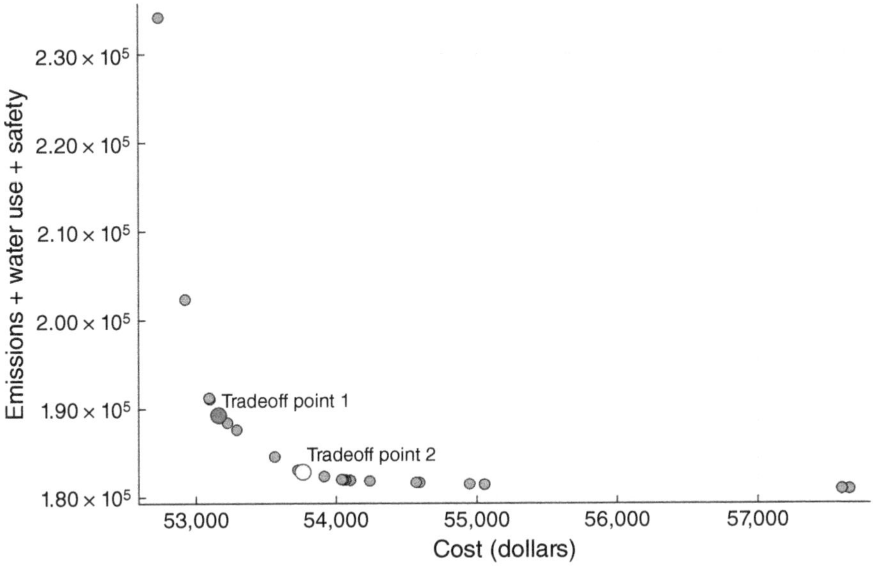

Figure 5.6 Pareto frontier based on objective reduction for optimal ammonia production at 6:00 AM August 16. Two trade-off points (knees of the curve) are highlighted for further analysis.

5.6 Case Study: Analysis of Planetary Boundary Objectives

As discussed in Section 5.2, the concept of planetary boundaries (PBs) has been introduced as a guiding framework for sustainable decision-making [26]. These nine boundaries delineate a safe operational space for humanity, highlighting critical processes essential for maintaining the stability and resilience of the Earth's systems. By including PBs in decision-making models and explicitly correlating model performance with their wide environmental impact, one can ultimately design a system that is, to some extent, more consistent with sustainable development goals [27–29]. In this case study, an adaptation of a previous work assessing planetary boundaries impacts for sustainable fuel supply chains in the United Kingdom [30] is considered. This work is particularly focused on hydrogen generation for vehicle use. It considers four hydrogen production technologies (steam reforming of methane, as well as wind-, solar-, and nuclear-powered electrolysis), two hydrogen storage technologies (compressed gas and liquefied gas), and two transportation mechanisms (truck and railcar). Hydrogen can be produced at or transported between 34 different locations, each with its own demand requirements. The supply chain optimization model decides where to build new production and storage facilities, and how to transport hydrogen within the supply chain. For full model details and parameter values, we refer the reader to the original work on this problem [30].

The original work assesses the impact of supply chain decisions on eight planetary boundary-related metrics that correspond to six of the nine overall planetary boundaries. Scenarios are considered where cost is minimized, and where the total transgression of all eight planetary boundary metrics is minimized. However, by lumping together all of the planetary boundary transgressions, one is unable to analyze the potential trade-off between specific planetary boundary outcomes. This case study instead considers the eight planetary boundary metrics each as their own objective, allowing for the use of the objective reduction tool to identify outcomes which are correlating versus competing. The result is a nine objective MaOP considering total cost (TC), atmospheric carbon dioxide concentration (PB1) and energy imbalance (PB2) corresponding to the climate change boundary; stratospheric ozone concentration (PB3) corresponding to the ozone depletion boundary; ocean carbonate ion concentration (PB4) corresponding to the ocean acidification boundary; phosphorous (PB5) and nitrogen flow (PB6) corresponding to the biogeochemical flow boundary; original forest cover remaining (PB7) corresponding to the land system change boundary; and blue water consumption (PB8) corresponding to the freshwater use boundary. Note that parameters for determining planetary boundary impact of different technologies were determined by proprietary LCA software and not made available in the original publication. As such, approximate values were generated for this case study which are reasonable for the given technologies and illustrative of the power of the objective reduction method.

The objective reduction tool described in this chapter is used to reduce the nine objective problem to a two objective problem. The objective reduction graph and resulting objective grouping is depicted in Figure 5.7. The first objective grouping consists of total cost, forest cover remaining, and blue water consumption, while all remaining planetary boundary outcomes are grouped in the second grouping. These results make a lot of intuitive sense when one considers the four candidate technologies for hydrogen production considered in this work. Steam reforming of methane, which is currently the most prevalent of the hydrogen production technologies used today, is the low cost option of the four. It also has a lower amount of water required per amount of hydrogen produced than any of the electrolysis technologies, each of which obtain all of the hydrogen from water. Moreover, installing additional wind or solar power to produce hydrogen via electrolysis has a large land requirement, resulting in a greater change in land use. However, in general, the fossil fuel-powered steam reforming process performs poorly when considering all of the other environmental outcomes.

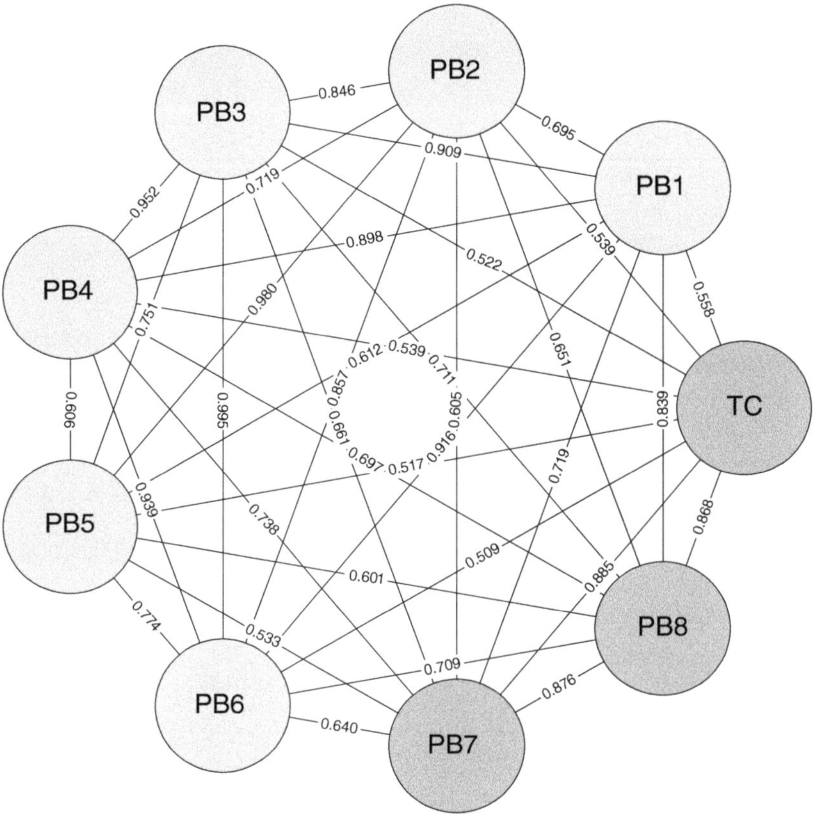

Figure 5.7 Weighted objective correlation graph and color-coded objective grouping of the nine-objective problem considered in this case study.

Systematically reducing the nine-objective problem to a two-objective problem allows for a more thorough assessment of the trade-offs encountered when attempting to optimize this system for sustainability. The result of using a weighted sum approach to generate the reduced space Pareto frontier is shown in Figure 5.8. In generating this plot, two "knee" points are encountered which denote promising compromise points between the two objective groups. Table 5.2 shows the resulting total cost and planetary boundary impacts for each of the two single group-optimal solutions, as well as the two highlighted knee points. In particular, while we see that completely minimizing the second planetary boundary group results in a significant degradation in the water use and land use metrics, choosing either of the knee points as the sustainable solution shows that it is possible to get most of the way to this point for much more reasonable values in these metrics. It is also important to note that although the objective grouping

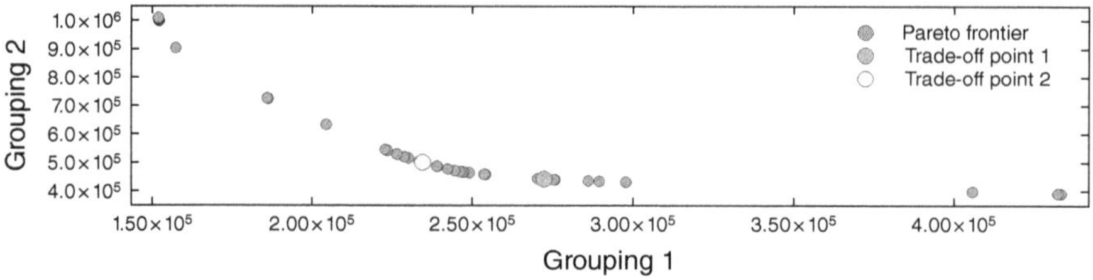

Figure 5.8 Pareto frontier between two grouped objectives with two knee points highlighted.

Table 5.2 Values of total cost (MM$) and eight planetary boundary metrics (a.u.) at two optimal and two trade-off points along Pareto.

	TC	***PB1***	**PB2**	***PB3***	***PB4***	**PB5**	***PB6***	***PB7***	***PB8***
Group 1 optimal	142,781	19,905	18,087	125,363	22,637	498,526	324,312	2673	6450
Trade-off Point 2	142,786	88,319	1881	96,901	96,811	17,502	199,162	53,118	38,662
Trade-off Point 1	142,790	106,019	1271	71,027	109,829	19,487	134,449	75,628	53,737
Group 2 optimal	143,870	25,864	4981	73,065	129,353	55,316	102,787	196,703	92,922

Group 1 objectives are bolded, while group 2 objectives are italicized.

preserves most trade-off information, there is always some information lost when reducing dimensionality due to the fact that none of the objectives are completely correlated. In these results, it is apparent that, for example, the PB2 and PB5 metrics have significantly lower values in the two trade-off solutions than in the solution that is optimal for group 2. Moreover, by just looking at the four provided points, it seems that PB4 is more strongly correlated with group 1 objectives (although a full analysis of all Pareto points confirms its strong correlation with PB3 and PB6). Nonetheless, the analysis shown here demonstrates the utility of the objective reduction tool to systematically assess trade-offs in MaOPs for sustainable systems in a way that would not be possible by, for example, just lumping all of the environmental planetary boundaries transgressions together.

5.7 Conclusion and Future Perspectives

Optimizing for a sustainable system is truly a wicked problem – there are many factors that one can consider, it is rarely clear how each factor interacts with the others, and there is almost never a single solution that performs best for every sustainability factor that one could consider. As scientists and engineers, while we would like to be able to generate a single, provably sustainable, decision, often the best we can do is provide an analysis of the trade-offs present between different sustainability goals. However, multi-objective and many-objective optimization provides a framework for performing this analysis. This chapter presented tools for systematically reducing high-dimensional MaOPs into tractable MOPs of two or three objectives, as well as methods for rigorously generating the Pareto frontier for such MOPs. Taken together, these tools promote better understanding of when sustainability objectives are correlating or competing and enable the efficient identification of potential compromise solutions which take into account all sustainability goals.

Ultimately, the methods described in this chapter are just one tool in the larger toolbox of sustainable decision-making. Truly designing a sustainable future requires an all-hands-on-deck effort involving engineers, organizational decision-makers, social scientists, community stakeholders, and policymakers. To this end, there is a strong need for engaging social and environmental scientists in order to better quantify the effect of different decisions on the various sustainability outcomes, particularly those where an intuitive relationship does not exist. While life cycle assessment tools have improved the ability to do this with environmental objectives, there is still a lot of room for improvement in this space, particularly when considering socially relevant objectives. Another area of future development entails working with community leaders for the generation of sustainability objective functions that are directly aligned with community concerns, in order to avoid solutions that exploit an underspecified region of the problem, or that appear strong given the goals considered today but ultimately develop a new set of problems down the road. Finally, the results of MaOP analysis need to be shared with policymakers in order to promote the development of sound, science-guided policy which constrains or incentivizes decision-makers to make sustainable decisions even when solely considering their own economic objective.

References

1 Garbie, I.H. (2015). Sustainability optimization in manufacturing enterprises. *Procedia CIRP* 26: 504–509.

2 Majewski, D.E., Wirtz, M., Lampe, M., and Bardow, A. (2017). Robust multi-objective optimization for sustainable design of distributed energy supply systems. *Computers and Chemical Engineering* 102: 26–39.

3 Beykal, B., Boukouvala, F., Floudas, C.A., and Pistikopoulos, E.N. (2018). Optimal design of energy systems using constrained grey-box multi-objective optimization. *Computers and Chemical Engineering* 116: 488–502.
4 Tian, X., Meyer, T., Lee, H., and You, F. (2019). Sustainable design of geothermal energy systems for electric power generation using life cycle optimization. *AIChE Journal* 66: e16898.
5 Roh, K., Al-Hunaidy, A.S., Imran, H., and Lee, J.H. (2019). Optimization-based identification of CO_2 capture and utilization processing paths for life cycle greenhouse gas reduction and economic benefits. *AIChE Journal* 65: e16580.
6 Zhao, X. and You, F. (2021). Waste high-density polyethylene recycling process systems for mitigating plastic pollution through a sustainable design and synthesis paradigm. *AIChE Journal* 67: e17127.
7 Gopalakrishnan, V. and Bakshi, B. (2018). Ecosystems as unit operations for local techno-ecological synergy: integrated process design with treatment wetlands. *AIChE Journal* 64: 2390–2407.
8 You, F., Tao, L., Graziano, D.J., and Snyder, S.W. (2012). Optimal design of sustainable cellulosic biofuel supply chains: multiobjective optimization coupled with life cycle assessment and input-output analysis. *AIChE Journal* 58: 1157–1180.
9 Allman, A., Tiffany, D., Kelley, S., and Daoutidis, P. (2017). A framework for ammonia supply chain optimization incorporating conventional and renewable generation. *AIChE Journal* 63: 4390–4402.
10 Liu, Z., Lim, M.Q., Kraft, M., and Wang, X. (2020). Simultaneous design and operation optimization of renewable combined cooling heating and power systems. *AIChE Journal* 66: e17039.
11 Saif, Y., Almansoori, A., and Elkamel, A. (2021). Sustainable optimization of waste management network over extended planning time horizon. *AIChE Journal* 67: e17256.
12 Mota, B., Gomes, M.I., Carvalho, A., and Barbosa-Pavoa, A.P. (2018). Sustainable supply chains: an integrated modeling approach under uncertainty. *Omega* 77: 32–57.
13 Duarte, I., Mota, B., Pinto-Varela, T., and Barbosa-Pavoa, A.P. (2022). Pharmaceutical industry supply chains: how to sustainably improve access to vaccines? *Chemical Engineering Research and Design* 182: 324–341.
14 Fonseca, J.D., Commenge, J.-M., Camargo, M. et al. (2021). Sustainability analysis for the design of distributed energy systems: a multi-objective optimization approach. *Applied Energy* 290: 116746.
15 Steffen, W., Richardson, K., Rockstron, J. et al. (2015). Planetary boundaries: guiding human development on a changing planet. *Science* 347: 1258955.
16 Wheeler, J., Paez, M.A., Guillen-Gosalbez, G., and Mele, F.D. (2018). Combining multi-attribute decision-making methods with multi-objective optimization

in the design of biomass supply chains. *Computers and Chemical Engineering* 113: 11–31.

17 United Nations (2024). *The 17 Goals*. United Nations.

18 Emmerich, A.H. and Deutz, M.T.M. (2018). A tutorial on multiobjective optimization: fundamentals and evolutionary methods. *Natural Computing* 17 (3): 585–609.

19 Wang, Z., Rangaia, G.P., and Wang, X. (2021). Preference ranking on the basis of ideal-average distance method for multi-criteria decision making. *Industrial and Engineering Chemistry Research* 60: 11216–11230.

20 Russell, J.M. and Allman, A. (2023). Sustainable decision making for chemical process systems via dimensionality reduction of many objective problems. *AIChE Journal* 69: e17962.

21 Newman, M.E.J. (2013). Spectral methods for community detection and graph partitioning. *Physical Review E* 88: 042822.

22 Blondel, V.D., Guilaume, J.-L., Lambiotte, R., and Lefebver, E. (2008). Fast unfolding of communities in large networks. *Journal of Statistical Mechanics: Theory and Experiment* 2008: P10008.

23 Traag, V.A., Waltman, L., and van Eck, N.J. (2019). From Louvain to Leiden: guaranteeing well-connected communities. *Scientific Reports* 9 (5233): 1–2.

24 Wang, H. and Allman, A. (2024). Analysis of the correlating or competing nature of cost-driven and emissions-driven demand response. *Computers and Chemical Engineering* 181: 108520.

25 Allman, A. and Daoutidis, P. (2018). Optimal scheduling for wind-powered ammonia generation: effects of key design parameters. *Chemical Engineering Research and Design* 131: 5–15.

26 Rockström, J., Steffen, W., Noone, K. et al. (2009). A safe operating space for humanity. *Nature* 461 (7263): 472–475.

27 Vázquez, D. and Guillén-Gosálbez, G. (2021). Process design within planetary boundaries: application to CO2 based methanol production. *Chemical Engineering Science* 246: 116891.

28 Algunaibet, I.M., Pozo, C., Galán-Martín, Á. et al. (2019). Powering sustainable development within planetary boundaries. *Energy & Environmental Science* 12 (6): 1890–1900.

29 Tulus, V., Pérez-Ramírez, J., and Guillén-Gosálbez, G. (2021). Planetary metrics for the absolute environmental sustainability assessment of chemicals. *Green Chemistry* 23 (24): 9881–9893.

30 Ehrenstein, M., Galán-Martín, Á., Tulus, V., and Guillén-Gosálbez, G. (2020). Optimising fuel supply chains within planetary boundaries: a case study of hydrogen for road transport in the UK. *Applied Energy* 276: 115486.

6

Optimization Models and Algorithms for Design and Planning of Sustainable Processes and Energy Systems

Seolhee Cho and Ignacio E. Grossmann

Department of Chemical Engineering, Carnegie Mellon University, Pittsburgh, PA, USA

6.1 Introduction

Sustainable development aims to meet social, economic, and technical goals without harming the environment [1]. As fossil fuel-based processes and energy systems have significantly impacted the environment, the interest in finding alternatives to solving current environmental issues is increasing. Alternatives receiving attention include hydrogen, methanol/ethanol, and biomass (algae) [2, 3]. Despite such interest, it is difficult for new and renewable resources to be utilized (or commercialized) due to multiple problems such as uncertain demand, insufficient infrastructure, stability/safety, and cost-efficiency [4–6]. As a result, various studies on sustainable development have been conducted, including the synthesis of new catalysts and chemicals [7, 8]; techno-economic and environmental analysis [9, 10], and infrastructure planning/operation using optimization techniques [11]. The importance of developing rigorous and efficient optimization models that can determine different decision levels is increasing for designing sustainable systems.

Optimization can be classified into continuous and discrete optimizations. While continuous optimization is expressed using constraints and continuous variables, discrete optimization includes constraints and discrete variables such as binary and/or integer [12]. This optimization model has been actively used to find the optimal process network, control strategy, and design/operation plans in sustainable processes and energy systems. Not only have the developed optimization models been used, but many algorithms have also been developed to effectively solve more complex and practical problems arising from the sustainable development area.

This chapter reviews optimization models and applications in sustainable processes and energy systems. In Section 6.2, we first review fundamental optimization models, such as linear programming (LP), nonlinear programming (NLP), mixed-integer programming (MIP), generalized disjunctive programming (GDP), and stochastic programming (SP) that are frequently used for sustainable processes and energy systems. Decomposition algorithms proposed to solve large-scale problems effectively are reviewed in Section 6.3. Lastly, in Section 6.4, successful applications of optimization models in sustainable processes and energy systems are explained, such as for hydrogen, biomass, methanol, and power systems.

6.2 Optimization Models

Figure 6.1 classifies the optimization problems depending on their basic properties. Specifically, there are four major types of optimization problems: continuous optimization, discrete optimization, logic-based optimization, and optimization under uncertainty. Each problem is addressed with different modeling techniques as seen in the figure.

The general formulation of optimization models consists of an objective function and a set of constraints, as shown in Eq. (6.1). The objective function $f(x, y)$ can be either a linear or a nonlinear function. The objective function is usually minimization, but maximization can also be used. The constraints are given by equalities $h(x, y) = 0$ and/or inequalities $g(x, y) \leq 0$. In the general notation below, x represents continuous variables, whereas y represents discrete variables, including binary and integer. While continuous optimization models have only continuous variables (i.e. no discrete variable y), pure discrete optimization

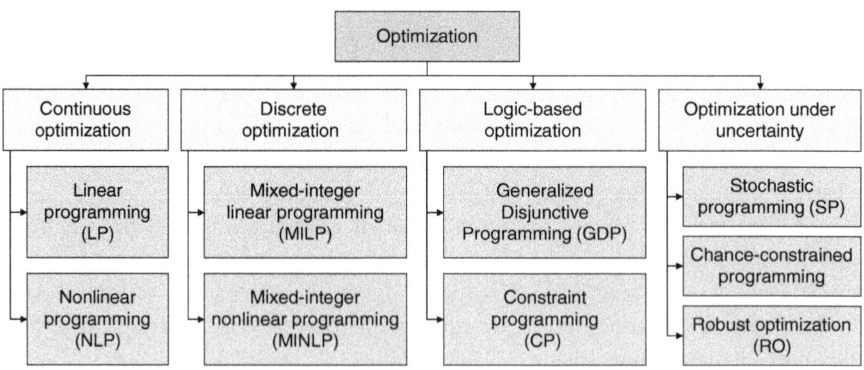

Figure 6.1 Classification of optimization problem.

models have only discrete variables (likewise, no continuous variable x).

$$\begin{aligned}\min_{x,y} \quad & f(x,y) \\ \text{s.t.} \quad & h(x,y) = 0 \\ & g(x,y) \leq 0 \\ & x \in \mathbb{R}^m \\ & y \in \mathbb{Z}^n.\end{aligned} \tag{6.1}$$

6.2.1 Continuous Optimization

Continuous optimization comprises linear programming (LP) and nonlinear programming (NLP). Both LP and NLP have only continuous variables. While LP involves exclusively linear equations, such as linear objective functions and linear inequalities/equalities, NLP incorporates at least one nonlinear equation. A particular instance of NLP is quadratic programming (QP), where the nonlinear functions are quadratic equations. There are two major methods to solve LPs: Simplex algorithm (also known as a pivoting method) and interior-point method (known as a barrier method) [13–15]. The simplex algorithm is known to be efficient for solving small-scale and medium-scale problems with several thousand variables and constraints but has exponential complexity. On the other hand, the interior-point method has polynomial complexity and performs best in very large-scale problems [16]. The major software packages for LPs include CPLEX [17], XPRESS [18], and Gurobi [19]. Algorithms for solving NLPs fall into three main categories: successive quadratic programming, reduced gradient method, and interior-point methods. A detailed explanation of each algorithm can be found in [20]. SNOPT [21] is a software for solving NLPs using the successive-quadratic programming (SQP) algorithm. It is known that SNOPT can effectively handle nonlinearities and requires relatively few iterations. MINOS [22] and CONOPT [23] are known as the best software for mostly linear constraints. CONOPT is more robust than MINOS as it tries to recover feasibility at each iteration. IPOPT [24] and KNITRO [25] implement the interior-point method to solve NLPs. It is known that they are effective in solving very large-scale problems.

6.2.2 Discrete Optimization

Discrete optimization can be classified into integer programming (IP) and mixed-integer programming (MIP), where IP only uses integer variables, but MIP uses both continuous and discrete variables. Depending on the types of constraints, MIP could be further divided into mixed-integer linear programming (MILP) and mixed-integer nonlinear programming (MINLP). When the objective function and all constraints are linear, the model becomes an MILP. On the other hand, when at least one equation is nonlinear, the model becomes a MINLP. There are three major algorithms to solve MILPs: branch-and-bound [26], cutting planes [27], and branch-and-cut [28]. The algorithms for MILP can also be used for MINLPs with removed nonlinearities via exact linearization or piece-wise linear approximations. The following algorithms, such as branch-and-bound, Generalized Benders Decomposition (GBD) [29], Outer-Approximation (OA) [30], and Extended Cutting Plane method (ECP) [31], address directly the solution of MINLP problems. The major software packages for MILP include CPLEX [17], XPRESS [18], and Gurobi [19], for MINLP they include DICOPT for OA [32], SBB for branch-and-bound [33], α-ECP for ECP [34], and finally BARON for global NLP and MINLP problems [35].

6.2.3 Logic-Based Optimization

Generalized disjunctive programming (GDP) is a mathematical optimization model that characterizes feasible regions using intersecting disjunctions of algebraic constraints involving continuous variables (both equality and inequality constraints). This model employs Boolean variables as indicators for each disjunct, activating the constraints associated with a disjunct when the Boolean value is true [36, 37]. The GDP is formulated as shown in Eq. (6.2).

$$
\begin{aligned}
\min_{x,Y} \quad & f(x) \\
\text{s.t.} \quad & r(x) \leq 0 \\
& \bigvee_{j \in \mathcal{J}_i} \begin{bmatrix} Y_{ij} \\ g_{ij}(x) \leq 0 \end{bmatrix} \quad \forall i \in \mathcal{I} \\
& \Xi(1, Y_{ij} \quad \forall j \in \mathcal{J}_i) \quad \forall i \in \mathcal{I} \\
& \Omega(Y) = \text{True} \\
& x^{\text{LB}} \leq x \leq x^{\text{UB}}, \quad x \in \mathbb{R}^m \\
& Y_{ij} \in \{\text{True}, \text{False}\} \quad \forall i \in \mathcal{I}, j \in \mathcal{J}_i,
\end{aligned}
\tag{6.2}
$$

where x is a set of continuous variables, which is bounded by x^{LB} and x^{UB}, Y_{ij} is a set of Boolean variables that take values of True or False. $f(x)$ is the objective function, $r(x) \leq 0$ is the set of global constraints, and $g_{ij}(x) \leq 0$ is the set of constraints applied when the Boolean indicator Y_{ij} is True for disjunct j in disjunction i. It is assumed that $f(x)$, $r(x)$, and $g_{ij}(x)$ are convex differentiable functions. $\Xi(1, Y_{ij} \quad \forall j \in \mathcal{J}_i)$ enforces that exactly 1 of the Boolean variables Y_{ij} should be True for disjunct j in disjunction i [38]. $\Omega(Y)$ specifies the logic constraints using propositional logic with a subset of Boolean variables. The GDP model can be reformulated into MILP/MINLP models with 0–1 binary variables using Big-M and Hull reformulations [37]. In both reformulations, disjunctions and logic propositions are translated into algebraic equations using binary and continuous variables. While the Big-M reformulation involves the introduction of a large parameter value (typically denoted as Big-M), the Hull reformulation disaggregates variables involved in each disjunction and reformulates logical constraints into algebraic equations in terms of binary variables. Theoretically, the Hull reformulation provides a tighter LP or NLP relaxation than the Big-M reformulation. The GDP model can also be solved using either disjunctive branch and bound or decomposition approaches, including logic-based outer approximation and logic-based generalized Benders decomposition [39–41]. *Pyomo.GDP* [42] and *DisjunctiveProgramming.jl* [43] are the software packages for GDP problems implemented in Pyomo and JuMP.

Constraint programming (CP) is a logic-based approach that can be applied to optimization [44]. It uses constraints to reduce the set of feasible solutions and applies search strategies to find the optimal solution. The CP algorithm uses advanced techniques, such as constraint propagation and domain reduction, instead of solving LP relaxation. The CP has been used actively in process scheduling by combining with MIP models. In detail, CP is used for the sequencing constraints, and MIP is used for the resource assignment constraints [45–47].

6.2.4 Optimization Under Uncertainty

Given many challenges facing sustainable systems, such as inaccurate forecasts of demand for alternative resources, difficulties in accurately predicting climate change, and new technologies in development, optimization under uncertainty is critical. To manage uncertainties in process systems, three mathematical frameworks, including stochastic programming, chance-constrained programming, and robust optimization, have been developed [48].

Stochastic programming (SP) aims to optimize the expected outcome based on a known probability distribution [49]. Two-stage stochastic programming is the most widely used case of stochastic programming. The decisions can be divided into stage 1 decisions, which are "here-and-now" in the sense that once selected, they cannot be changed or adjusted, and stage 2 decisions, which are "wait-and-see" in that they can be adjusted according to the realization of the uncertain parameters. Equation (6.3) represents the mathematical formulation of the two-stage SP model.

$$\begin{aligned}\min_{x,y_\omega} \quad & \sum_{\omega\in\Omega} f(x, y_\omega) \\ \text{s.t.} \quad & g(x, y_\omega) \le 0 \\ & x \in X \\ & y_\omega \in Y_\omega,\ \forall \omega \in \Omega,\end{aligned} \tag{6.3}$$

where x represents the first-stage decisions, whereas y_ω represents the second-stage decisions in scenario ω (it is also known as recourse variables). The two-stage stochastic programming problem can be extended to a multistage stochastic programming problem. Multistage stochastic programming makes decisions sequentially over several stages, revealing new information about uncertain parameters at each stage. A scenario tree is usually used, where each branch represents a possible realization of uncertainties over time [48, 49].

Chance-constrained programming deals with uncertainty by ensuring that certain constraints are satisfied with a specified probability [50]. Instead of requiring constraints to be met under all scenarios, it allows for a controlled level of risk, meaning constraints involving uncertain parameters are only required to hold in a certain percentage of cases. On the other hand, robust optimization (RO) focuses on optimizing for the "worst-case" scenario within a predefined uncertainty set [51]. Robust optimization ensures solutions remain feasible across all possible scenarios within a predefined uncertainty set. Instead of relying on probabilistic distributions, it optimizes the worst-case scenario to guarantee performance under the most extreme conditions.

6.3 Solution Strategies

Decomposition in optimization is a computational approach that decomposes a large and complex optimization problem into smaller and more manageable subproblems. This can be helpful when the original problem is too large to solve efficiently. Major decomposition methods include Benders decomposition, Lagrangean decomposition, and bilevel decomposition.

6.3.1 Benders Decomposition

Benders decomposition is used for problems with "complicating variables," which make it difficult to decompose the original problem into subproblems [16]. Figure 6.2(a) shows the high-level view of Benders decomposition. Equation (6.4) shows an LP problem suitable for Benders decomposition, in which y represents the complicating variables. When the complicating variable (i.e. y) is fixed, the original problem can be decomposed into N subproblems in terms of x_i, which can be quickly and independently solved.

$$\begin{aligned} \max \quad & a^T y + \sum_{i=1}^{N} c_i^T x_i \\ \text{s.t.} \quad & Ay + D_i x_i = d_i \quad \forall i = 1, \dots, N \\ & y \geq 0, x_i \geq 0 \quad \forall i = 1, \dots, N. \end{aligned} \tag{6.4}$$

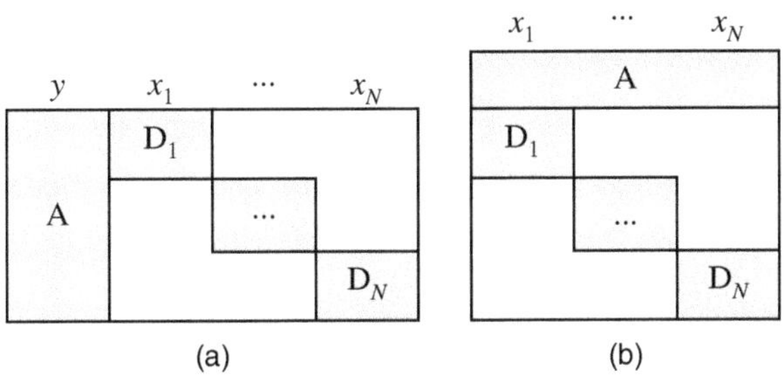

Figure 6.2 Structure of (a) Benders decomposition and (b) Lagrangean decomposition.
Source: Reproduced from Modified from [16] / Cambridge University Press.

Benders decomposition has been used in two-stage stochastic programming problems [48]. Specifically, the first-stage variables are regarded as "complicating variables." When the first-stage variables are fixed, the rest of the problem can be decomposed by scenario and solved in parallel. Benders decomposition algorithm starts by defining a master problem with only the complicating variables. After the master problem with the complicating variables is solved, the first-stage variables in the original problem are fixed. Then, the original problem with fixed values for the first-stage variables can be decomposed into subproblems depending on the scenarios. After obtaining the optimal solution from the subproblems, valid Benders cuts are added to the master problem, and the process continues until it converges. Benders decomposition can converge in a finite number of iterations when the subproblems have only continuous variables and linear constraints, meaning there will be a dual gap between upper and lower bounds if the subproblems have discrete variables.

6.3.2 Lagrangean Decomposition

Lagrangean decomposition is used for problems with "complicating constraints" [16]. Figure 6.2(b) shows the high-level view of Lagrangean decomposition. Equation (6.5) shows an LP problem suitable for Lagrangean decomposition. If the complicating constraint ($Ax = b$) is removed by dualizing and adding it to the objective function, the problem can be decomposed into N subproblems.

$$\begin{aligned} \max \quad & c^T x \\ \text{s.t.} \quad & Ax = b \\ & D_i x_i = d_i \quad \forall i = 1, \dots, N \\ & x_i \geq 0 \quad \forall i = 1, \dots, N. \end{aligned} \tag{6.5}$$

Lagrangean decomposition has been used in multistage stochastic programming problems [48]. Contrary to Benders decomposition, in which the first-stage decisions are fixed, Lagrangean decomposition copies the first-stage decisions for all scenarios. Lagrangean decomposition algorithm starts by relaxing complicating constraints. The original problem is relaxed by adding the complicating constraints to the objective function using Lagrange multipliers. The relaxed problems are then decomposed into several subproblems, which can be solved independently. In minimization problems, the subproblems yield a lower bound, while the original problem provides an upper bound. The Lagrange multipliers are updated until the algorithm converges. Contrary to Benders decomposition, where Benders cuts are added to the master problem, the key step in Lagrangean decomposition is to properly update the Lagrange multipliers. The subgradient or the cutting plane methods are generally used to update the multipliers.

6.3.3 Bilevel Decomposition

Bilevel decomposition is a tailored decomposition technique that exploits the hierarchical structure of problems [52, 53]. For example, the problem that optimizes both planning and scheduling can be effectively solved by using bilevel decomposition. The bilevel decomposition decomposes the original problem into an upper-level problem, which is a planning problem with relaxed scheduling decisions, and a lower-level problem, which is a scheduling problem with fixed planning decisions. While the upper-level problem provides a lower bound, the lower-level problem provides an upper bound in minimization problems. To converge the algorithm, tailored logic cuts, such as design cut, superset cut, and subset cut, are added to the upper level as well as integer cut. Design cuts are for expanding the capacity, whereas superset and subset cuts exclude previously obtained configurations. Integer cuts also exclude a specific solution previously revealed as the optimal solution.

6.4 Algebraic Modeling Languages

Algebraic Modeling Languages (AMLs) are high-level programming languages designed to represent and solve mathematical optimization problems. AMLs provide a structured way of expressing optimization models, making it easier to write and understand complex mathematical formulations. These languages typically include features for defining variables, constraints, objectives, and data in a way that closely resembles the mathematical notation used in optimization theory [54].

Table 6.1 shows the summary of computational features of currently available AMLs, such as AMPL, AIMMS, GAMS, Pyomo, and JuMP. AMPL (A Mathematical Programming Language) is a high-level language designed for modeling and solving optimization problems, such as LP, NLP, and MILP/MINLP.

Table 6.1 Summary of computational features of AMLs [55, 56].

	AMPL	AIMMS	GAMS	Pyomo	JuMP
Solver access	Easy	Easy	Easy	Difficult[a)]	Difficult[a)]
Easy of use	Easy	Easy	Hard[b)]	Hard[b)]	Hard[b)]
Visualization	Limited	Advanced	Limited	Limited[c)]	Limited[c)]
License	Commercial	Commercial	Commercial	Open-source	Open-source

a) Need installation to use solvers.
b) Need coding knowledge.
c) Can be extended by using Python or Julia libraries.

AMPL separates the model, data, and solver, allowing users to switch between solvers like Gurobi or CPLEX without altering the model. Its interactive environment supports rapid debugging of optimization models [57]. AIMMS (Advanced Interactive Multidimensional Modeling System) is an optimization modeling tool that offers a user-friendly graphical interface as well as a mathematical modeling environment. It supports the development of both optimization models and applications, making it suitable for decision-support systems. A key feature of AIMMS is its built-in capability to develop customized graphical user interfaces (GUIs) for models, allowing nonexperts to interact with the optimization tools [58]. GAMS (General Algebraic Modeling System) is a high-level modeling system designed for mathematical programming and optimization. GAMS is known for its capability to handle large and complex optimization problems in a wide variety of applications, including LP, NLP, MILP/MINLP, and SP. GAMS allows models to be structured with reusable components, which is especially beneficial for managing complex optimization models with shared sub-models [59]. GAMS has a shortcoming of lacking visualization capabilities, but to solve this, GAMS Development Corporation launched GAMS MIRO, an interface for deployment of GAMS models and results, in 2020 [60]. Pyomo is a Python-based open-source optimization modeling language that provides flexibility for defining and solving various optimization problems. It is integrated into Python, allowing users to leverage Python's extensive libraries. A unique feature of Pyomo is its ability to support both algebraic modeling and abstract modeling, making it adaptable to different problem structures [61]. JuMP is a domain-specific modeling language for mathematical optimization that is embedded within the Julia programming language. It offers a high-level interface for defining optimization models while utilizing Julia's high-performance computing capabilities. One of the advantages of JuMP is its speed and efficiency, as it is designed to provide near-interactive performance, which is ideal for large-scale optimization and real-time applications [62].

6.5 Applications in Sustainable Process and Energy Systems

This section reviews previous works on sustainable processes and energy systems, including hydrogen, biomass, methanol, and power systems.

6.5.1 Hydrogen

Hydrogen has been regarded as one of the promising energy carriers for sustainable energy systems due to its lower environmental impact than fossil fuels. However, the lack of appropriate infrastructure has made it difficult to realize the hydrogen economy, and to solve this, the Process Systems Engineering (PSE) community has conducted research specifically related to the planning of hydrogen infrastructure. Almansoori and Shah [63, 64] first propose snapshot and multi-period MILP models for the design and operation of the hydrogen supply chain, which considers hydrogen production, distribution, and storage. Multiple hydrogen production facilities, such as steam methane reforming and gasification, and hydrogen storage facilities, such as liquid and compressed hydrogen storage, are considered. They [65] then extend the deterministic model to a stochastic model by considering uncertainty in the hydrogen demand. Hugo et al. [66] propose a multi-objective MILP model that addresses the challenges of developing energy-efficient, environmentally friendly, and cost-effective hydrogen pathways. The model evaluates various technological options for hydrogen production, storage, distribution, and dispensing under the cost and environmental impact minimization environment. Kim and Moon [67] present a multi-objective MILP model to identify cost-efficient and safety-focused infrastructure of hydrogen systems. Using multiobjective optimization techniques, the model provides Pareto optimal solutions that balance cost and safety. Additionally, it assesses the impact of demand uncertainty by comparing deterministic and stochastic solutions.

Green hydrogen refers to the hydrogen that is produced by water electrolysis using renewable electricity. Unlike traditional hydrogen production, which is often derived from fossil fuels such as natural gas, green hydrogen is considered a more sustainable option as it does not emit greenhouse gas while producing [68]. Figure 6.3 shows the pathway for green hydrogen from renewable sources such as wind, solar, and biomass. Some works have been reported to suggest the optimal infrastructure and operation strategies for green hydrogen. Won et al. [69] propose a new methodology to identify the optimal configuration and operation of renewable source-based hydrogen supply systems. They formulate a MILP model, which can integrate multiple resources and various technologies. Kim and Kim [70] propose a snapshot optimization model for a wind-based hydrogen supply chain, whereas Cho et al. [71] develop an optimization model for a biomass-based hydrogen infrastructure. Table 6.2 summarizes the previous works performed.

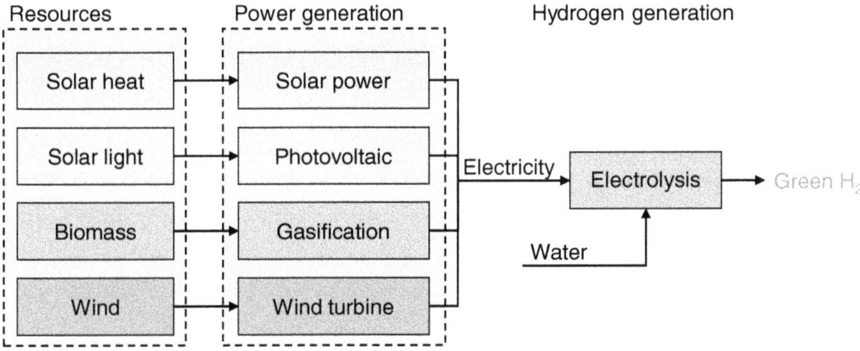

Figure 6.3 Green hydrogen pathway.

Table 6.2 Summary of works on hydrogen infrastructure planning.

Authors	Method	AML	Period	Uncertainty	Multi-objective	Green H_2	Refs.
Almansoori et al.	MILP	GAMS	No	No	No	No	[63]
Almansoori et al.	MILP	GAMS	Year	No	No	No	[64]
Almansoori et al.	MILP	GAMS	Year	Yes	No	No	[65]
Hugo et al.	MILP	—	Year	No	Emission	No	[66]
Kim et al.	MILP	GAMS	Year	Yes	Safety	No	[67]
Won et al.	MILP	GAMS	No	No	No	Yes	[69]
Kim et al.	MILP	GAMS	No	No	No	Yes	[70]
Cho et al.	MILP	GAMS	No	No	No	Yes	[71]
Cho et al.	MILP	GAMS	Year	No	No	Yes	[72]
Woo et al.	MILP	GAMS	Month	No	No	Yes	[73]
Dautel et al.	MILP	Pyomo	Hour	No	No	Yes	[74]

6.5.2 Biomass

Biomass is a promising alternative for sustainable energy systems because it offers significant environmental and technical advantages. As a renewable energy source derived from organic materials, biomass has a lower environmental impact than fossil fuels, reducing greenhouse gas emissions. One key application of biomass is in producing biofuels, such as biodiesel and bioethanol. Bioethanol is commonly blended with gasoline to reduce carbon emissions, whereas biodiesel is blended with diesel fuel. Many researchers in the PSE community have conducted various optimization studies. Yue et al. [75] explore the challenges and opportunities in optimizing biomass-to-bioenergy supply chains, focusing on energy pathways from biomass to biofuels and power. It highlights key components of bioenergy supply chains, offers a comprehensive review of existing optimization efforts, and identifies future research directions in multi-scale modeling, sustainability, and uncertainty management. Kim et al. [76] present a framework for identifying and evaluating biomass-to-fuel production strategies using a technology superstructure that includes various conversion technologies, feedstocks, intermediates, and final products. Table 6.3 summarizes previous works in this area.

6.5.3 Methanol

Methanol is a promising alternative solution for sustainable energy systems due to its numerous advantages. Methanol can be produced from various renewable resources, including biomass, CO_2, and natural gas, and reduces reliance on fossil fuels. Its properties allow for efficient combustion in existing engines and infrastructure with minimal modifications, making it an attractive option for transportation and energy generation. Methanol also serves as a valuable feedstock in the production of chemicals, plastics, and pharmaceuticals, while supporting a circular economy. In addition to optimization for the design of methanol-based energy systems, research on process simulation or process synthesis is also being conducted. For example, Cho et al. [9] simulate two methanol synthesis processes, that is syngas-to-methanol and direct CO_2-to-methanol processes and evaluate their techno-economic and environmental capabilities. Cho et al. [84] develop a knowledge-based assessment platform for the early-stage screening of CO_2 direct hydrogenation catalysts for

Table 6.3 Summary of works on biomass-based energy systems.

Authors	Method	AML	Uncertainty	Multi-objective	Feedstock	Products	Refs.
Kim et al.	MILP	GAMS	No	Yes	Biomass	Chemicals	[76]
Allman et al.	MILP	JuMP	Yes	No	Waste	Electricity	[77]
Čuček et al.	MILP	GAMS	No	No	Crops, residue	Biofuel	[78]
You et al.	MILP	GAMS	No	Emission	Cellulosic biomass.	Biofuel	[79]
O'Neill et al.	MILP	GAMS	Yes	No	Energy crop	Biofuel	[80]
Martın et al.	MINLP	GAMS	No	No	Switchgrass	Hydrogen	[81]
Ng et al.	MINLP	GAMS	No	Emission	Cellulosic biomass.	Biofuel	[82]
Lee et al.	MILP	GAMS	No	No	Crops, residue	Bioethanol	[83]

methanol. Studies on strategic planning and operation of methanol-based energy systems have been reported. Wolday et al. [85] model a syngas-to-methanol synthesis plant and optimize its total annual gas emissions, system utility, and waste. Hernández-Camacho et al. [86] propose a multi-objective optimization model to determine the optimal infrastructure of the methanol supply chain in terms of the economic, environmental, and social points of view. Xu et al. [87] propose a multi-period MINLP model for the methanol supply chain that utilizes modular production lines powered by renewable energy. The model determines optimal production sites, the number of production lines, and supply chain scenarios. Table 6.4 summarizes the previous works performed.

6.5.4 Power Systems

As interest in electrification and decarbonization increases, various alternatives are emerging in power systems to reduce carbon emissions, such as adopting renewable generation technologies or adding carbon capture systems to existing fossil fuel-based power generation facilities. However, proper expansion planning is necessary to expand and modify the structure of the existing power system. Capacity expansion planning of power systems focuses on determining the optimal investment decisions such as the number, size, location, and time of generators and transmission lines required to satisfy the load demand [88]. The PSE community has been recently optimizing the infrastructure and operation strategies of power systems with high penetration of renewable generation and carbon capture systems. Lara et al. [89] propose a MILP model that optimizes long-term investment decisions and hourly operation decisions, which minimize the overall costs. A nested Benders decomposition is also proposed to effectively solve the large-scale multiperiod MILP problems. The authors [90] then extend

Table 6.4 Summary of works on methanol energy systems.

Authors	Type	Method	AML	Area	Refs.
Cho et al.	Simul.	—	Aspen Plus	Synthesis	[9]
Cho et al.	Simul., Opt.	LP	Aspen Plus, GAMS	Synthesis	[84]
Wolday et al.	Simul., Opt.	MILP	Aspen Plus, MATLAB	Synthesis	[85]
Hernández-Camacho et al.	Opt.	MILP	GAMS	Planning	[86]
Xu et al.	Opt.	MILP	GAMS	Planning	[87]

Source: Simul., Simulation; Opt., Optimization.

the model by considering uncertainty in renewable source availability, load demand, carbon tax, and fuel price. An advanced decomposition called stochastic dual dynamic integer programming is also applied. Li et al. [91] propose an MILP model for expansion planning of generation and transmission lines by extending the model proposed in [89] and including direct current (DC) power flow constraints.

Reliability refers to the ability of power systems to supply uninterrupted power to satisfy the load demand [92]. However, meeting the fluctuating demand is not always feasible due to the stochastic factors that impact power systems, such as random equipment failures and supply disruptions caused by extreme weather. Additionally, accurately predicting the power output from variable generation technologies like wind turbines and solar panels is difficult due to the intermittent nature of renewable energy sources. Reliability-constrained expansion planning models have been reported. Ahmadi et al. [93] introduce a robust linear expansion planning model that incorporates uncertainty in forecasted load while addressing reliability, fuel, and emissions constraints. Moreira et al. [94] present a two-stage min–max–min model to co-optimize transmission and renewable generation expansion, addressing renewable uncertainty and security criteria while balancing the costs of reserve resources and new transmission lines. Dehghan et al. [95] propose a reliability-constrained robust expansion planning model and a tri-level decomposition algorithm that addresses uncertainties in electricity demand, wind power generation, and equipment availability. Cho et al. [96] develop a GDP model for long-term capacity expansion planning of reliable power generation systems. In particular, the analytical method using the probability of equipment failures and capacity failure states is used to rigorously estimate the power system reliability. Table 6.5 summarizes the previous works performed in this area.

Table 6.5 Summary of works on expansion planning of power systems.

Authors	Method	AML	Uncertainty	Reliability	CO_2 emission	Refs.
Lara et al.	MILP	GAMS	No	No	Yes	[89]
Lara et al.	MILP,SP	Pyomo	Yes	No	Yes	[90]
Li et al.	MILP	Pyomo	No	No	Yes	[91]
Ahmadi et al.	MILP	GAMS	Yes	Yes	Yes	[93]
Moreira et al.	MILP	GAMS	Yes	Yes	No	[94]
Dehghan et al.	MILP	GAMS	Yes	Yes	No	[95]
Cho et al.	GDP	GAMS	No	Yes	Yes	[96]
Aghaei et al.	MILP	GAMS	No	Yes	No	[97]

6.6 Conclusions

In conclusion, this paper provides a comprehensive review of computational models and techniques used in discrete and continuous optimization, including LP, NLP, MIP, GDP, and SP. The discussion of decomposition methods, such as Benders and Lagrangean decomposition, highlights their potential effectiveness for solving large-scale optimization problems. Additionally, the review of AML like GAMS, JuMP, and Pyomo illustrates their important role in optimization models implementations. This paper also reviews the successful applications of these optimization models in sustainable energy systems such as hydrogen, biomass, methanol, and power systems.

References

1 Mensah, J. (2019). Sustainable development: meaning, history, principles, pillars, and implications for human action: literature review. *Cogent Social Sciences* 5 (1): 1653531. https://doi.org/10.1080/23311886.2019.1653531.

2 Chai, W.S., Chew, C.H., Munawaroh, H.S.H. et al. (2021). Microalgae and ammonia: a review on inter-relationship. *Fuel* 303: 121303. https://doi.org/10.1016/j.fuel.2021.121303.

3 Dunn, S. (2002). Hydrogen futures: toward a sustainable energy system. *International Journal of Hydrogen Energy* 27 (3): 235–264. https://doi.org/10.1016/S0360-3199(01)00131-8.

4 Ball, M. and Wietschel, M. (2009). The future of hydrogen - opportunities and challenges. *International Journal of Hydrogen Energy* 34 (2): 615–627. https://doi.org/10.1016/j.ijhydene.2008.11.014.

5 Erdemir, D. and Dincer, I. (2021). A perspective on the use of ammonia as a clean fuel: challenges and solutions. *International Journal of Energy Research* 45 (4): 4827–4834. https://doi.org/10.1002/er.6232.

6 Ahmed, G.B.L., Abdelaziz, E.M., and Hallenbeck, P.C. (2013). Addressing the challenges for sustainable production of algal biofuels: I. algal strains and nutrient supply. *Environmental Technology* 34 (13–14): 1783–1805. https://doi.org/10.1080/09593330.2013.827748.

7 Tian, F., Zhou, N., Chen, W. et al. (2024). Progress in green ammonia synthesis technology: catalytic behavior of ammonia synthesis catalysts. *Advanced Sustainable Systems* 8 (8): 2300618. https://doi.org/10.1002/adsu.202300618.

8 Kiss, A.A., Pragt, J., Vos, H. et al. (2016). Novel efficient process for methanol synthesis by CO_2 hydrogenation. *Chemical Engineering Journal* 284: 260–269. https://doi.org/10.1016/j.cej.2015.08.101.

9 Cho, S., Do, T.N., and Kim, J. (2023). Advanced design and comparative analysis of methanol production routes from CO_2 and renewable H_2: via syngas vs. direct hydrogenation processes. *International Journal of Energy Research* 2023: 6270858. https://doi.org/10.1155/2023/6270858.

10 Do, T.N., You, C., and Kim, J. (2022). A CO_2 utilization framework for liquid fuels and chemical production: techno-economic and environmental analysis. *Energy Environmental Science* 15: 169–184. https://doi.org/10.1039/D1EE01444G.

11 Cho, S. and Kim, J. (2019). Multi-site and multi-period optimization model for strategic planning of a renewable hydrogen energy network from biomass waste and energy crops. *Energy* 185: 527–540. https://doi.org/10.1016/j.energy.2019.07.053.

12 Perez, H.D. and Grossmann, I.E. (2021). Recent advances in computational models for the discrete and continuous optimization of industrial process systems. In: *Advances on Links Between Mathematics and Industry* (ed. P.Q. Estévez, B. Coll, R.M. Crujeiras, et al.), 1–31. Springer International Publishing.

13 Nash, J.C. (2000). The (Dantzig) simplex method for linear programming. *Computing in Science & Engineering* 2 (01): 29–31. https://doi.org/10.1109/5992.814654.

14 Illés, T. and Terlaky, T. (2002). Pivot versus interior point methods: pros and cons. *European Journal of Operational Research* 140 (2): 170–190. https://doi.org/10.1016/S0377-2217(02)00061-9.

15 Tomlin, J.A. (1989). A note on comparing simplex and interior methods for linear programming. In: *Progress in Mathematical Programming: Interior-Point and Related Methods* (ed. N. Megiddo), 91–103. Springer Science & Business Media https://doi.org/10.1007/978-1-4613-9617-8˙6.

16 Grossmann, I.E. (2021). *Advanced Optimization for Process Systems Engineering, Cambridge Series in Chemical Engineering*. Cambridge University Press.

17 IBM (2022). Ibm ilog cplex optimization studio. https://www.ibm.com/products/ilog-cplex-optimization-studio (accessed 25 September 2024).

18 Cross-platform software (2022). FICO Xpress optimizer. https://www.fico.com/en/products/fico-xpress-optimization (accessed 25 September 2024).

19 Gurobi Optimization (2023). Gurobi Optimizer Reference Manual. https://www.gurobi.com (accessed 25 September 2024).

20 Biegler, L.T. (2010). *Nonlinear Programming*. Society for Industrial and Applied Mathematics https://doi.org/10.1137/1.9780898719383.

21 UCSD Optimization Software (2005). Sparse Nonlinear OPTimizer. https:/ccom.ucsd.edu/~optimizers/solvers/snopt/ (accessed 25 September 2024).

22 Stanford Business Software Inc. (2021). Modular in-core nonlinear optimization system. http://www.sbsi-sol-optimize.com/asp/sol_products_minos_desc.htm (accessed 25 September 2024).

23 ARKI Consulting & Development A/S (2021). CONOPT. https://www.conopt.com/ (accessed 25 September 2024).

24 Wächter, A. and Laird, C. (2024). Interior Point OPTimizer. https://github.com/coin-or/Ipopt (accessed 25 September 2024).

25 Artelys (2024). Nonlinear interior point trust region optimization. https://www.artelys.com/solvers/knitro/ (accessed 25 September 2024).

26 Dakin, R.J. (1965). A tree-search algorithm for mixed integer programming problems. *The Computer Journal* 8 (3): 250–255. https://doi.org/10.1093/comjnl/8.3.250.

27 Balas, E., Ceria, S., and Cornuéjols, G. (1993). A lift-and-project cutting plane algorithm for mixed 0–1 programs. *Mathematical Programming* 58 (1): 295–324. https://doi.org/10.1007/BF01581273.

28 Johnson, E.L., Nemhauser, G.L., and Savelsbergh, M.W. (2000). Progress in linear programming-based algorithms for integer programming: an exposition. *INFORMS Journal on Computing* 12 (1): 2–23. https://doi.org/10.1287/ijoc.12.1.2.11900.

29 Geoffrion, A.M. (1972). Generalized benders decomposition. *Journal of Optimization Theory and Applications* 10 (4): 237–260. https://doi.org/10.1007/BF00934810.

30 Duran, M.A. and Grossmann, I.E. (1986). An outer-approximation algorithm for a class of mixed-integer nonlinear programs. *Mathematical Programming* 36 (3): 307–339. https://doi.org/10.1007/BF02592064.

31 Westerlund, T. and Pettersson, F. (1995). An extended cutting plane method for solving convex MINLP problems. *Computers & Chemical Engineering* 19: 131–136. https://doi.org/10.1016/0098-1354(95)87027-X.

32 Grossmann, I. E., Viswanathan, J., Vecchietti, A., et al. (2002). GAMS/DICOPT: A discrete continuous optimization package.

33 Drud, A. (2001). SBB: A new solver for mixed integer nonlinear programming.

34 Lastusilta, T., Bussieck, M., and Westerlund, T. (2009). https://doi.org/10.1021/ie801378n). An experimental study of the GAMS/AlphaECP MINLP solver. *Industrial & Engineering Chemistry Research* 48 (07): 7337–7345.

35 The Optimization Firm (n.d.). BARON Solver. https://minlp.com/ (accessed 25 September 2024).

36 Raman, R. and Grossmann, I. (1994). Modelling and computational techniques for logic based integer programming. *Computers & Chemical Engineering* 18 (7): 563–578. https://doi.org/10.1016/0098-1354(93)E0010-7.

37 Grossmann, I.E. and Trespalacios, F. (2013). Systematic modeling of discrete-continuous optimization models through Generalized Disjunctive

Programming. *AICHE Journal* 59 (9): 3276–3295. https://doi.org/10.1002/aic.14088.

38 Perez, H.D. and Grossmann, I.E. (2024). Extensions to Generalized Disjunctive Programming: hierarchical structures and first-order logic. *Optimization and Engineering* 25: 959–998. https://doi.org/10.1007/s11081-023-09831-x.

39 Türkay, M. and Grossmann, I.E. (1996). Logic-based MINLP algorithms for the optimal synthesis of process networks. *Computers & Chemical Engineering* 20 (8): 959–978. https://doi.org/10.1016/0098-1354(95)00219-7.

40 Trespalacios, F. and Grossmann, I.E. (2015). Improved Big-M reformulation for generalized disjunctive programs. *Computers & Chemical Engineering* 76: 98–103. https://doi.org/10.1016/j.compchemeng.2015.02.013.

41 Lee, S. and Grossmann, I.E. (2000). New algorithms for nonlinear generalized disjunctive programming. *Computers & Chemical Engineering* 24 (9): 2125–2141. https://doi.org/10.1016/S0098-1354(00)00581-0.

42 Chen, Q., Johnson, E.S., Bernal, D.E. et al. (2022). Pyomo.GDP: an ecosystem for logic based modeling and optimization development. *Optimization and Engineering* 23 (1): 607–642. https://doi.org/10.1007/s11081-021-09601-7.

43 Perez, H.D., Joshi, S., and Grossmann, I.E. (2023). Disjunctiveprogramming.jl: generalized disjunctive programming models and algorithms for JuMP. *Proceedings of the JuliaCon Conferences* 1 (1): 117.

44 Hooker, J.N. and van Hoeve, W.-J. (2018). Constraint programming and operations research. *Constraints* 23: 172–195. https://doi.org/10.1007/s10601-017-9280-3.

45 Jain, V. and Grossmann, I.E. (2001). Algorithms for hybrid MILP/CP models for a class of optimization problems. *INFORMS Journal on Computing* 13 (4): 258–276. https://doi.org/10.1287/ijoc.13.4.258.9733.

46 Maravelias, C.T. and Grossmann, I.E. (2004). A hybrid MILP/CP decomposition approach for the continuous time scheduling of multipurpose batch plants. *Computers & Chemical Engineering* 28 (10): 1921–1949. https://doi.org/10.1016/j.compchemeng.2004.03.016.

47 Castro, P.M., Grossmann, I.E., and Rousseau, L.-M. (2011). *Decomposition Techniques for Hybrid MILP/CP Models Applied to Scheduling and Routing Problems*, 135–167. New York: Springer https://doi.org/10.1007/978-1-4419-1644-0˙4.

48 Li, C. and Grossmann, I.E. (2021). A review of stochastic programming methods for optimization of process systems under uncertainty. *Frontiers in Chemical Engineering* 2: 622241. https://doi.org/10.3389/fceng.2020.622241.

49 Birge, J.R. and Louveaux, F. (2011). *Introduction to Stochastic Programming*. New York, NY: Springer.

50 Li, P., Arellano-Garcia, H., and Wozny, G. (2008). Chance constrained programming approach to process optimization under uncertainty.

Computers & Chemical Engineering 32 (1): 25–45. https://doi.org/10.1016/j.compchemeng.2007.05.009.

51 Lappas, N.H. and Gounaris, C.E. (2016). Multi-stage adjustable robust optimization for process scheduling under uncertainty. *AICHE Journal* 62 (5): 1646–1667. https://doi.org/10.1002/aic.15183.

52 Iyer, R.R. and Grossmann, I.E. (1998). A bilevel decomposition algorithm for long-range planning of process networks. *Industrial & Engineering Chemistry Research* 37 (2): 474–481. https://doi.org/10.1021/ie970383i.

53 Dogan, M.E. and Grossmann, I.E. (2006). A decomposition method for the simultaneous planning and scheduling of single-stage continuous multiproduct plants. *Industrial & Engineering Chemistry Research* 45 (1): 299–315. https://doi.org/10.1021/ie050778z.

54 Kallrath, J. (2004). *Mathematical Optimization and the Role of Modeling Languages*, 3–24. Boston, MA: Springer US https://doi.org/10.1007/978-1-4613-0215-51.

55 Jusevičius, V., Oberdieck, R., and Paulavičius, R. (2021). Experimental analysis of algebraic modelling languages for mathematical optimization. *Informatica* 32: 283–304. https://doi.org/10.15388/21-INFOR447.

56 Karbowski, A. and Wyskiel, K. (2021). Comparative study of AMPL, Pyomo and JuMP optimization modeling languages on a network linear programming problem example. *Pomiary Automatyka Robotyka* 25: 23–30. https://doi.org/10.14313/PAR˙241/23.

57 Fourer, R., Gay, D.M., and Kernighan, B.W. (2003). *AMPL: A Modeling Language for Mathematical Programming*. Duxbury Pr.

58 Roelofs, M. and Bisschop, J. (2013). *AIMMS –The Language Reference*. Paragon Decision Technology.

59 GAMS Development Corporation (1987). General algebraic modeling system. https://www.gams.com/ (accessed 25 September 2024).

60 GAMS Development Corporation (2020). General algebraic modeling system–MIRO. https://www.gams.com/miro/index.html (accessed 25 September 2024).

61 Bynum, M.L., Hackebeil, G.A., Hart, W.E. et al. (2021). *Pyomo –Optimization Modeling in Python*. Cham: Springer.

62 Lubin, M., Dowson, O., Dias Garcia, J. et al. (2023). JuMP 1.0: recent improvements to a modeling language for mathematical optimization. *Mathematical Programming Computation* 15: 581–589. https://doi.org/10.1007/s12532-023-00239-3.

63 Almansoori, A. and Shah, N. (2006). Design and operation of a future hydrogen supply chain: snapshot model. *Chemical Engineering Research and Design* 84 (6): 423–438. https://doi.org/10.1205/cherd.05193.

64 Almansoori, A. and Shah, N. (2009). Design and operation of a future hydrogen supply chain: multi-period model. *International Journal of Hydrogen Energy* 34 (19): 7883–7897. https://doi.org/10.1016/j.ijhydene.2009.07.109.

65 Almansoori, A. and Shah, N. (2012). Design and operation of a stochastic hydrogen supply chain network under demand uncertainty. *International Journal of Hydrogen Energy* 37 (5): 3965–3977. https://doi.org/10.1016/j.ijhydene.2011.11.091.

66 Hugo, A., Rutter, P., Pistikopoulos, S. et al. (2005). Hydrogen infrastructure strategic planning using multi-objective optimization. *International Journal of Hydrogen Energy* 30 (15): 1523–1534. https://doi.org/10.1016/j.ijhydene.2005.04.017.

67 Kim, J. and Moon, I. (2008). Strategic design of hydrogen infrastructure considering cost and safety using multiobjective optimization. *International Journal of Hydrogen Energy* 33 (21): 5887–5896. https://doi.org/10.1016/j.ijhydene.2008.07.028.

68 Hassan, Q., Algburi, S., Sameen, A.Z. et al. (2024). Green hydrogen: a pathway to a sustainable energy future. *International Journal of Hydrogen Energy* 50: 310–333. https://doi.org/10.1016/j.ijhydene.2023.08.321.

69 Won, W., Kwon, H., Han, J.-H., and Kim, J. (2017). Design and operation of renewable energy sources based hydrogen supply system: technology integration and optimization. *Renewable Energy* 103: 226–238. https://doi.org/10.1016/j.renene.2016.11.038.

70 Kim, M. and Kim, J. (2017). An integrated decision support model for design and operation of a wind-based hydrogen supply system. *International Journal of Hydrogen Energy* 42 (7): 3899–3915. https://doi.org/10.1016/j.ijhydene.2016.10.129.

71 Cho, S., Woo, Y.B., Kim, B.S., and Kim, J. (2016). Optimization-based planning of a biomass to hydrogen (B_2H_2) system using dedicated energy crops and waste biomass. *Biomass and Bioenergy* 87: 144–155. https://doi.org/10.1016/j.biombioe.2016.02.025.

72 Cho, S. and Kim, J. (2016). An optimization-based planning of investment strategies for a renewable energy supply system from biomass utilization. *Korean Journal of Chemical Engineering* 33: 2808–2819. https://doi.org/10.1007/s11814-016-0209-0.

73 Woo, Y.B., Cho, S., Kim, J., and Kim, B.S. (2016). Optimization-based approach for strategic design and operation of a biomass-to-hydrogen supply chain. *International Journal of Hydrogen Energy* 41 (12): 5405–5418. https://doi.org/10.1016/j.ijhydene.2016.01.153.

74 Dautel, J.L., Thakur, J., and Elberry, A.M. (2024). Enabling industrial decarbonization: a MILP optimization model for low-carbon hydrogen supply chains. *International Journal of Hydrogen Energy* 77: 863–891. https://doi.org/10.1016/j.ijhydene.2024.06.050.

75 Yue, D., You, F., and Snyder, S.W. (2014). Biomass-to-bioenergy and biofuel supply chain optimization: overview, key issues and challenges. *Computers & Chemical Engineering* 66: 36–56. https://doi.org/10.1016/j.compchemeng.2013.11.016.

76 Kim, J., Sen, S.M., and Maravelias, C.T. (2013). An optimization-based assessment framework for biomass-to-fuel conversion strategies. *Energy & Environmental Science* 6: 1093–1104. https://doi.org/10.1039/C3EE24243A.

77 Allman, A., Lee, C., Martın, M., and Zhang, Q. (2021). Biomass waste-to-energy supply chain optimization with mobile production modules. *Computers & Chemical Engineering* 150: 107326. https://doi.org/10.1016/j.compchemeng.2021.107326.

78 Čuček, L., Martın, M., Grossmann, I.E., and Kravanja, Z. (2014). Multi-period synthesis of optimally integrated biomass and bioenergy supply network. *Computers & Chemical Engineering* 66: 57–70. https://doi.org/10.1016/j.compchemeng.2014.02.020.

79 You, F. and Wang, B. (2011). Life cycle optimization of biomass-to-liquid supply chains with distributed-centralized processing networks. *Industrial & Engineering Chemistry Research* 50 (17): 10102–10127. https://doi.org/10.1021/ie200850t.

80 O'Neill, E.G., Martinez-Feria, R.A., Basso, B., and Maravelias, C.T. (2022). Integrated spatially explicit landscape and cellulosic biofuel supply chain optimization under biomass yield uncertainty. *Computers & Chemical Engineering* 160: 107724. https://doi.org/10.1016/j.compchemeng.2022.107724.

81 Martın, M. and Grossmann, I.E. (2011). Energy optimization of hydrogen production from lignocellulosic biomass. *Computers & Chemical Engineering* 35 (9): 1798–1806. https://doi.org/10.1016/j.compchemeng.2011.03.002.

82 Ng, R.T.L. and Maravelias, C.T. (2016). Design of cellulosic ethanol supply chains with regional depots. *Industrial & Engineering Chemistry Research* 55 (12): 3420–3432. https://doi.org/10.1021/acs.iecr.5b03677.

83 Lee, M., Cho, S., and Kim, J. (2017). A comprehensive model for design and analysis of bioethanol production and supply strategies from lignocellulosic biomass. *Renewable Energy* 112: 247–259. https://doi.org/10.1016/j.renene.2017.05.040.

84 Cho, S., Kim, C., and Kim, J. (2021). Techno-economic assessment and early-stage screening of CO_2 direct hydrogenation catalysts for methanol production using knowledge-based surrogate modeling. *Energy Conversion and Management* 244: 114477. doi: 10.1016/j.enconman.2021.114477.

85 Wolday, A.K., Gujarathi, A.M., and Ramteke, M. (2023). Multi-objective optimization of methanol production for energy efficiency and environmental sustainability. *Computers & Chemical Engineering* 179: 108426. https://doi.org/10.1016/j.compchemeng.2023.108426.

86 Hernández-Camacho, N.V., Gómez-Castro, F.I., Ponce-Ortega, J.M., and Martın, M. (2024). Production of methanol from renewable sources in Mexico: supply chain optimization. *Computers & Chemical Engineering* 188: 108780. https://doi.org/10.1016/j.compchemeng.2024.108780.

87 Xu, W., Li, Y., Xiang, B. et al. (2024). Multi-period supply chain design and optimization for modular methanol production driven by renewable energy. *Energy Conversion and Management* 319: 118868. https://doi.org/10.1016/j.enconman.2024.118868.

88 Conejo, A.J., Baringo, L., Kazempour, S.J., and Sissiqui, A.S. (2016). *Investment in Electricity Generation and Transmission –Decision Making under Uncertainty*. Springer International Publishing https://doi.org/10.1007/978-3-319-29501-5.

89 Lara, C.L., Mallapragada, D.S., Papageorgiou, D.J. et al. (2018). Deterministic electric power infrastructure planning: mixed-integer programming model and nested decomposition algorithm. *European Journal of Operational Research* 271 (3): 1037–1054. https://doi.org/10.1016/j.ejor.2018.05.039.

90 Lara, C.L., Siirola, J.D., and Grossmann, I.E. (2020). Electric power infrastructure planning under uncertainty: stochastic dual dynamic integer programming (SDDiP) and parallelization scheme. *Optimization and Engineering* 21 (4): 1243–1281. https://doi.org/10.1007/s11081-019-09471-0.

91 Li, C., Conejo, A.J., Liu, P. et al. (2021). Mixed-integer linear programming models and algorithms for generation and transmission expansion planning of power systems. *European Journal of Operational Research* 297 (3): 1071–1082. https://doi.org/10.1016/j.ejor.2021.06.024.

92 Cho, S., Li, C., and Grossmann, I.E. (2022). Recent advances and challenges in optimization models for expansion planning of power systems and reliability optimization. *Computers & Chemical Engineering* 165: 107924. https://doi.org/10.1016/j.compchemeng.2022.107924.

93 Ahmadi, A., Mavalizadeh, H., Nezhad, A.E. et al. (2020). A robust model for generation and transmission expansion planning with emission constraints. *SIMULATION* 96 (7): 605–621. https://doi.org/10.1177/0037549720915773.

94 Moreira, A., Pozo, D., Street, A., and Sauma, E. (2017). Reliable renewable generation and transmission expansion planning: co-optimizing system's resources for meeting renewable targets. *IEEE Transactions on Power Systems* 32 (4): 3246–3257. https://doi.org/10.1109/TPWRS.2016.2631450.

95 Dehghan, S., Amjady, N., and Conejo, A.J. (2016). Reliability-constrained robust power system expansion planning. *IEEE Transactions on Power Apparatus and Systems* 31 (3): 2383–2392. https://doi.org/10.1109/TPWRS.2015.2464274.

96 Cho, S., Tovar-Facio, J., and Grossmann, I.E. (2023). Disjunctive optimization model and algorithm for long-term capacity expansion planning of reliable power generation systems. *Computers & Chemical Engineering* 174: 108243. https://doi.org/10.1016/j.compchemeng.2023.108243.

97 Aghaei, J., Amjady, N., Baharvandi, A., and Akbari, M.-A. (2014). Generation and transmission expansion planning: MILP-based probabilistic model. *IEEE Transactions on Power Systems* 29 (4): 1592–1601. https://doi.org/10.1109/TPWRS.2013.2296352.

7

Multiscale Modeling and Optimization of Carbon Capture Processes

Kyeongjun Seo[1,2], Mark A. Stadtherr[1], and Michael Baldea[1]

[1] *McKetta Department of Chemical Engineering, The University of Texas at Austin, Austin, TX, USA*
[2] *Department of Chemical Engineering and Biotechnology, Tech University of Korea, Siheung-Si, Republic of Korea*

7.1 Introduction

Anthropogenic CO_2 emissions have had a significant planetary impact due to the amplification of the greenhouse effect, leading to global heating and its consequences. Global fossil CO_2 emissions from energy and industrial sources (including cement production) were estimated [1] as 37.1 billion metric tons (Gt) CO_2 in 2022 and 37.5 Gt CO_2 in 2023, including a US contribution of 5.06 Gt CO_2 in 2022 and 4.90 Gt CO_2 in 2023. Total CO_2 emissions from large facilities reporting to the US Environmental Protection Agency were 2.48 Gt CO_2 in 2022 [2]. Power plants were the largest point sources, emitting 1.58 Gt CO_2, while refineries and chemical plants were the leading sources in the manufacturing sector, releasing 0.162 and 0.167 Gt CO_2, respectively [2].

Mitigating (and eventually eliminating) CO_2 emissions from the power generation and manufacturing sectors requires a concerted and multi-faceted effort [3]. This includes replacing fossil-fuel/combustion-based power generation with renewable electricity sources, decarbonizing process heating (including heating electrification), identifying and implementing improvements in process efficiency, and developing and deploying new technologies involving low- or no-carbon emission processing pathways. All these approaches have limitations; for example, renewable energy generation rates fluctuate during the day, and these resources are generally not dispatchable without some form of storage [4]. Efficiency gains are intrinsically limited by the laws of thermodynamics and kinetics and tend to be small for established processes. Developing completely new processing pathways is fraught with technical difficulties and economic challenges when competing against incumbent technologies.

Against this backdrop, carbon capture technology is expected to play an important complementary role in significantly decarbonizing the power generation and manufacturing sectors in the near to medium term [3, 5–7]. Depending on the application considered, carbon capture can be deployed on its own or in conjunction with any or all of the above decarbonization measures. However, there is no "one-size-fits-all" carbon capture solution, and the design of capture processes from point sources in these sectors must abide by design practices followed with all other chemical processes. That is, the plant must be adapted to each site, accounting for the flow rate of the CO_2-rich stream to be processed, its CO_2 concentration, the presence of other components in the gas stream (including water vapor, NO_x/SO_x), utilities available locally, local weather conditions, etc.

In this context, the techno-economic optimization of the process design is essential, and mathematical modeling and flowsheet optimization are central to this effort. Carbon capture systems present specific modeling and optimization requirements, and the purpose of this chapter is to delineate these requirements along with emerging challenges and the opportunities that they present. In particular, we will highlight the multiple length and time scales of the modeling and optimization problems, the need to account for uncertainty and flexibility in the temporal domain, and the computational barriers associated with these concerns. We set our discussion in general terms, followed by the presentation of several research vignettes reflecting recent results derived by the authors and other workers in the field. To focus our discussion, we concentrate on solvent-based capture from point sources.

7.2 Modeling of Carbon Capture Processes

7.2.1 Process Structure and Operation

The structure of a solvent-based carbon capture process is shown in Figure 7.1. A CO_2-rich gas stream enters the absorber column, which is typically filled with structured or random packing. The gas stream (which is at pressure and temperature slightly above ambient) comes into contact with a capture solvent that is low in CO_2 ("lean solvent"). CO_2 is absorbed physically and/or chemically in the solvent, which leaves at the bottom of the absorber as the "rich solvent." The target capture rate is typically 90% (i.e. 90% of the CO_2 present in the feed gas stream is retained in the solvent), although higher rates are being considered. After processing, the gas stream is released to the atmosphere. Owing to the exothermic nature of the absorption process, cooling is provided to maintain a favorable CO_2 capture equilibrium. Heat exchange between the lean and rich solvent streams occurs in the cross heat exchanger. There, the hot lean solvent is used to preheat the rich

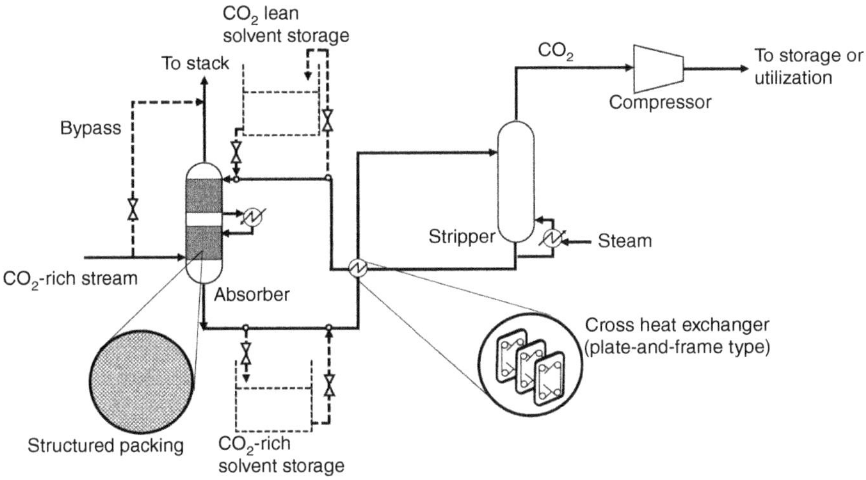

Figure 7.1 Schematic diagram of a solvent-based process for carbon capture from point sources. The core process units are represented in black lines, while flexibility-enabling features are highlighted in dashed lines.

solvent, which enters the thermal regeneration unit. Heat is used here to release CO_2 from the solvent, resulting in a pure CO_2 stream that is compressed for transportation, storage, or further processing.

The process is continuous and is typically designed for a nominal capacity (in terms of the flow rate and composition of the CO_2-rich feed gas stream, and target capture rate), which in turn dictates the heat duty required for solvent regeneration. Traditionally, operation at or close to steady state has been preferred (as is, in fact, the case with most continuous chemical processes). Nevertheless, in the context of including carbon capture in a portfolio of CO_2 emission mitigation technologies as described above, the operation of capture plants such as the one described here is expected to be increasingly variable [8–11]. Variability in operation is dictated by externalities/exogenous factors, which include process (e.g. feed disturbances) and market conditions (e.g. power demand). These factors can directly impact CO_2 capture strategies, as the goal is to adjust operations to optimize capture costs in response to these variations. Variability is reflected both in (significant) fluctuations in processing capacity (i.e. in the flow rate of the feed gas stream) and in capture rate. Variable operation is likely to become the norm, rather than the exception, for capture plants attached to both manufacturing facilities (whose emissions vary over time as production rates change – a process condition) and for plants attached to power generation facilities (whose emission rates fluctuates as demand for electricity changes daily and seasonally – a market condition). While the emission rates may change significantly, the CO_2

concentration in the flue gas is expected to remain relatively stable, as the same fuel sources are being used.

In addition to bringing the challenge of variable operation, these externalities – which are largely unpredictable – also present an economic opportunity. Consider the case of a carbon capture process attached to a fossil-fuel-based power plant. In this case, the heat required for solvent regeneration is provided by steam diverted from the power plant, which reduces power generation rates (this reduction is substantial, in the order of 30% of the nameplate capacity [12]). Since electricity prices change constantly (and often substantially) in the day-ahead and real-time electricity markets, the economic value of the diverted steam stream is not constant in time. It reflects a cost of (lost) opportunity – the steam used to regenerate the capture solvent is not used to generate saleable power – and is much higher when electricity prices are high than when they are low. In light of these facts, a variable operation strategy for the CO_2 capture system that decouples and desynchronizes power generation rates from carbon capture rates is likely to be economically advantageous, lowering the overall cost of carbon capture for power generation. Somewhat counterintuitively, this entails lowering capture rates when power demand (and CO_2 generation rates) are high, and conversely increasing the capture rates when power demand is low, such that *on average over a given period of time* (typically 24 hours), the desired capture rate is achieved. This then makes more steam available for power generation when electricity prices are high, increasing revenue from power plant operation.

Such variable operation and the decoupling of the operating rates of the CO_2 source and the CO_2 capture plant call for increasing the flexibility of the capture process, which we define as its ability to maintain reasonable performance at off-design conditions. Specifically, a solvent-based carbon capture process is flexible if it is able to accommodate significant fluctuations (particularly decreases) in the feed gas stream flowrate while ensuring, among other conditions, that no weeping or flooding occurs in the absorber, that mass transfer coefficients are sufficiently large to reach the target capture rate, and that there are not heat transfer limitations (pinches) in the heat exchangers.

Several design modifications can be implemented to this end; these are illustrated by dashed lines in Figure 7.1. A bypass stream can be used to divert a portion of the feed gas stream straight to the atmosphere. This lowers the overall capture rate; however, it also has the effect of lowering the flow rate of the lean solvent required to reach the capture rate for the portion of the feed gas stream that does enter the absorber. In turn, this reduces the flow rate of the rich solvent and the amount of energy (and steam) needed for solvent regeneration, thereby freeing up steam for power generation. Another means for decoupling power generation from solvent regeneration is the use of solvent storage. Two tanks, one for rich solvent and one for lean solvent, are used. With these ancillary units, rich solvent

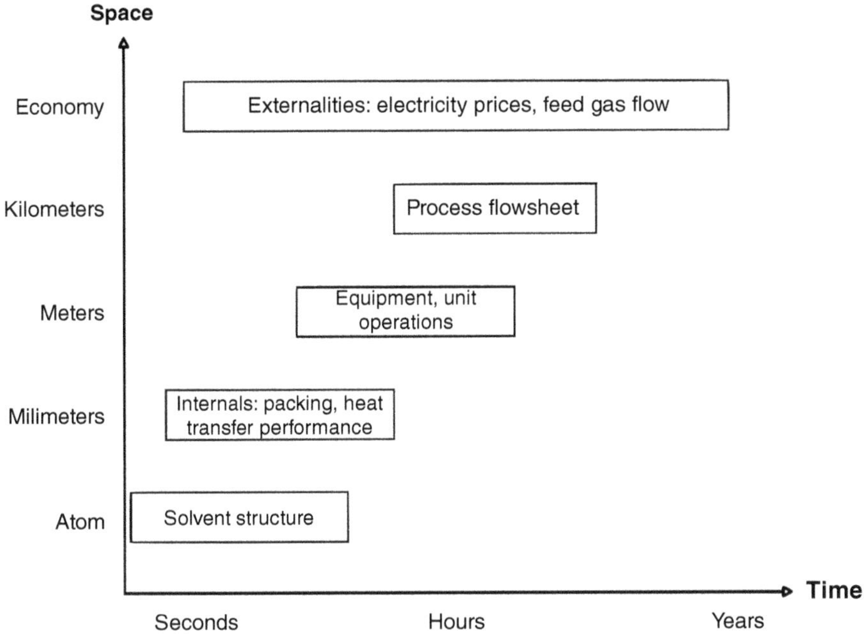

Figure 7.2 Spatial and temporal scales involved in multiscale modeling of solvent-based carbon capture plants.

can be stored during peak power demand times to lower the solvent regeneration rate (and heat duty). A target level of CO_2 capture can be maintained by supplying the absorber with lean solvent from the corresponding tank. The rich solvent can be regenerated during periods of low electricity demand/prices and stored in the lean solvent tank. Implementing solvent storage does carry the cost of purchasing an amount of solvent that exceeds what is needed for steady state operation, and the cost of the tanks.

7.2.2 Mathematical Modeling

Capturing the intricacies of the design and operation of the class of carbon capture processes described in the previous section requires models that span multiple temporal and spatial time scales, as illustrated in Figure 7.2. We describe these below; acronyms are assigned in the interest of the subsequent discussion on optimization problem formulations.

- Fluctuations in the feed gas stream flow rate occur over relatively short time spans (minutes–hours), but may trend in the upwards or downwards direction over much longer time horizons (seasonally). It is also possible that longer-term trends (over years/decades) become apparent. Likewise, electricity prices/power demand fluctuates over short time spans (seconds/minutes), over the course of a day and seasonally. These trends are driven by regional/national economic considerations, and hence reflect phenomena occurring on a significant spatial time scale. Importantly, neither feed gas flow rates nor economic factors are easily predicted with accuracy. Overall trends can be discerned, but accurate real-time predictions are in general not available (or possible). Therefore, modeling of feed gas (FG) fluctuations and electricity price (EP) fluctuations likely captures the statistical characteristics of these variables.

- The time constant of a chemical plant is in the order of hours, and carbon capture plants are no exception. Thus, at the flowsheet (FS) level (where the physical scale is in the hundreds of meters/kilometers), models must be capable to reflect process dynamics over the minute/hour time scale. Here, it is important to capture the relevant energy and material balances.
- Unit operations (columns, heat exchangers, etc.) within the carbon capture process flowsheet have time constants in the order of minutes (or up to several hours) and geometric dimensions in the order of tens of meters. In addition to accurately representing the material and energy balances of each unit, unit operation (UO) models should capture other important safety and performance characteristics (such as column flooding/weeping, compressor surge/choke, and heat exchanger pinch).
- Reflecting the performance characteristics of the unit operations as mentioned above requires an understanding of the performance of internal components (INT).
  - Packed columns are particularly well-suited for CO_2 absorption processes, as their packing materials significantly increase the surface area for vapor-liquid contact, enhancing mass transfer and reducing pressure drop, while tray columns are typically more appropriate for distillation processes [13, 14]. The performance of the absorber column is strongly influenced by the choice of packing (typically structured packing) used to enhance gas–liquid contact. The relevant length scale for representing packing performance is in the order of millimeters/centimeters. The phase equilibria and/or chemical reactions involved in absorbing and releasing the CO_2 into and from the solvent, respectively, should be described. These phenomena occur over the second/minute time scale. The absorber column cannot be assumed to be at equilibrium; mass transfer rates between the vapor and liquid phases are important and should be reflected in the model. For example, Afkhamipour and Mofarahi [15] showed that, in the context of post-combustion carbon capture, a rate-based model provided better predictions of absorber performance than an equilibrium-based model.

- Modern capture process designs make use of intricate heat exchanger networks (see, e.g. [16]) in the solvent recovery section, with some heat exchangers handling phase change and two-phase streams. These heat exchangers are of the plate-and-frame type, and accurately capturing their performance (particularly when two-phase flow is involved) requires spatially distributed models and an appropriate representation of the performance (heat transfer coefficient, pressure drop) of the heat transfer plates.

- Modeling the solvent is crucial.
  - For a solvent (e.g. aqueous amine, ionic liquid [17–21]) of **known** molecular structure, in addition to the properties mentioned earlier, models should accurately predict the heat of CO_2 absorption, as well as the density, viscosity, and heat capacity of the solvent as a function of CO_2 loading and temperature. We refer to such models as a physical property package (PP).
  - The design space for solvents (both amines and ionic liquids) is vast, and it has been hypothesized by several authors [22–25] that the solvent molecule (or mixture thereof) can be tailored to the application. Equivalently, a solvent structure can be selected from several known options, or a **new** molecule can be designed/identified such that the **overall design** (process and solvent) is economically optimal for capturing CO_2 from a feed stream with a specific flow rate and composition, given any site-specific constraints.
  - If a new molecular design is pursued, models should be able to reflect structure–property relations (i.e. be able to predict macroscopic properties of the solvent based on the structure of the molecule). Such models are referred to as quantitative structure property relations (QSPRs).
  - Solvent degradation (SDEG) over the relevant time scales should also be captured.
- Economic modeling should reflect the cost of equipment (EC) as a function of size/capacity and operating conditions, the cost of the solvent (initial charge and replacement due to degradation) (SC), and the cost of utilities (UC) as a function of demand. These costs may not be known accurately and may vary over time.

7.2.3 Multiscale Optimization

The mathematical models described above serve as the basis for formulating and solving simulation and optimization problems for carbon capture plants, concerning both process design and process operations. We outline the corresponding problem formulations below; Table 7.1 summarizes the information used in each problem. In general, the objective function is a combination of annualized capital cost (CAPEX) and annual operating cost (OPEX). CAPEX includes the cost of equipment, determined from the EC model and amortized over a chosen time horizon and the initial cost of the solvent (SC), while OPEX may include the cost

Table 7.1 Tableau of carbon capture plant optimization problem formulations.

	P:DO				
	P:DO1	**P:DO2**	**P:DO3**	**P:DO4**	**P:OP**
FG	Nominal	Nominal	Nominal	Scenarios/time series	
EP	Nominal	Nominal	Nominal	Scenarios/time series	✓
FS	✓	✓	✓	✓	✓
UO	✓	✓	✓	✓	✓
INT	•	•	•	•	✓
PP	✓	✓ (multiple)			✓
SDEG	•	• (multiple)			
QSPR			✓		
EC	✓	✓	✓	✓	•
UC	✓	✓	✓	✓	
SC	✓	✓			

A checkmark (✓) denotes that a model is a necessary part of the optimization problem constraints, while a bullet (•) denotes an optional component. FG: feed gas, EP: electricity price, FS: flowsheet, UO: unit operation, INT: internal components, PP: property package, QSPR: quantitative structure property relations, SDEG: solvent degradation, EC: cost of equipment, UC: cost of utilities, SC: cost of solvent.

of utilities (UC), solvent replacement (SC), and may account for revenue from tax incentives for carbon capture. The cost of personnel, downtime, etc. may be accounted for as well.

- Problem: Design Optimization (P:DO) refers to identifying the equipment capacities/physical sizes, operating conditions (pressures, temperatures, and flow rates and compositions of all streams and utilities) that are optimal with respect to the chosen objective function.
  - P:DO1: At a basic level, design optimization considers nominal values of the feed gas flow rate and electricity rates/utility costs, and a single given solvent. The solution comprises the equipment dimensions, nominal stream flow rates and compositions, all relevant temperatures and pressures, and corresponding costs. This problem is typically formulated as a large-scale nonlinear program (NLP).
  - P:DO2: Design optimization can be extended to consider solvent selection. In this case, multiple solvents are represented via their respective physical property (PP), solvent cost (SC), and solvent degradation (SDEG) models. This multiple length-scale problem is formulated as a large-scale mixed-integer nonlinear program (MINLP), with the integer decisions corresponding to the

selection of the solvent. The solution is the same as P:DO1, but includes the selection of the optimal solvent.

- P:DO3: In a similar vein, P:DO1 can be reformulated as a multiscale problem involving solvent design/discovery. In this case, QSPRs are used to model the solvent. Depending on the type of QSPR used, the problem can be formulated either as an NLP or MINLP, and the solution includes a new molecular structure in addition to the variables described above. In this case, it is unlikely that the solvent cost (SC) can be estimated with any degree of accuracy.
- Another class of design problems (P:DO4) accounts for uncertainty in the feed gas characteristics and energy prices. This is quite frequently done by representing price and feed gas data as time series extracted from historical data. Some degree of aggregation/selection is required to limit problem size and ensure that the optimization problem remains computationally tractable.

In this context, Problems P:DO1–P:DO3 can be posed as the corresponding problems under uncertainty, while also expanding the model to account for the process dynamics. By incorporating the dynamics of the process, the optimization problems lend themselves naturally to a two-stage formulation, whereby the upper-level decision variables pertain to process/equipment design (e.g. time-invariant size of each unit operation), and the lower-level variables pertain to process operations (e.g. time-varying values of flow rates, compositions, pressures, and temperatures of each process unit and stream, including any variables related to flexible operation, such as feed bypass). Scenario-based representations can be employed. Roughly speaking, this involves creating multiple instances – scenarios – of the same optimization problem, one for each data set, and computing the objective function value as a weighted average of the values of the objective function for each scenario. The values of the equipment design (upper-level) decisions are the same for all scenarios, while the values of the lower-level decision variables are specific to each scenario. Note that in order to study/exploit process flexibility, it is necessary to account for the process dynamics explicitly.

- Problem: Operational Optimization (P:OP) refers to identifying operating conditions that minimize the instantaneous/short-term operating cost. Here the design decisions (equipment sizes, solvent storage tank sizes) and the solvent selection are fixed.

7.3 Multiscale Modeling and Optimization Results

We review recent instances of solving problem P:DO in a multiscale setting that involves simultaneous optimization of the carbon capture process and the capture solvent.

Seo et al. [28] addressed P:DO4. They utilized a stochastic programming approach to optimize flexible carbon capture systems connected to a fossil-fueled power plant, addressing uncertainties in electricity demand and prices. A rate-based dynamic model was used for the absorption column. A scenario-based stochastic problem was formulated to simultaneously optimize both process design and operation. The scenarios considered reflected electricity prices on three separate days at the Wolf Hollow II power station in Granbury, TX (Figure 7.3). Data for an entire year were considered, and scenario weights were assigned according to the Euclidean distance between the electricity prices of each selected day and those of all other days in the year. For high, moderate, and low variability cases, the corresponding scenario weights were 0.0082, 0.3224, and 0.6694, respectively, based on ERCOT electricity price data from 2020. Key constraints included a time-average CO_2 removal rate of at least 90% and ensuring the conservation of inventory levels in solvent storage tanks.

This work confirmed that optimal flexible operation effectively decouples CO_2 absorption and solvent regeneration/CO_2 compression. This allows the carbon

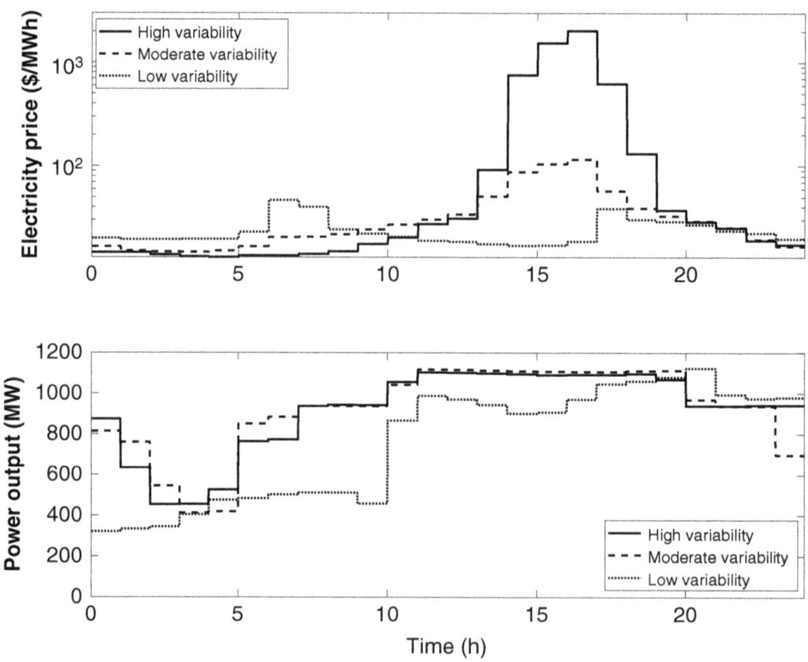

Figure 7.3 Day ahead electricity prices [26] and power output from Wolf Hollow II power station [27]. Three selected days in 2020 were chosen as representation of high, moderate, and low variability scenarios in electricity prices. *Source:* Data are obtained from Seo et al. [28].

capture system to adapt varying load patterns and avoid the high costs associated with regeneration and compression during peak hours. By solving the two-stage optimization problem, the optimal process design variables were determined, and the optimal operating conditions were identified for different scenarios, as shown in Figure 7.4. The results showed that the instantaneous capture rate is minimized during the peak electricity price hours. During periods of lower electricity cost, the instantaneous capture rate increases to compensate, ensuring an overall capture rate of 90%. Similarly, the solvent regeneration rate is slowed by storing the CO_2-rich solvent during high-price periods. The stored solvent is regenerated when electricity prices are lower and power plant load is reduced.

Also within the context of Problem P:DO4, Zantye et al. [29] examined the synergistic integration of renewable power generation and flexible carbon capture associated with an fossil-fueled power plant. Renewable generation provides clean energy for the carbon capture system, while the flexible carbon capture

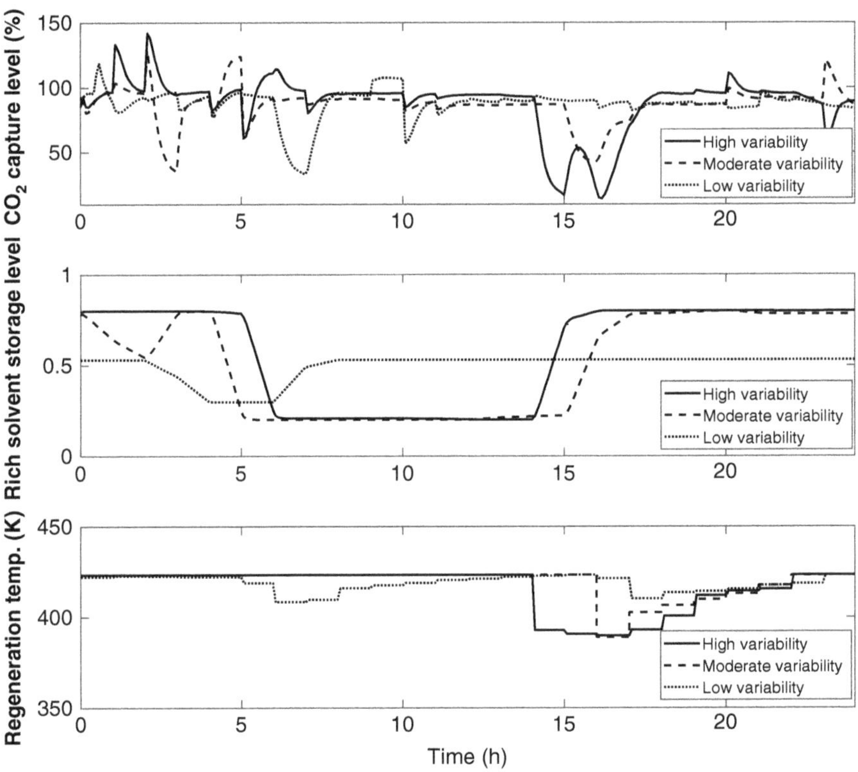

Figure 7.4 Optimized instantaneous CO_2 capture level, rich solvent storage level, and regeneration temperature for high, moderate, and low variability scenarios in electricity prices. *Source:* Seo et al., 2023 / with permission of Elsevier.

process acts as a form of energy storage to counter the intermittency of renewable generation. Specifically, the integration of renewable power sources in this work aimed to supply the steam/electricity required for carbon capture systems and to thus improve the overall economics by using renewable energy to offset some of the operational costs.

The framework used involves a two-stage modeling approach to handle the complexity of multiscale variability in electricity prices, solar power and wind power generation. The first stage of the optimization addresses long-term design decisions, such as the retrofit of the coal plant with a CO_2 capture system, the installation of renewable energy technologies and their capacities, and the capacity of an electric boiler and the corresponding investment. The second stage focuses on hourly operational decisions, including the power flows among the system components, and flexible carbon capture rates (the rates of CO_2 absorption, desorption, and compression). Their model used deterministic scenarios to represent temporal variability in renewable energy and electricity prices, and aimed to maximize the net present value of the integrated system. Key expenses included investments in electric boilers, and CO_2 capture systems, and operating costs related to fuel, maintenance, and CO_2 storage. Revenue sources included long-term contracts, spot market sales, and CO_2 sales. The approach integrated renewable energy to supply both grid electricity and the energy required for carbon capture, improving overall system economics and flexibility.

This work employed hourly time-varying profiles to capture fluctuations in electricity prices and renewable availability. While this accurately describes these variations, it significantly increases the computational complexity in simultaneously optimizing system design and operation. To manage this complexity, a temporal clustering approach was used. This approach groups time periods with similar parameter profiles, reducing the number of scenarios needed while adequately representing the variability of key parameters. Time periods with similar profiles are aggregated, resulting in a simplified scenario tree that captures key variations in the parameters. Then the process design decisions are determined using the simplified scenarios, while the operating decisions are found based on the full set of original scenarios. This approach balances computational efficiency with solution accuracy, enabling effective handling of large-scale, two-stage stochastic optimization problems.

Zantye at al. applied their two-stage optimization framework to both statewide and nationwide case studies across the United States. The findings suggested that when there is a carbon tax exceeding US$80 per ton and renewable energy prices drop below US$300 per kW, integrating CO_2 capture with all coal plants becomes advantageous. Among renewable options, solar energy is selected as a more favorable choice than wind, with nearly one-third of the coal plants across the country being paired with co-located solar PV farms. The levelized cost of the integrated system is found to be lower than that of constructing a new natural gas combined cycle plant, while achieving CO_2 emission reductions between 87.5% and 91%. In conclusion, under a scenario where carbon emissions are properly taxed, coal plant operators would benefit more from investing in a hybrid system combining CO_2 capture with renewable energy, rather than replacing coal plants with cleaner natural gas combined cycle units. This integrated approach offers significant potential for cost-effective, low-carbon energy production, facilitating the shift from a fossil fuel-based power sector to one led by renewable energy sources.

Zhang et al. [30] addressed Problem P:DO3. They employed machine-learning-based surrogate models to reduce the complexity of process models. The application is precombustion carbon capture using physically absorbing ionic liquid solvents. The design variables include both molecular structures of the ionic liquid solvent and process operating conditions, which collectively influence absorption kinetics and thermodynamics. A rate-based absorption process model was also incorporated to capture these effects. This multiscale design problem was formulated as an MINLP problem, with the objective of minimizing process costs while satisfying both process-level and molecular-level constraints. Thermodynamic properties, such as equilibrium gas solubility, are traditionally calculated using group-contribution-based physical models such as UNIFAC. In this work, machine-learning-based models were developed to predict gas solubility based on experimental data. This approach is mathematically simple but reliable, reducing the complexity of the integrated problem.

Both physics-based and data-driven models were employed (hybrid modeling approach) for predicting the physical and kinetic properties of the capture solvents, including group contribution methods to estimate molecular weight, melting point, boiling point, and molar volume. An artificial neural network based group contribution model was used for predicting viscosity, while correlation models estimate density and heat capacity. For CO_2 absorption, the mass transfer coefficient was calculated using an empirical correlation model, and CO_2 solubility was predicted using machine learning models. Structural constraints were also imposed to ensure a feasible molecule and to limit the use of certain substituent groups. The optimized process and solvent molecule demonstrated a 14.8% cost reduction compared to the conventional Selexol process, mainly due to lower electricity consumption, making it a more economical option for precombustion carbon capture.

Another strategy to address the numerical challenge of Problem P:DO3 is through continuous molecular targeting [25, 31, 32]. This approach involves considering continuous molecular parameters within physics-based equations-of-state, such as perturbed-chain statistical associating fluid theory (PC-SAFT), as adjustable variables in the optimization problem. With this approach, the optimal molecule and process design can be identified within a continuous decision space, circumventing the need for discrete variables.

Seo et al. [33] recently introduced a continuous formulation for Problem P:DO3 for carbon capture using physically absorbing ionic liquid solvents. The approach integrates structure–property prediction models into macroscopic process models, employing the PC-SAFT model to estimate the thermophysical properties of potential solvents. Continuous PC-SAFT parameters are used as molecular descriptors, providing reliable predictions for properties such as density, heat capacity, viscosity, and CO_2 solubility. Importantly, the model captures viscosity changes in the potential ionic liquid solvents upon CO_2 absorption, which is necessary for accurate mass transfer and liquid holdup predictions. In addition, the model further improves viscosity predictions by categorizing solvents based on their anion characteristics, ensuring accurate predictions across varying temperatures and process conditions.

The objective of the P:DO3 problem solved by Seo et al. [33] was to minimize annualized capture costs, including both CAPEX and OPEX. The decision variables included process operating conditions, molecular descriptors, and design parameters, while the problem was constrained by macroscopic process constraints (such as CO_2 capture efficiency and temperature limits) and molecular constraints derived from the PC-SAFT model. The search space for the molecular descriptors (PC-SAFT parameters) was confined to the convex polytope formed by the descriptors of the known cation and anion moieties in the database, as shown in Figure 7.5. While QSPRs are described using three molecular descriptors for each cation and anion, these constraints were relaxed for solvent discovery to explore broader possibilities in an expanded molecular design space. Interested

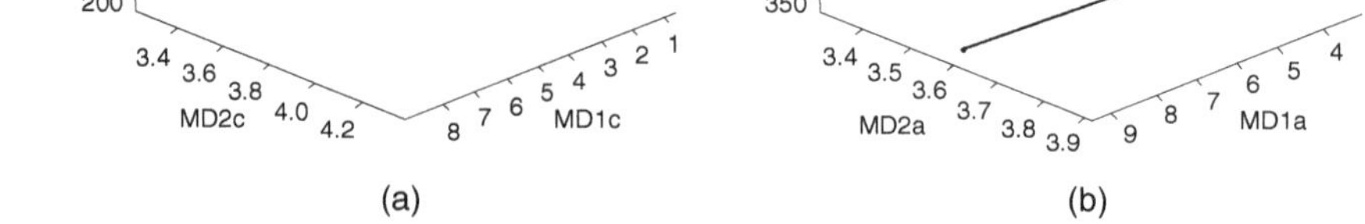

Figure 7.5 Convex polytope created by ionic liquid (a) cations and (b) anions. The ranges of PC-SAFT parameters of the solvent molecule are restricted by the convex polytope. MD stands for the molecular descriptor; c and a denote cation and anion, respectively. *Source:* Seo et al., 2024 / with permission of American Chemical Society.

readers are referred to Seo et al. [33] for more details. This multiscale approach demonstrated significant cost savings compared to the optimal process design using a base-case solvent. It was noted that the optimal (and other highly ranked) solvents provided a good balance among the properties of CO_2 capacity, heat capacity, and molecular volume, all of which collectively impact overall process efficiency. Optimization in the expanded molecular design space suggested that new solvents could be developed with improved properties, but experimental validation is required to confirm their chemical feasibility.

Lee et al. [24] developed a robust integrated molecule and process design framework for the simultaneous design of optimal aqueous chemical solvents and post-combustion CO_2 capture processes (P:DO3). The framework incorporates feasibility tests within an outer approximation (OA) algorithm, which decomposes the problem into a nonlinear programming (NLP) subproblem and a mixed-integer linear programming (MILP) master problem. This approach can handle nonconvex problems by equality relaxation and augmented penalty techniques. The feasibility tests, based on physical properties and thermodynamic behavior, filter out infeasible solvents and process conditions before solving complex NLPs, improving efficiency and focusing on physically attainable molecules. The solvent design space was defined based on parameters in the SAFT-γ Mie group contribution approach. Additional constraints were used to ensure the consistency and feasibility of the solvent structures, including limits on the number and type of functional groups, and the molecular design space was further refined to address isomer degeneracy.

Lee et al. applied this approach to three distinct design cases, each with varying constraints on amine functionalities, to identify optimal aqueous solvents for CO_2 absorption processes. The proposed algorithm utilized multiple starting points to enhance the likelihood of finding globally optimal solutions. The use of the feasibility tests was found to successfully narrow down the design space for the capture solvent, improving the accuracy and efficiency of the optimizations, thus enabling the identification of high-performance solvents and while reducing the computational cost. The final solvents identified showed significant improvements over the benchmark solvent monoethanolamine (MEA), with reduced total annualized costs and energy consumption.

7.4 Conclusions

Future implementations of carbon capture from point sources are likely to require significant operational flexibility driven by uncertainties in the flow rate of the CO_2-rich feed gas, as well as uncertain electricity prices and generation rates. Process designs that are tailored to each source/site lead to the lowest capture costs. Design optimization should consider all the relevant spatial and temporal scales of the process, from economic externalities to the choice of solvent molecule. This in turn calls for developing accurate models across multiple scales, and formulating and solving the resulting multiscale optimization problems under uncertainty.

In this chapter, we highlighted recent advances in multiscale modeling and optimization of point-source carbon capture systems, focusing on solvent-based processes. We began with defining a generic process structure and several design modifications that enhance flexibility. We inventoried the relevant phenomena, models and the corresponding time and space scales. A classification of the design and operational optimization problems was also provided. Finally, a review of recent literature results on multiscale design under uncertainty was provided.

In general, the design of flexible processes under uncertainty is approached as a two-stage optimization problem, whereby first-stage decisions consider a long time horizon and involve equipment design, while second-stage decisions are operational and consider shorter time spans. In spite of existing progress, challenges remain; these are related to (i) properly capturing uncertainties (existing results rely on different types of aggregation of, for example, feed gas and electricity price data to generate a small number of representative scenarios); (ii) dealing with the curse of dimensionality related to modeling multiple time and space horizons; and (iii) creating tractable representations of the structure–property relations of solvent molecules for the purpose of molecular design/discovery. In the latter realm, machine learning tools have been successfully used for developing reduced-order models. Nevertheless, a further challenge, (iv) is related to experimental validation of the properties of the discovered molecules. Additionally, (v) improving the accuracy of equipment cost models and broader techno-economic analysis data is crucial to obtaining practically relevant results.

Acknowledgments

Partial support from the National Science Foundation (Award 2133543) and the US Department of Energy (Award DE-EE0010725) is acknowledged with gratitude.

Disclaimer

This chapter was prepared as an account of work sponsored by an agency of the United States Government. Neither the United States Government nor any agency thereof, nor any of their employees, makes any warranty, express or implied, or assumes any legal liability or responsibility for the accuracy, completeness, or usefulness of any information, apparatus, product, or process disclosed, or represents that its use would not infringe privately owned rights. Reference herein to any specific commercial product, process, or service by trade name, trademark, manufacturer, or otherwise does not necessarily constitute or imply its endorsement, recommendation, or favoring by the United States Government or any agency thereof. The views and opinions of authors expressed herein do not necessarily state or reflect those of the United States Government or any agency thereof.

References

1 Global Carbon Project (2023). Global carbon budget. https://www.globalcarbonproject.org/carbonbudget (accessed 02 December 2025).

2 U.S. Environmental Protection Agency (2023). 2022 greenhouse gas emissions from large facilities. https://ghgdata.epa.gov/flight (accessed 02 December 2025).

3 Cresko, J., Rightor, E., Carpenter, A. et al. (2022). Industrial decarbonization roadmap. US Department of Energy DOE/EE-2635. https://www.energy.gov/eere/doe-industrial-decarbonization-roadmap/ (accessed 02 December 2025).

4 Gonzalez-Salazar, M.A., Kirsten, T., and Prchlik, L. (2018). Review of the operational flexibility and emissions of gas- and coal-fired power plants in a future with growing renewables. *Renewable and Sustainable Energy Reviews* 82: 1497–1513. https://doi.org/10.1016/j.rser.2017.05.278.

5 Chao, C., Deng, Y., Dewil, R. et al. (2021). Post-combustion carbon capture. *Renewable and Sustainable Energy Reviews* 138: 110490. https://doi.org/10.1016/j.rser.2020.110490.

6 Osman, A.I., Mahmoud Hefny, M.I.A., Maksoud, A. et al. (2021). Recent advances in carbon capture storage and utilisation technologies: a review. *Environmental Chemistry Letters* 19: 797–849. https://doi.org/10.1007/s10311-020-01133-3.

7 Wilberforce, T., Olabi, A.G., Sayed, E.T. et al. (2021). Progress in carbon capture technologies. *Science of the Total Environment* 761: 143203. https://doi.org/10.1016/j.scitotenv.2020.143203.

8 Cheng, F., Patankar, N., Chakrabarti, S., and Jenkins, J.D. (2022). Modeling the operational flexibility of natural gas combined cycle power plants coupled with flexible carbon capture and storage via solvent storage and flexible regeneration. *International Journal of Greenhouse Gas Control* 118: 103686. https://doi.org/10.1016/j.ijggc.2022.103686.

9 Cohen, S.M., Webber, M.E., and Rochelle, G.T. (2012). The impact of electricity market conditions on the value of flexible CO_2 capture. *ASME International*

Mechanical Engineering Congress and Exposition 6: 581–593. https://doi.org/10.1115/IMECE2012-88119.

10 Rúa, J., Bui, M., Nord, L.O., and Dowell, N.M. (2020). Does CCS reduce power generation flexibility? A dynamic study of combined cycles with post-combustion CO_2 capture. *International Journal of Greenhouse Gas Control* 95: 102984. https://doi.org/10.1016/j.ijggc.2020.102984.

11 Wu, X., Wang, M., Liao, P. et al. (2020). Solvent-based post-combustion CO_2 capture for power plants: a critical review and perspective on dynamic modelling, system identification, process control and flexible operation. *Applied Energy* 257: 113941. https://doi.org/10.1016/j.apenergy.2019.113941.

12 Kang, C.A., Brandt, A.R., and Durlofsky, L.J. (2014). Optimizing heat integration in a flexible coal–natural gas power station with CO_2 capture. *International Journal of Greenhouse Gas Control* 31: 138–152. https://doi.org/10.1016/j.ijggc.2014.09.019.

13 Salem, A. and Amanpour Reyhani, F. (2015). Applied aspects for enhanced CO_2 capture from reformer gas: comparison between the performance of valve tray absorber and packed column, part I. *International Journal of Greenhouse Gas Control* 42: 237–245. https://doi.org/10.1016/j.ijggc.2015.07.028.

14 Yeh, J.T., Pennline, H.W., and Resnik, K.P. (2001). Study of CO_2 absorption and desorption in a packed column. *Energy & Fuels* 15 (2): 274–278. https://doi.org/10.1021/ef0002389.

15 Afkhamipour, M. and Mofarahi, M. (2013). Comparison of rate-based and equilibrium-stage models of a packed column for post-combustion CO_2 capture using 2-amino-2-methyl-1-propanol (AMP) solution. *International Journal of Greenhouse Gas Control* 15: 186–199. https://doi.org/10.1016/j.ijggc.2013.02.022.

16 Tsay, C., Pattison, R.C., Zhang, Y. et al. (2019). Rate-based modeling and economic optimization of next-generation amine-based carbon capture plants. *Applied Energy* 252: 113379. https://doi.org/10.1016/j.apenergy.2019.113379.

17 Lian, S., Song, C., Liu, Q. et al. (2021). Recent advances in ionic liquids-based hybrid processes for CO_2 capture and utilization. *Journal of Environmental Sciences* 99: 281–295. https://doi.org/10.1016/j.jes.2020.06.034.

18 Meng, F., Meng, Y., Tongyao, J. et al. (2022). Research progress of aqueous amine solution for CO_2 capture: a review. *Renewable and Sustainable Energy Reviews* 168: 112902. https://doi.org/10.1016/j.rser.2022.112902.

19 Rochelle, G., Chen, E., Freeman, S. et al. (2011). Aqueous piperazine as the new standard for CO_2 capture technology. *Chemical Engineering Journal* 171 (3): 725–733. https://doi.org/10.1016/j.cej.2011.02.011.

20 Shama, V.M., Swami, A.R., Aniruddha, R. et al. (2021). Process and engineering aspects of carbon capture by ionic liquids. *Journal of CO2 Utilization* 48: 101507. https://doi.org/10.1016/j.jcou.2021.101507.

21 Zhou, T., Shi, H., Ding, X., and Zhou, Y. (2021). Thermodynamic modeling and rational design of ionic liquids for pre-combustion carbon capture. *Chemical Engineering Science* 229: 116076. https://doi.org/10.1016/j.ces.2020.116076.

22 Borhani, T.N. and Wang, M. (2019). Role of solvents in CO_2 capture processes: the review of selection and design methods. *Renewable and Sustainable Energy Reviews* 114: 109299. https://doi.org/10.1016/j.rser.2019.109299.

23 Chai, S., Song, Z., Zhou, T. et al. (2022). Computer-aided molecular design of solvents for chemical separation processes. *Current Opinion in Chemical Engineering* 35: 100732. https://doi.org/10.1016/j.coche.2021.100732.

24 Lee, Y.S., Galindo, A., Jackson, G., and Adjiman, C.S. (2023). Enabling the direct solution of challenging computer-aided molecular and process design problems: chemical absorption of carbon dioxide. *Computers & Chemical Engineering* 174: 108204. https://doi.org/10.1016/j.compchemeng.2023.108204.

25 Stavrou, M., Lampe, M., Bardow, A., and Gross, J. (2014). Continuous molecular targeting–computer-aided molecular design (CoMT–CAMD) for simultaneous process and solvent design for CO_2 capture. *Industrial & Engineering Chemistry Research* 53 (46): 18029–18041. https://doi.org/10.1021/ie502924h.

26 Electric Reliability Council of Texas (2022). Day-ahead price. http://www.ercot.com/ (accessed 02 December 2025).

27 United States Environmental Protection Agency (2022). Air markets program data. https://campd.epa.gov (accessed 02 December 2025).

28 Seo, K., Retnanto, A.P., Martorell, J.L. et al. (2023). Simultaneous design and operational optimization for flexible carbon capture process using ionic liquids. *Computers & Chemical Engineering* 178: 108344. https://doi.org/10.1016/j.compchemeng.2023.108344.

29 Zantye, M.S., Arora, A., and Faruque Hasan, M.M. (2021). Renewable-integrated flexible carbon capture: a synergistic path forward to clean energy future. *Energy Environmental Science* 14: 3986–4008. https://doi.org/10.1039/D0EE03946B.

30 Zhang, X., Ding, X., Song, Z. et al. (2021). Integrated ionic liquid and rate-based absorption process design for gas separation: global optimization using hybrid models. *AIChE Journal* 67: e17340. https://doi.org/10.1002/aic.17340.

31 Bardow, A., Steur, K., and Gross, J. (2010). Continuous-molecular targeting for integrated solvent and process design. *Industrial & Engineering Chemistry Research* 49 (6): 2834–2840. https://doi.org/10.1021/ie901281w.

32 Schilling, J., Lampe, M., Gross, J., and Bardow, A. (2017). 1-stage CoMT-CAMD: an approach for integrated design of ORC process and working fluid using PC-SAFT. *Chemical Engineering Science* 159: 217–230. https://doi.org/10.1016/j.ces.2016.04.048.

33 Seo, K., Brennecke, J.F., Edgar, T.F. et al. (2024). Multiscale design of ionic liquid solvents and CO_2 capture processes for high-pressure point sources. *ACS Sustainable Chemistry & Engineering* 12 (2): 706–715. https://doi.org/10.1021/acssuschemeng.3c03697.

8

Integrated Design and Operability Optimization of Sustainable Process Intensification Systems

Yuhe Tian[1], Rahul Bindlish[2], and Efstratios N. Pistikopoulos[3,4]

[1]*Department of Chemical and Biomedical Engineering, West Virginia University, Morgantown, WV, USA*
[2]*Technical Expertise and Support Technology Center, The Dow Chemical Company, Houston, TX, USA*
[3]*Artie McFerrin Department of Chemical Engineering, Texas A&M University, College Station, TX, USA*
[4]*Texas A&M Energy Institute, Texas A&M University, College Station, TX, USA*

8.1 Introduction

It is essential to develop innovative chemical process technologies which can achieve step-change improvements in all facets of sustainability, efficiency, and economic competitivity [1, 2]. Toward this direction, process intensification (PI) is regarded as a promising class of novel process technologies, having demonstrated the potential to achieve 20–80% energy and capital expenditure savings [3, 4]. The statistics survey presented in Figure 8.1 also illustrates the increasingly important role of PI in shaping the transitions of modern chemical process industry toward sustainability, decarbonization, digitalization, etc.

PI realizes substantial process performance improvements by exploiting the fundamental principles to intensify physicochemical phenomena [5]. To name a few examples of PI principles: (i) leveraging the synergy between multifunctional phenomena to overcome equilibrium limits (e.g. combined reaction and separation), (ii) enhancing mass and heat transfer rates (e.g. miniaturization), (iii) adopting periodic operations (e.g. pressure swing adsorption), and (iv) utilizing alternative energy sources (e.g. microwave heating). In this way, process intensification opens up the opportunity to develop novel out-of-the-box unit operations and process schemes. More than 72 intensified technologies have been developed to date [6]. Eight representative PI technologies are presented in Figure 8.2 including reactive distillation, microreactors, membrane distillation, rotating packed bed, dividing wall column, simulated moving bed, membrane reactors, and microwave reactors. A literature survey is also performed to illustrate the research development of each technology in the past decade (2015–2023)

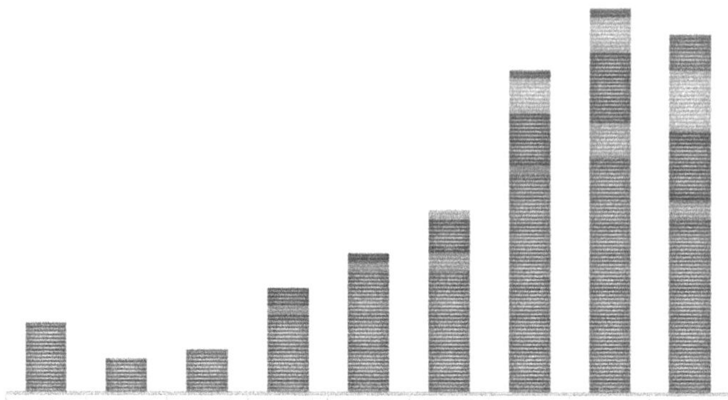

	2015	2016	2017	2018	2019	2020	2021	2022	2023
Decarbonization							1	1	4
Electrification						1	4	4	7
Circular economy				2	1	4	6	8	8
Digitalization				1	1	2	1	4	2
Sustainability	8	4	5	9	14	14	25	27	20

Figure 8.1 Literature statistics on the role of process intensification to chemical industry transitions. *Source:* Citation database: Web of Science Core Collection, Update: September 2024.

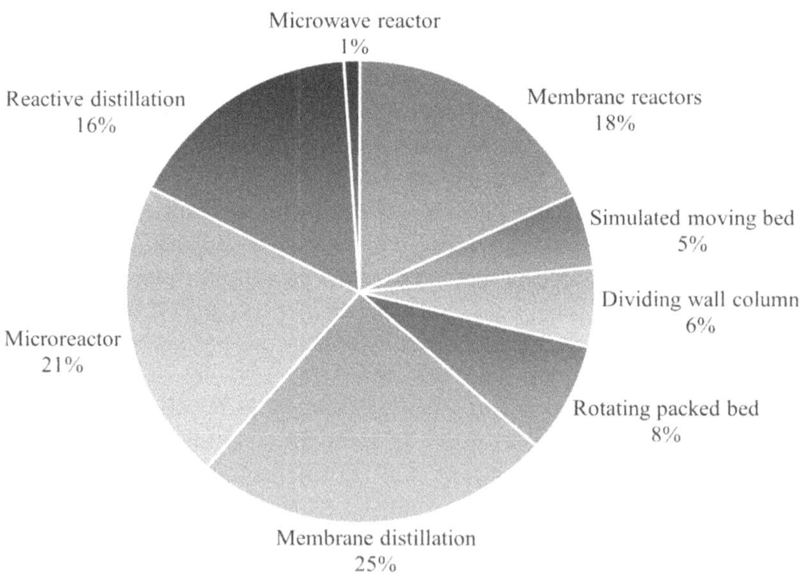

Figure 8.2 Literature statistics on modeling of process intensification technologies. *Source:* Citation database: Web of Science Core Collection, Update: September 2024.

with a particular focus on process modeling. A summary of the literature statistics for earlier years can be found in our prior work [7]. As can be noticed, intensified distillation-based systems still represent a majority of PI modeling efforts. Significant research has also been done on the modeling of membrane reactors and microreactors. Indicative review articles is given in Table 8.1 to elucidate the recent advancements of process intensification modeling, design, and optimization from both algorithmic and application aspects.

Despite the extensive research efforts, the industrial deployments of intensified process technologies are yet limited. As the most commercialized PI technique, more than 150 industrial installations have been reported for reactive distillation out of more than 40,000 distillation columns being operated worldwide [20, 21]. The design complexity of intensified equipment remains one of the key open challenges. As the models for most PI technologies are not yet supported in commercial chemical process simulators, the quantitative analyses and comparison

Table 8.1 Latest review articles on PI modeling, design, and optimization – an indicative list.

References	Highlight	Year
Simkoff et al. [8]	Relationship between process design, intensification, control, and energy efficiency	2020
Asprion [9]	Simulation of distillation column, including intensified structures	2020
Biegler [10]	Nonlinear optimization strategies with application to process intensification	2020
Yao et al. [11]	Modeling of hydrodynamics and mass transfer in microchannels	2021
Rivera et al. [12]	Computational fluid dynamics simulation for electrochemical reactors	2021
Ng et al. [13]	Modeling and simulation of hollow fiber membranes for gas separation	2021
Goyal et al. [14]	Multiscale modeling and simulation of microwave multiphase reactors	2022
Chen et al. [15]	Computational fluid dynamics simulation for rotating packed bed	2022
Shu et al. [16]	Reactive distillation design and optimization	2022
Pasha et al. [17]	Hydrodynamics and mass transfer modeling in microreactors for CO_2 absorption	2022
Singh et al. [18]	Modeling of rotating packed beds for carbon capture	2023
Kiss et al. [19]	Process synthesis of reactive distillation systems with operability considerations	2024

with their conventional counterparts are typically excluded at the conceptual design stage [22]. At the flowsheet level, there also lacks a well-accepted protocol to systematically identify process intensification opportunities as well as to optimally integrate the new PI equipment with existing process infrastructure [23]. Another barrier lies in the operability and control, as process intensification may shrink the feasible operating window, reduce the available degrees of freedom, and render faster dynamics [24, 25]. These processes are thus more vulnerable to disturbances. Through rigorous model-based analyses, our prior work [26] has highlighted the need of simultaneous process design, intensification, and operability optimization to make sure that the final optimal process configuration will be operable under disturbances.

To address these challenges, this chapter introduces a systematic framework to synthesize optimal, intensified, and operable process systems with enhanced sustainability. The remainder of the chapter is organized as follows: Section 8.2 details the phenomena-based methodological framework using the Generalized Modular Representation Framework (GMF) and its integration with model-based operability approaches. Section 8.3 demonstrates the framework using two case studies: (i) MMA purification using dividing wall column, and (ii) methyl tert-butyl ether (MTBE) production via combined reaction and separation. Section 8.4 presents concluding remarks.

8.2 Methodology Framework

8.2.1 Prelude: Phenomena-Based Process Synthesis

Optimization-based process synthesis utilizes mathematical programming to identify the best design solution based on a superstructure which contains all the plausible flowsheet structural alternatives [27, 28]. As shown in Figure 8.3, conventional process synthesis methods employ unit operation as the elementary building blocks. In this way, the incorporation of PI considerations requires pre-postulating intensified unit operations into the initial superstructure. Such computational analyses have mostly been studied to examine the viability to adopt a specific intensified unit operation within a pre-specified flowsheet structure [29–31]. However, process synthesis based on unit operations may not allow to discover innovative unit/flowsheet solutions, which are nonintuitive to the existing design practice.

Given this, process synthesis using phenomena building blocks has received recent impetus as a promising way toward systematic process intensification [32–34]. As shown in Figure 8.3, phenomena-based process synthesis builds on the idea that, despite the variety of unit operations, they can be represented with

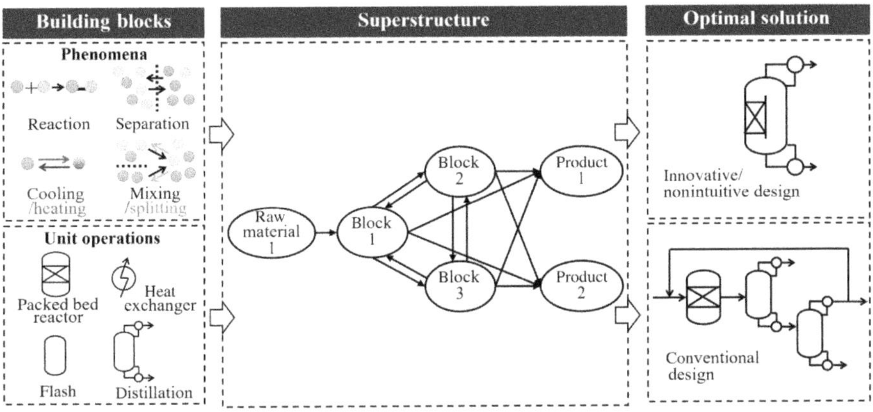

Figure 8.3 Optimization-based process synthesis: unit operation-based versus phenomena-based. *Source:* Reproduced from [27] / Annual Reviews / CC BY 4.0.

a more unified set of physicochemical phenomena (e.g. reaction, separation, heating, cooling). New unit operation designs can be generated by flexibly integrating different phenomena to maximize multifunctional synergies without being restricted by pre-postulated unit/flowsheet structures.

Motivated by this, in what follows we introduce the GMF for phenomena-based process synthesis, integration, and intensification [35]. With particular interest, we will showcase how GMF can effectively address several key research questions for phenomena-based process synthesis:

- **Representation Capability:** It remains an open question on how to properly define and mathematically model phenomena building blocks in order to achieve maximum design creativity and performance enhancement.
- **Combinatorial Design Space:** Phenomena-based synthesis presents a substantially larger design space from a lower aggregated level than unit operations.
- **Integrated Design with Operability:** With very limited equipment information, open challenges lie in how to quantify the operability of phenomenological configurations and how to generate optimal designs with desired operability.

8.2.2 Generalized Modular Representation Framework

Our central hypothesis is that many chemical processes can be described utilizing an aggregation of mass/heat transfer phenomena. On this basis, GMF involves the following phenomena building blocks (Figure 8.4a):

i) **Mass/Heat (M/H) Exchange Module:** The inlet streams may perform mass and heat exchange driven by the Gibbs free energy-based driving force

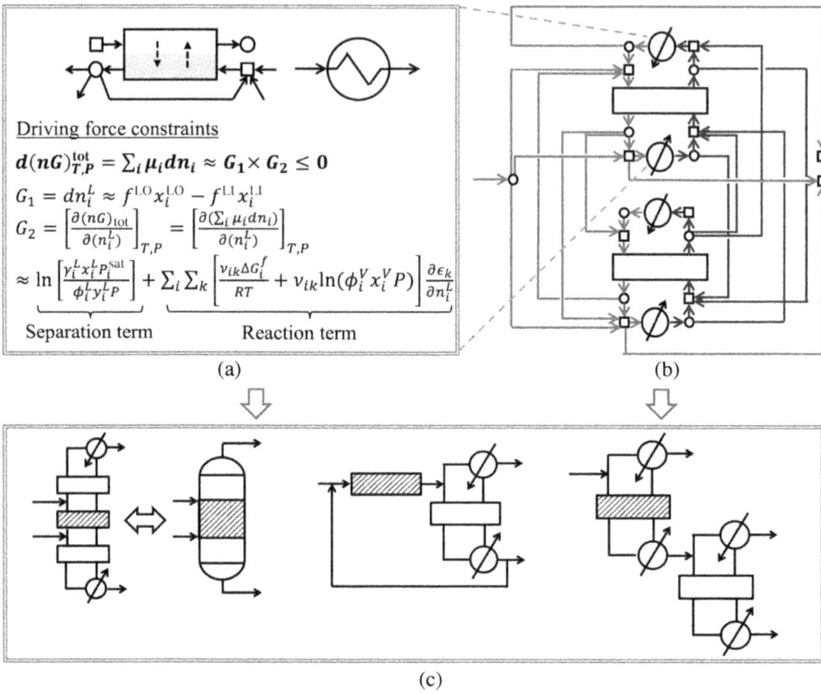

Figure 8.4 A schematic of process synthesis via GMF. (a) Phenomena building blocks, (b) superstructure optimization, and (c) process representation.

constraints (to be detailed later). The mass exchange can take place in the form of reaction, separation, reactive separation, etc.

ii) **Pure Heat Exchange (HE) Module:** The inlet streams may perform heat exchange driven by the temperature gradient. The heat exchange can take place in the form of heat integration between two process streams, heating, or cooling between a utility stream and a process stream.

Using the mass/heat exchange module as an example, the driving force constraints (Figure 8.4a) orient from the change of total Gibbs free energy, that is, $d(nG)^{\text{tot}}_{T,P} = \sum_i \mu_i dn_i \leq 0$, to determine mass transfer feasibility while intensifying mass transfer toward thermodynamic limit. Note that the constraints only perform mass transfer feasibility check rather than enforcing any equilibrium conditions. For a liquid–vapor system driven by phase separation (such as that involved in distillation), the driving force constraints are proven to be the sum of a separation term (analogous to liquid–vapor separation constant) and a reaction term (as a function of reaction extent). The mathematical formulations are highlighted in Figure 8.4a. Auxiliary binary (0–1) variables can also be used to dictate the existence (or not) of the reaction and/or separation term. Detailed derivations can be found in our prior work [36].

This enables GMF to systematically determine the synergy (or not) of multifunctional phenomena, that is, reaction and separation, based on thermodynamic driving forces. The optimal "task" to be performed by this generic "phenomena" M/H module can be automatedly identified as intensified reactive separation, conventional separation, or reaction. Our recent work has also extended the driving force constraints to capture membrane reactors, using the chemical potential-based formulation to characterize mass transfer feasibility across membranes. This compact physical representation feature of GMF also contributes to substantially reduce the numerical complexity for superstructure optimization (Figure 8.4b).

The resulting GMF synthesis model features a mixed-integer nonlinear optimization problem including an objective function, a physical model (e.g. mass/energy balances, phase defining, kinetics), and a structural model (e.g. stream and module interconnections). Table 8.2 provides indicative modeling constraints. The Generalized Benders Decomposition algorithm is used to solve this mixed-integer nonlinear optimization problem [37]. The resulting optimal GMF phenomena-based design solution(s) are further translated to unit operation-based process and validated with unit operation-based simulation (Figure 8.4c).

8.2.3 Integrated Synthesis and Operability Optimization

The process solutions generated via GMF are optimal with respect to the nominal design conditions. A key question thus lies in how to ensure these processes are safely operable to deliver the expected functions under disturbances and uncertainties. To this purpose, we have developed an integrated approach to incorporate model-based safety and flexibility analyses with GMF synthesis as given in Figure 8.5. This leads to an iterative algorithm featuring multiperiod process synthesis formulation to systematically identify optimal and intensified process designs with guaranteed safety and flexibility performances.

8.2.3.1 Safety Considerations via Risk Analysis

To evaluate the safety of phenomena-based GMF designs, we leverage quantitative risk analysis as shown in Figure 8.6. The step-wise procedure to determine the risk of each GMF heat or mass/heat exchange module is detailed below:

Step 1. **Determine Failure Frequency:** Note that the GMF phenomena-based representation does not pre-postulate the type of equipment. To obtain an initial evaluation of the failure frequency associated with various tasks, we

Table 8.2 GMF process synthesis model – an indicative list.

Objective function minimize cost	$\min \text{ Obj} = \sum C_n f_l^n + \sum C_{cu} Qc_e + \sum C_{hu} Qh_e + \sum C_e y_e$
Physical model (for each module)	
Mass and energy balances	$f^{\mathrm{LI}} x_i^{\mathrm{LI}} + f^{\mathrm{VI}} x_i^{VI} - f^{\mathrm{LO}} x_i^{\mathrm{LO}} - f^{\mathrm{VO}} x_i^{\mathrm{VO}} + \sum \nu_i \times r = 0$ $f^{\mathrm{LI}} h^{\mathrm{LI}} + f^{\mathrm{VI}} h^{\mathrm{VI}} - f^{\mathrm{LO}} h^{\mathrm{LO}} - f^{\mathrm{VO}} h^{\mathrm{VO}} + \sum \Delta H \times r = 0$
Driving force constraints	$G_1 \times G_2 \leq 0, G_1 = f^{\mathrm{LO}} x_i^{LO} - f^{\mathrm{LI}} x_i^{\mathrm{LI}}$ $G_2 = \ln\left[\frac{\gamma_i^L x_i^L P_i^{\mathrm{sat},L}}{\phi_i^V x_i^V P_{tot}}\right] +$ $\sum_i \sum_k \left[\frac{\nu_{ik} \Delta G_i^f}{RT_e^L} + \nu_{ik} \ \ln(\phi_i^V x_i^V P_{tot})\right] \frac{\partial \epsilon_k}{\partial n_i^L}$
Phase defining constraints	Liquid: $\sum K_{\mathrm{eq},i} x_i \leq 1$, vapor: $\sum x_i / K_{\mathrm{eq},i} \leq 1$, $K_{\mathrm{eq},i} = \gamma_i P_i^{\mathrm{sat}} / P$
Physical properties	$\gamma_i = \gamma(x_i, T, P)$ [Non-Random Two-Liquid Model (NRTL), UNIQUAC, etc.]
Reaction kinetics, etc.	$r = k[\Pi a^\alpha - \frac{1}{Ka} a^\beta]$, $k = A \exp(-\frac{-E_a}{RT})$, $Ka = Ka(T)$
Structural model (for superstructure)	
Module/stream existence	$y_{e+1} - y_e \leq 0, f^{\mathrm{LL}} - y^{LL} \times f_{\max} \leq 0$
Maximum module availability	$\sum y_e - N_{\max} \leq 0$
Connection rules, etc.	$y - \left[\sum y^{\mathrm{IL}} + \sum y^{\mathrm{LL}} + \sum y^{CL}\right] \leq 0$

Nomenclature: i is the set for components, e is the set for modules, L and V are liquid and vapor phases,

LO and LI are outlet and inlet streams, f is the flowrate, x is the molar fraction, γ is activity coefficient,

h is the enthalpy, P^{sat} is the saturated vapor pressure, ν is the stoichiometric coefficient, G is the Gibbs free energy,

n is the molar amount, μ is the chemical potential, ϕ is the fugacity coefficient, R is the gas constant, ϵ is the reaction extent,

T is the pressure, P is the pressure, and y is the binary (0–1) variable.

approximate the HE modules as heat exchangers and the M/H modules as process vessels (if only separation takes place) or reactor vessels (if any reactive conversion takes place). The average values of failure frequency are adopted from Stoffen [38] based on historic data, as given in Table 8.3. Modularization can be systematically considered in GMF to improve process safety by utilizing smaller equipment. The use of backup/redundant units to mitigate potential failure has not yet been incorporated.

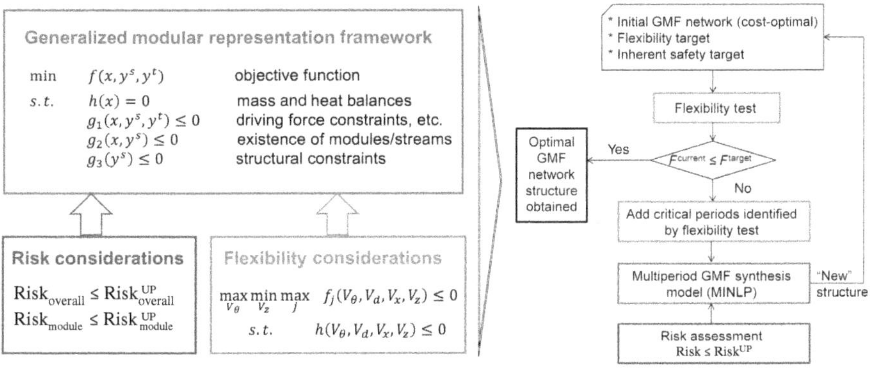

Figure 8.5 Integrated process synthesis with flexibility and safety considerations.

Figure 8.6 Risk analysis.

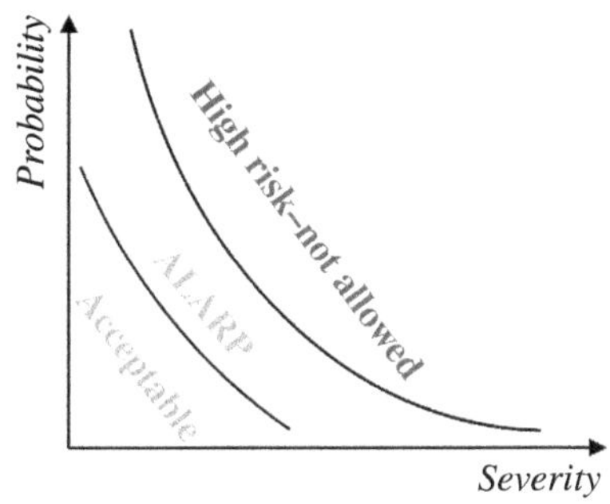

Table 8.3 Failure frequencies.

	Heat exchanger	Process vessels	Reactor vessels
Frequency (per year)	5×10^{-5}	5×10^{-6}	5×10^{-6}

Step 2. **Determine Consequence Severity:** The consequence severity can be estimated using the indication number (A) as defined in Eq. (8.1) [39]:

$$A = \frac{W \times O^1 \times O^2 \times O^3}{S}, \tag{8.1}$$

where W stands for the substance quantity in the module, O^i are factors reflecting the process conditions. Specifically, O^1 represents the installation type, that is, storage or processing. Since GMF modules will all take processing tasks, $O^1 = 1$. O^2 represents the positioning, that is, outdoor or indoor. This work assumes outdoor placement with $O^2 = 1$. O^3 quantifies the substance amount in gas phase after release, accounting for the process conditions such as phase states and process temperature. S is the limiting value factors corresponding to various types of risks, such as toxicity, flammability, or explosiveness. As an example, for explosiveness risk, S is the amount of substance (in kilogram) that can release the same energy with 1 ton trinitrotoluene.

Step 3. **Calculate Risk:** The risk of each module can then be computed following Eq. (8.2). Overall, the risk calculation takes into consideration the total amount of hazardous substances and the indicative process conditions. This serves to provide an initial estimation to penalize the hazardous risks inherently to the process design, as early as at the phenomena-based synthesis stage.

$$\text{Risk} = \text{Faulty Probability} \times \text{Consequence Severity}. \tag{8.2}$$

Step 4. **Define Maximum Risk Tolerance:** Inequality constraints can be set to define the maximum risk allowed in the process as per Eq. (8.3). These constraints can be explicitly incorporated with the GMF process synthesis model discussed in Section 8.2.2 to obtain design solutions at the desired level of process safety. This constraint can be posed to limit the risk of an individual GMF module and/or the entire process as will be showcased via the case study.

$$\text{Risk} \le \text{Risk}^{\text{UP}}. \tag{8.3}$$

8.2.3.2 Design Under Uncertainty via Flexibility Analysis

Flexibility measures the process capability to have feasible operation under uncertainties [40]. If we recast the GMF process synthesis model using the general format of Eq. (8.4), the flexibility test problem [41] can be formulated as Eq. (8.5). An assumption here is that the dimension of x is the same as the dimension of h, which however can be easily relaxed. Equation (8.5) tests if the current process design is feasible (or not) to operate under the entire uncertainty range. Namely, for a given design d, if there always exists a control variable z which makes the set of inequality constraints J_f satisfied, for all the uncertainties in the expected range of $U(\theta)$.

General model formulation

$$\begin{aligned} \min_{d,x,z} \quad & \text{obj}(d,x,z,\theta) \\ \text{s.t.} \quad & h(d,x,z,\theta) = 0 \\ & g(d,x,z,\theta) \le 0 \qquad \forall \theta \in U(\theta), \end{aligned} \tag{8.4}$$

where x is the vector of state variables, d is the vector of design variables, z is the vector of control variables, θ is the vector of uncertain parameters, $U(\theta)$ gives the expected range of uncertainties, h and g gives the equality and inequality constraints, respectively, and *obj* defines the objective function.

The tri-level optimization problem is given in Eq. (8.5). f denotes the new set of inequality constraints assuming that the state variables can be eliminated via the equality constraints. Namely, assume $x = x(d, z, \theta)$ can be solved from $h(d, x, z, \theta) = 0$, which yields $g(d, x(d, z, \theta), z, \theta) = f(d, z, \theta) \leq 0$. This assumption can be easily generalized for the cases in which state variables cannot be eliminated. Equation (8.5) can be further transformed to a single-level mixed-integer optimization problem following the active set strategy based on Karush–Kuhn–Tucker (KKT) conditions [42]. The mathematical formulation is presented in Eq. (8.6), the solution of which yields the decision if the current design is flexible over the expected uncertainties. In the case of not flexible, the solution is also able to identify a critical uncertainty point which renders the largest infeasibility (i.e. constraint violation). However, given the nonconvexity of GMF model, the solution via KKT reformulation cannot theoretically guarantee flexibility. The resulting design solution should be validated against the entire uncertainty range (as that shown in Figure 8.11).

Flexibility test – Tri-level optimization

$$\max_{\theta \in U(\theta)} \min_{z} \max_{j \in J_f} \quad f_j(\theta, d, z) \leq 0. \tag{8.5}$$

Flexibility test – Active set strategy

$$\begin{aligned}
&\max_{\theta, u, z, \lambda_j, s_j, y_j} \quad u \\
&\text{s.t.} \quad 1 - \sum_{j \in J_f} \lambda_j = 0, \qquad \sum_{j \in J_f} \lambda_j \frac{\partial f_j}{\partial z} = 0 \\
&\qquad s_j + f_j(d, z, \theta) - u = 0, \qquad s_j - U(1 - y_j) \leq 0 \\
&\qquad \lambda_j - y_j \leq 0 \qquad \sum_{j \in J} y_j = n + 1 \\
&\qquad y_j = 0, 1 \quad s_j \geq 0, \quad \lambda_j \geq 0 \quad j \in J_f, \quad \theta \in T.
\end{aligned} \tag{8.6}$$

The above flexibility test problem considers a given design d. The synthesis of flexible designs requires to simultaneously determine a design d which is optimal and flexible under uncertainties. To this purpose, a multiperiod GMF process synthesis approach is developed featuring the following iterative algorithm highlighted in Figure 8.5:

Step 1. **Start:** The algorithm starts with an initial GMF design and an expected range of uncertainties. The initial design can be obtained via GMF synthesis without operability considerations, or via simulation of known unit operation configurations.

Step 2. **Flexibility Test:** Flexibility test is performed against the current design using Eq. (8.6). If the design passes flexibility test, a flexible design solution is obtained and the algorithm stops. If the design fails flexibility test, the critical uncertainty point is recorded for use in the next step.

Step 3. **Multiperiod GMF Process Synthesis:** The critical point is added to formulate a multiperiod synthesis model [43]. In this way, the most critical uncertainty condition is explicitly accounted for by GMF synthesis in tandem with the original nominal conditions. The solution of this multiperiod synthesis problem will yield a new GMF design which is guaranteed to be flexible under the critical point. The algorithm then iterates back to Step 2 to check the flexibility over the entire uncertainty range.

8.3 Case Studies

We present two case studies to showcase the framework: (i) Design of an industrial-scale process for MMA purification, (ii) Synthesis of safely operable reaction/separation process for MTBE production. The case studies and numerical results are adapted from [36, 44].

8.3.1 MMA Purification

8.3.1.1 Process Description

MMA serves as the basic monomer to commercially produce poly-methyl methacrylate plastics. A dividing wall column (DWC, Figure 8.7) has been designed in the patent by Jewell et al. [45] to purify the mixture effluent from MMA reactors. On the basis of the same energy consumption and the same number of column stages, this DWC design could achieve 7.6% higher MMA recovery compared to its conventional counterpart using two sequential distillation columns. In what follows, we investigate this MMA purification process via GMF [44] with the patent design as base case.

The reactor effluent mixture includes 175.259 kmol/h MMA, 15.7643 kmol/h water (H_2O), 9.60129 kmol/h MMA oligomers (MMAOLG), and 0.332871 kmol/h methanol (MeOH). A key challenge for this quaternary separation is the vapor–liquid and liquid–liquid phase equilibriums that may be exhibited by this mixture. Azeotropes may also be formed between MeOH-MMA or H_2O-MMA at various temperatures and pressures. To capture the complex mixture behavior, we adopt the UNIQUAC model and parameters [46] which have been verified against data from experimental investigations. The purification targets are set as per the patent [45] and the Dow Sales Specifications [47]: (i) MMA product

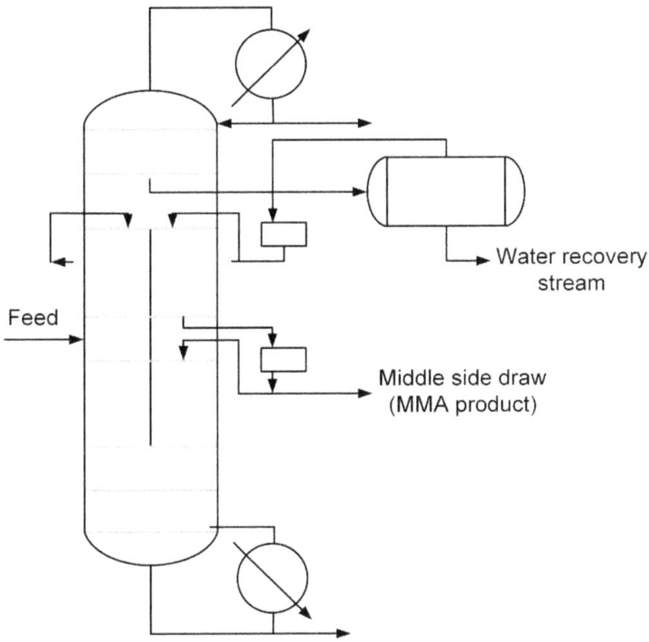

Figure 8.7 MMA dividing wall column – patent design. *Source:* Jewell et al. [45] / American Chemical Society / CC BY 4.0.

flowrate ≥ 172 kmol/h, (ii) MMA product purity more than ≥ 99.80 wt% and ≥ 99.8%, and (iii) purity of water recovery stream ≥ 99.0 mol%. The process synthesis objective is to identify a design at the optimal cost and the corresponding operating parameters, while simultaneously considering both conventional and intensified process alternatives.

8.3.1.2 GMF Simulation of Base Case Design

We first verify if GMF can effectively represent and simulate the base case DWC design presented in the patent. To this purpose, a unit operation-based simulation is developed to model the DWC using Aspen Plus Petlyuk column (Figure 8.8a). The main column comprises 17 stages. The prefractionator comprises 5 stages with the reactor effluent fed onto the 3rd stage. Both columns are operated at 0.06 atm for temperature control to prevent MMA polymerization. An upper side draw stream is extracted from one stage above the dividing section. The stream is then directed to a water decanter in which liquid–liquid phase separation occurs. The decanter generates an aqueous stream (namely the water recovery stream) and an organic stream (to be recycle to the column). The MMA product stream is obtained as the middle side draw with 172.0 kmol/h flowrate and 99.8%

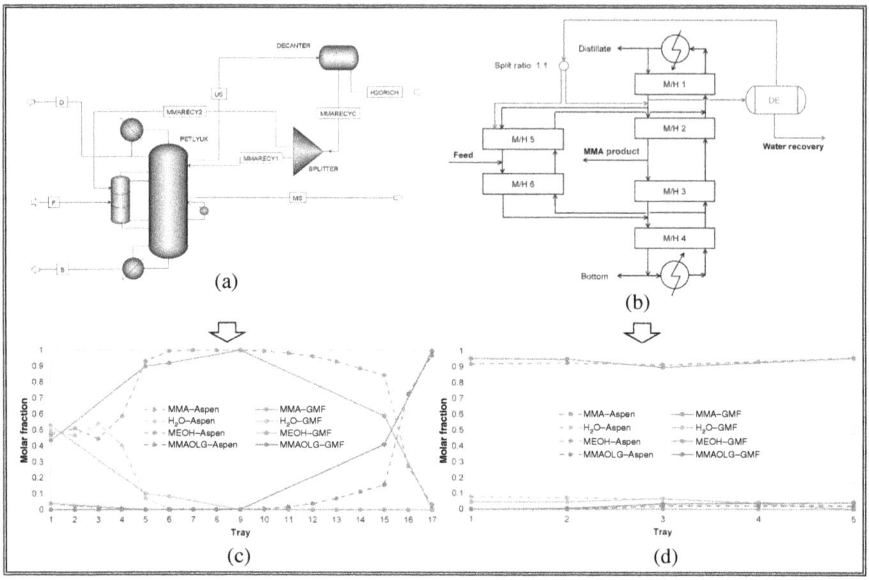

Figure 8.8 MMA dividing wall column – GMF base case design representation. (a) Aspen simulation, (b) GMF representation, (c) validation – liquid molar fraction in main column, and (d) validation – liquid molar fraction in prefractionator.

MMA purity. The heating duty is calculated as 5035.5 kW and the cooling duty as 5560.4 kW.

As depicted in Figure 8.8b, the GMF representation of this base case DWC comprises 6 M/H modules in total (i.e. M/H 1–M/H 4 for main column and M/H 5–M/H 6 for prefractionator). Another M/H module is employed to represent the water decanter. Two HE modules are used specifically as the column reboiler and condenser. The feed/product locations and interconnecting stream placements follow that suggested by the Aspen simulation. The reactor effluent is fed onto M/H 6. The upper side draw is extracted from M/H 1 outlet liquid stream. MMA product is extracted from M/H 2 outlet liquid stream. The simulation results of this GMF-based design are validated against that of Aspen simulation. Figure 8.8c,d compares the GMF and Aspen liquid molar fraction profiles respectively in main column and in prefractionator. GMF can well capture the major mass transfer trends in the column, which is critical to the next step predictive process design optimization.

8.3.1.3 GMF Synthesis for Grassroots Design

To synthesize the optimal design, the GMF superstructure (Figure 8.9a) is posed with a maximum of HE and M/H modules that can be used. The Mixed-Integer

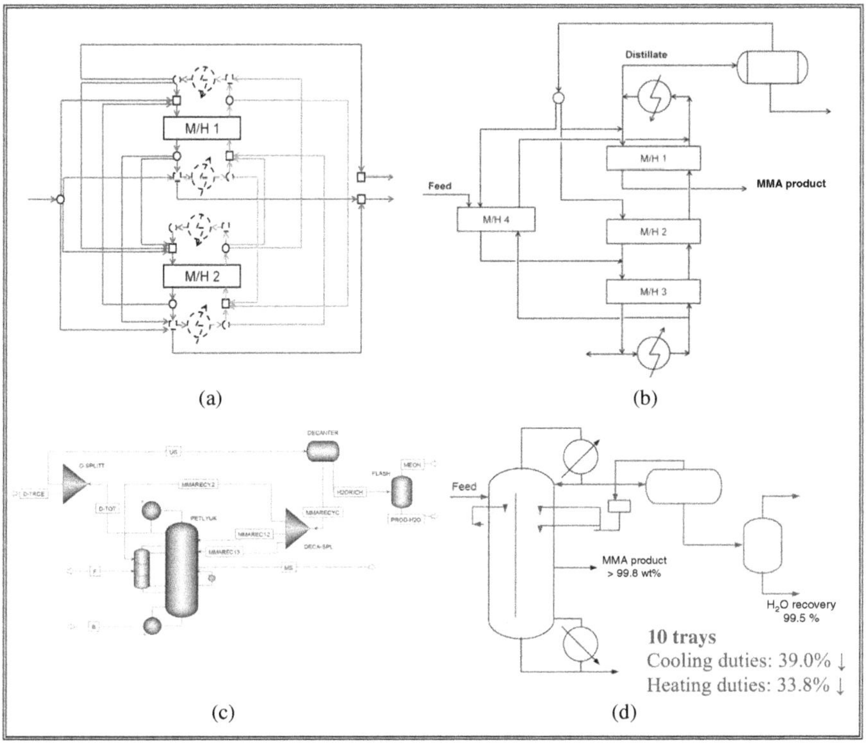

Figure 8.9 MMA dividing wall column – GMF grassroots design. (a) GMF superstructure representation, (b) optimal GMF design solution, (c) validation with unit operation-based simulation, and (d) final process design.

Nonlinear Programming (MINLP)-based process synthesis model is set up in General Algebraic Modeling System (GAMS), which provides a key feature of dynamic sets to only include the activated modules in algorithmic solution. The Generalized Benders Decomposition (GBD) algorithm converges after 25 iterations. As the GMF synthesis model is highly nonlinear and highly nonconvex, GBD is utilized to identify local optimal solutions. The optimal design solution is identified as Figure 8.9b. Compared to the base case, several major structural and operating changes of the optimal design include: (i) Four M/H modules are used in the DWC design which indicates a potentially smaller equipment; (ii) the distillate is directly sent to decanter which thus replaces the need of the original upper side draw stream; and (iii) the dewatered organic stream from decanter is split into three streams and fed onto various M/H modules.

As shown in Figure 8.9c, the optimal GMF design is then translated to unit operation-based design and validated using Aspen Plus. It gives a Petlyuk column integrated with a water decanter. M/H 1–3 are translated to 2, 3, and 3 column stages in the main column. M/H 4 is translated to a 3-stage prefractionator. The two HE modules correspond to the reboiler and condenser of the main column. For stream connections, following the GMF design, the first stage of prefractionator takes in the feed stream and approximately 10% of the dewatered organic stream from decanter. The remaining dewatered organic stream is split equality and sent to the third/fifth stage in the main column. In the decanter, liquid–liquid phase separation occurs at 323.15 K and 0.4 atm. However, the resulting aqueous stream from decanter contains around 2% methanol. A flash drum is further added to achieve 99.5 mol% purity of the water recovery stream. Figure 8.9d shows the equivalent DWC configuration, in which the dividing wall should vertically extend from the 2nd to 10th stage. The unit operation-based simulation in Aspen Plus confirms that the GMF optimal design can result in significant utility savings (37.4% for heating and 39.0 % for cooling) and a reduced number of column stages. In this way, a better design solution has been identified for this MMA purification process with enhanced cost, energy, and sustainability performances.

8.3.2 MTBE Production

8.3.2.1 Process Description

In this second case study, we consider the synthesis of a combined reaction/separation process and showcase how operability considerations can affect the optimal design configuration. We investigate the case study for MTBE production from isobutylene (IB4) and methanol (MeOH) [36, 48]. MTBE is a type of fuel additive to gasoline. The reaction rate is given by Eq. (8.7) using Amberlyst 15 as the catalyst resin. UNIQUAC model is used to describe the nonideal liquid–vapor mixture behavior between MTBE, IB4, MeOH, and 1-butene (NB4, inert).

$$r = k\left[\frac{a_{\mathrm{IB4}}}{a_{\mathrm{MeOH}}} - \frac{1}{Ka}\frac{a_{\mathrm{MTBE}}}{a_{\mathrm{MeOH}}^2}\right] \quad \mathrm{kmol/(h \cdot kg\ cat)}, \tag{8.7}$$

where r is the reaction rate, k is the rate constant for which the temperature-dependent expression can be found in [49], Ka is the equilibrium constant for which the temperature-dependent expression can be found in [50], and a is the activity.

The raw materials comprise two feed streams as shown in Table 8.4. The desired product is MTBE in liquid stream (197 mol/s flowrate, 98 mol% purity). However, uncertainty exists in the liquid feed flowrate (±10 mol/s), which should be accounted for during process design. There are also concerns regarding the safety performance of this process with respect to toxicity, flammability, and explosiveness risks. The hazardous properties are given in Table 8.5. In this

Table 8.4 Raw materials for MTBE production.

	Temperature (K)	Pressure (atm)	Flowrate (mol/s)	Molar fractions
Liquid feed	320	11	215.5	$x_{MeOH} = 1$
Vapor feed	350	11	545	$x_{IB4} = 0.3578, x_{NB4} = 0.6422$

Table 8.5 Hazardous property data.

	LC_{50} (rat, 1 h, in h)/mg	Flammable?	Trinitrotoluene equivalence/kg
MeOH	65.6	Yes	4.62
IB4	155	Yes	2.05
NB4	164.5	Yes	2.03
MTBE	21.3	Yes	2.62

context, the synthesis objective for this process is to identify the cost-optimal process design solution(s), which should also ensure safe and flexible operations under uncertainty.

8.3.2.2 Integrated GMF Synthesis and Operability Optimization

Following the integrated strategy presented in Section 8.2.3, three GMF design alternatives (Figure 8.10) are obtained with respect to various levels of safety and flexibility performances.

a) **Cost-Optimal Design (Figure** 8.10**a):** The objective is minimizing the total annualized cost, which includes pseudo-capital cost (estimated based on key process variables such as M/H vapor flowrates [48]) and utility cost (as a function of HE heat duties). The resulting optimal design comprises 6 M/H modules featuring a reactive distillation configuration. Specifically, 1 M/H module takes the role of rectifying section with pure separation, 2 M/H modules take the role of reaction zone with reactive separation, and another 1 M/H module takes the role of stripping section with pure separation. The liquid and vapor feeds are, respectively, fed above and below the reaction zone. Two HE modules are employed for the column reboiler and condenser. This design results in a total annualized cost of US$ 1.71 × 10^6 dollars per year.
b) **Operable Design 1 (Figure** 8.10**b):** The above cost-optimal design considers the liquid feed flowrate at the nominal condition without uncertainty.

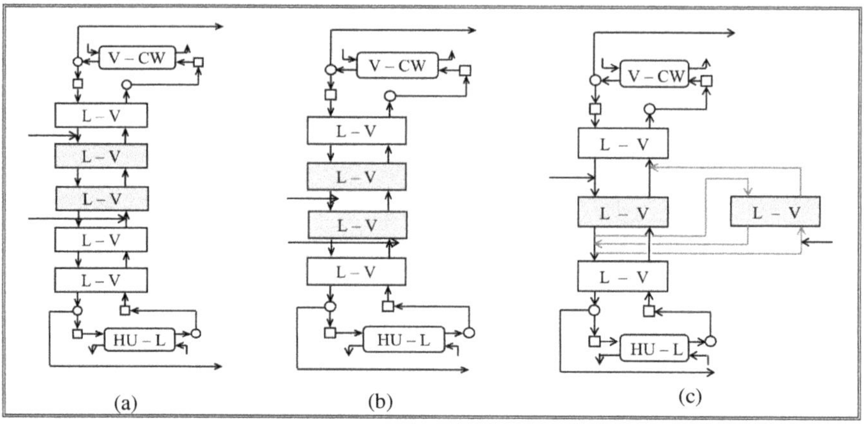

Figure 8.10 MTBE production – GMF design with operability. (a) Cost-optimal, (b) +10% feasible operation range −20% overall process risk, (c) +10% feasible operation range −20% overall process risk module risk <30% process risk.

Operable design 1 is obtained by adding on: (i) flexibility requirement under uncertain feed flowrate; and (ii) improved safety by reducing all toxicity, flammability, and explosiveness risks by 20% in the overall process design. Optimized against the total annualized cost, the new process design solution includes 4 M/H modules and 2 HE modules. Notably, the M/H modules feature larger module diameters which allows to tolerate feed flowrate uncertainties. This is also consistent with the observations of other research works [51]. On the other hand, the reduced number of total modules may indicate a smaller reactive distillation column in response to the safety constraints, that is, smaller is safer. This design results in a total annualized cost of US$ 1.82×10^6 dollars per year.

c) **Operable Design 2 (Figure** 8.10**c):** This design further requires the risk of each individual module to be no more than 30% of the risk of the entire process. This avoids any module being a risk-critical component to the process, for example, due to heavier mass/heat transfer loads as that observed in the top reactive separation M/H module in operable design 1. The resulting optimal design still comprises 4 M/H modules and 2 HE modules; however, the stream connections feature a reactive Petlyuk column configuration rather than a single reactive distillation column as previous. In other words, the thermal coupling between two reactive separation M/H modules contribute to better distribute the mass/heat transfer load. The new interconnecting streams also introduce additional degrees of freedoms to improve process operability. The total annualized cost of this design is US$ 1.90×10^6 dollars per year.

These three GMF design alternatives are translated to unit operation-based process. Aspen Plus simulation is used for validation, following the same strategy discussed in Section 8.3.1.3. The GMF designs respectively correspond to a 15-stage reactive distillation (cost-optimal design), a 13-stage reactive distillation (operable design 1), and a 12-stage reactive Petlyuk column (operable design 2). The validation results are summarized in Figure 8.11. In particular, we have verified: (i) all the unit operation-based design can satisfy the production target

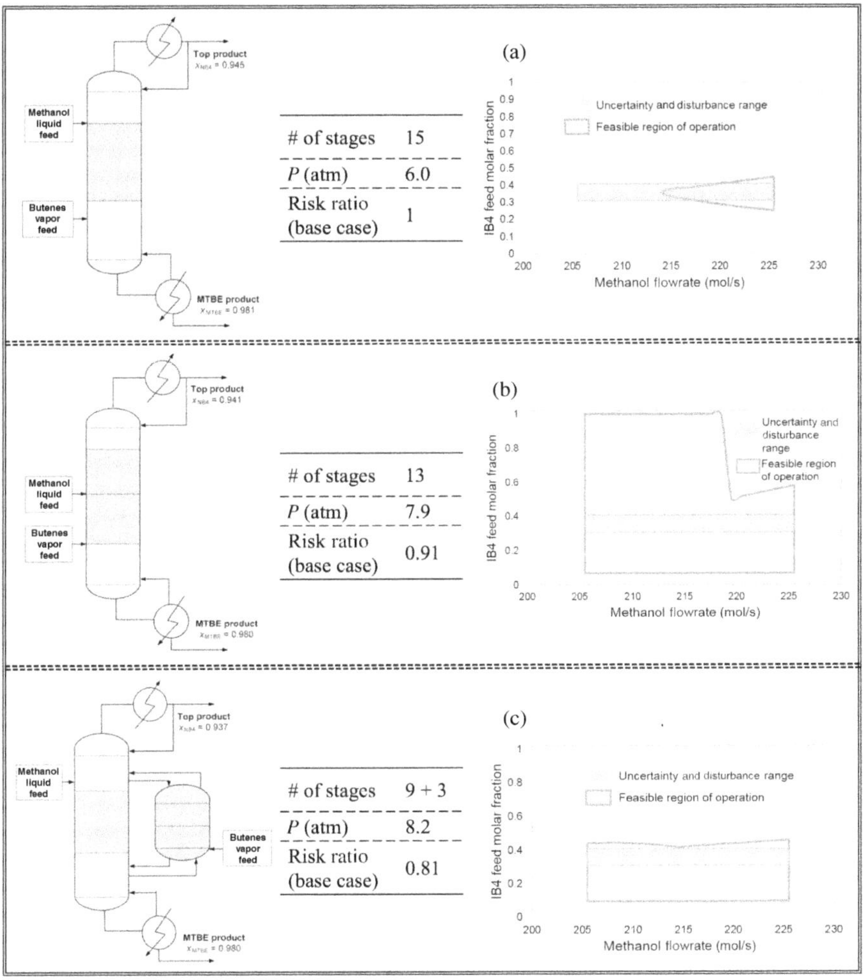

Figure 8.11 MTBE production – process design validation. (a) Cost-optimal, (b) +10% feasible operation range −20% overall process risk, (c) +10% feasible operation range −20% overall process risk module risk <30% process risk.

(i.e. 197 mol/s MTBE product at 98.0 mol% purity), (ii) the operable design are indeed safer than the cost-optimal design using the quantitative risk metric, and (iii) the operable designs are feasible under liquid feed flowrate uncertainties. Specifically to verify the flexibility, the feasible operating space is rigorously mapped for the three unit operation-based designs. It can be seen that, for the uncertainty range, the cost-optimal design is not flexible while the two operable designs are. This thus confirms that the operability promise of phenomenological process designs via GMF can be well retained in the final unit operation-based designs.

8.4 Concluding Remarks

In this chapter, we have introduced the GMF, which utilizes mass/heat exchange building blocks for process intensification synthesis. To ensure safe and feasible operations under uncertainty, GMF is integrated with model-based flexibility and safety analyses to systematically generate optimal, operable, and intensified design solutions. Two case studies are provided to showcase the methodology: (i) Industrial MMA purification using dividing wall column, and (ii) MTBE production via combined reaction/separation. Significant energy and cost savings are achieved via GMF process synthesis, leading to more sustainable process design solutions.

Acknowledgment

Financial support from NSF GOALI project (CBET-2401564), ACS PRF (67606-DNI9), Texas A&M Energy Institute, and West Virginia University is gratefully acknowledged.

References

1 Energy Information Administration (2023). *Annual Energy Outlook*. Energy Information Administration.
2 Cresko, J., Rightor, E., Carpenter, A. et al. (2022). *US Department of Energy's Industrial Decarbonization Roadmap*, https://doi.org/10.2172/1961393.
3 Bielenberg, J. and Palou-Rivera, I. (2019). The RAPID manufacturing institute–reenergizing US efforts in process intensification and modular chemical processing. *Chemical Engineering and Processing - Process Intensification* 138: 49–54.

4 Abdulrahman, I., Máš˘ a, V., and Teng, S.Y. (2021). Process intensification in the oil and gas industry: a technological framework. *Chemical Engineering and Processing-Process Intensification* 159: 108208.

5 Van Gerven, T. and Stankiewicz, A. (2009). Structure, energy, synergy, time –the fundamentals of process intensification. *Industrial & Engineering Chemistry Research* 48 (5): 2465–2474.

6 Creative Energy (2017). *European Roadmap for Process Intensification*, https://efce.info/efce_media/-p-531.pdf.

7 Tian, Y., Demirel, S.E., Hasan, M.F., and Pistikopoulos, E.N. (2018). An overview of process systems engineering approaches for process intensification: state of the art. *Chemical Engineering and Processing - Process Intensification* 133: 160–210.

8 Simkoff, J.M., Lejarza, F., Kelley, M.T. et al. (2020). Process control and energy efficiency. *Annual Review of Chemical and Biomolecular Engineering* 11 (1): 423–445.

9 Asprion, N. (2020). Modeling, simulation, and optimization 4.0 for a distillation column. *Chemie Ingenieur Technik* 92 (7): 879–889.

10 Biegler, L.T. (2020). Nonlinear optimization strategies for process separations and process intensification. *Chemie Ingenieur Technik* 92 (7): 867–878.

11 Yao, C., Zhao, Y., Ma, H. et al. (2021). Two-phase flow and mass transfer in microchannels: a review from local mechanism to global models. *Chemical Engineering Science* 229: 116017.

12 Rivera, F.F., Pérez, T., Castaeda, L.F., and Nava, J.L. (2021). Mathematical modeling and simulation of electrochemical reactors: a critical review. *Chemical Engineering Science* 239: 116622.

13 Ng, E., Lau, K., Lau, W., and Ahmad, F. (2021). Holistic review on the recent development in mathematical modelling and process simulation of hollow fiber membrane contactor for gas separation process. *Journal of Industrial and Engineering Chemistry* 104: 231–257.

14 Goyal, H., Chen, T.Y., Chen, W., and Vlachos, D.G. (2022). A review of microwave-assisted process intensified multiphase reactors. *Chemical Engineering Journal* 430: 133183.

15 Chen, W.C., Fan, Y.W., Zhang, L.L. et al. (2022). Computational fluid dynamic simulation of gas-liquid flow in rotating packed bed: a review. *Chinese Journal of Chemical Engineering* 41: 85–108.

16 Shu, C., Li, X., Li, H., and Gao, X. (2022). Design and optimization of reactive distillation: a review. *Frontiers of Chemical Science and Engineering* 16 (6): 799–818.

17 Pasha, M., Liu, S., Zhang, J. et al. (2022). Recent advancements on hydrodynamics and mass transfer characteristics for CO_2 absorption in microreactors. *Industrial & Engineering Chemistry Research* 61 (34): 12249–12268.

18 Singh, M.P., Alatyar, A.M., Berrouk, A.S., and Saeed, M. (2023). Numerical modelling of rotating packed beds used for CO_2 capture processes: a review. *The Canadian Journal of Chemical Engineering* 101 (11): 6170–6202.

19 Kiss, A.A., Muthia, R., Pazmio-Mayorga, I. et al. (2024). Conceptual methods for synthesis of reactive distillation processes: recent developments and perspectives. *Journal of Chemical Technology & Biotechnology* 99 (6): 1263–1290.

20 Harmsen, G.J. (2007). Reactive distillation: the front-runner of industrial process intensification: a full review of commercial applications, research, scale-up, design and operation. *Chemical Engineering and Processing: Process Intensification* 46 (9): 774–780.

21 Weinfeld, J.A., Owens, S.A., and Eldridge, R.B. (2018). Reactive dividing wall columns: a comprehensive review. *Chemical Engineering and Processing-Process Intensification* 123: 20–33.

22 Tula, A.K., Eden, M.R., and Gani, R. (2019). Computer-aided process intensification: challenges, trends and opportunities. *AIChE Journal* 66 (1): 16819.

23 Li, Q., Somoza-Tornos, A., Grievink, J., and Kiss, A.A. (2024). Challenges and opportunities for process intensification in Europe from a process systems engineering perspective. *Frontiers in Energy Research* 12: 1340635.

24 Baldea, M. (2015). From process integration to process intensification. *Computers & Chemical Engineering* 81: 104–114.

25 Nikačević, N.M., Huesman, A.E., Van den Hof, P.M., and Stankiewicz, A.I. (2012). Opportunities and challenges for process control in process intensification. *Chemical Engineering and Processing: Process Intensification* 52: 1–15.

26 Pistikopoulos, E.N., Tian, Y., and Bindlish, R. (2021). Operability and control in process intensification and modular design: challenges and opportunities. *AIChE Journal* 67 (5): e17204.

27 Pistikopoulos, E.N. and Tian, Y. (2024). Advanced modeling and optimization strategies for process synthesis. *Annual Review of Chemical and Biomolecular Engineering* 15: 81–103.

28 Mencarelli, L., Chen, Q., Pagot, A., and Grossmann, I.E. (2020). A review on superstructure optimization approaches in process system engineering. *Computers & Chemical Engineering* 136: 106808.

29 Onel, O., Niziolek, A.M., Butcher, H. et al. (2017). Multi-scale approaches for gas-to-liquids process intensification: CFD modeling, process synthesis, and global optimization. *Computers & Chemical Engineering* 105: 276–296.

30 Qian, Z., Chen, Q., and Grossmann, I.E. (2017). Optimal synthesis of rotating packed bed reactor. *Computers & Chemical Engineering* 105: 152–160.

31 Kim, S.H., Landa, H.O.R., Ravutla, S. et al. (2022). Data-driven simultaneous process optimization and adsorbent selection for vacuum pressure swing adsorption. *Chemical Engineering Research and Design* 188: 1013–1028.

32 Garg, N., Kontogeorgis, G.M., Gani, R., and Woodley, J.M. (2020). A process synthesis-intensification method for generation of novel and intensified solutions. *Chemical Engineering and Processing-Process Intensification* 156: 108103.

33 Demirel, S.E., Li, J., and Hasan, M.F. (2019). Systematic process intensification. *Current Opinion in Chemical Engineering* 25: 108–113.

34 Skiborowski, M. (2023). Synthesis and design methods for energy-efficient distillation processes. *Current Opinion in Chemical Engineering* 42: 100985.

35 Pistikopoulos, E.N. and Tian, Y. (2022). *Synthesis and Operability Strategies for Computer-Aided Modular Process Intensification*. Elsevier.

36 Tian, Y. and Pistikopoulos, E.N. (2018). Synthesis of operable process intensification systems-steady-state design with safety and operability considerations. *Industrial & Engineering Chemistry Research* 58 (15): 6049–6068.

37 Geoffrion, A.M. (1972). Generalized benders decomposition. *Journal of Optimization Theory and Applications* 10: 237–260.

38 Stoffen, P.G. (2005). Guidelines for quantitative risk assessment. *Ministerie van Volkshuisvesting Ruimtelijke Ordening en Milieu*. CPR E; 18.

39 Freeman, R.A. (1990). CCPS guidelines for chemical process quantitative risk analysis. *Plant/Operations Progress* 9 (4): 231–235.

40 Li, C. and Grossmann, I.E. (2021). A review of stochastic programming methods for optimization of process systems under uncertainty. *Frontiers in Chemical Engineering* 2: 622241.

41 Swaney, R.E. and Grossmann, I.E. (1985). An index for operational flexibility in chemical process design. Part II: computational algorithms. *AIChE Journal* 31 (4): 631–641.

42 Grossmann, I.E. and Floudas, C.A. (1987). Active constraint strategy for flexibility analysis in chemical processes. *Computers & Chemical Engineering* 11 (6): 675–693.

43 Papalexandri, K. and Pistikopoulos, E. (1994). A multiperiod MINLP model for the synthesis of flexible heat and mass exchange networks. *Computers & Chemical Engineering* 18 (11–12): 1125–1139.

44 Tian, Y., Meduri, V., Bindlish, R., and Pistikopoulos, E.N. (2022). A Process Intensification synthesis framework for the design of dividing wall column systems. *Computers & Chemical Engineering* 160: 107679.

45 Jewell, D.W., Pendergast, J.G., Worley, W.G. et al. (2019). Process for purification of methyl methacrylate. US Patent 10,392,337.

46 Wu, Y.C., Hsu, C., Huang, H.P., and Chien, I.L. (2011). Design and control of a methyl methacrylate separation process with a middle decanter. *Industrial & Engineering Chemistry Research* 50 (8): 4595–4607.

47 Dow 2018. The Dow Chemical Company sales specification on methyl methacrylate (accessed 19 September 2024).

48 Ismail, S.R., Proios, P., and Pistikopoulos, E.N. (2001). Modular synthesis framework for combined separation/reaction systems. *AIChE Journal* 47 (3): 629–649.

49 Rehfinger, A. and Hoffmann, U. (1990). Kinetics of methyl tertiary butyl ether liquid phase synthesis catalyzed by ion exchange resin–I. Intrinsic rate expression in liquid phase activities. *Chemical Engineering Science* 45 (6): 1605–1617.

50 Colombo, F., Cori, L., Dalloro, L., and Delogu, P. (1983). Equilibrium constant for the methyl tert-butyl ether liquid-phase synthesis using UNIFAC. *Industrial & Engineering Chemistry Fundamentals* 22 (2): 219–223.

51 Di Pretoro, A., Montastruc, L., Manenti, F., and Joulia, X. (2019). Flexibility analysis of a distillation column: indexes comparison and economic assessment. *Computers & Chemical Engineering* 124: 93–108.

9

Circular Economy Assessment Tools for Process Systems

Paola Munoz-Briones, Kenneth Martinez, Javiera Vergara-Zambrano, and Styliani Avraamidou

Department of Chemical and Biological Engineering, University of Wisconsin-Madison, Madison, WI, USA

9.1 Introduction

Quantifiable metrics are essential for sustainable process design and new technology development as they provide a measure of progress, guide informed decision-making, and ensure alignment with established sustainability goals. By evaluating performance and identifying critical areas of improvement, metrics ensure that efforts are directed toward the most impactful domains, driving development and refinement. Sustainability metrics enable companies and organizations not only to comply with environmental regulations but also to optimize material and energy use and reduce waste, potentially leading to cost savings and improved efficiencies. Circular Economy (CE) metrics can enhance competitiveness by enabling performance benchmarking, driving innovation in sustainability, and distinguishing companies in the market [1]. Moreover, they can promote long-term success by encouraging the use of renewable energy and recycled materials, facilitating the transition from a linear to a CE [2].

The traditional linear model "take-make-waste," where natural resources are extracted, transformed into products, used, and then discarded as waste, imposes a significant burden to natural resources and results in significant negative environmental and socioeconomic impacts. With the rising population and increasing demand to meet social needs, this economic model has become increasingly unsustainable. The CE concept emerges as a promising solution, aiming to minimize environmental pollution and promote the closed-loop handling of resources, minimizing losses, waste, and the consumption of natural resources, increasing the use of renewable resources, and preserving the value of materials and products for a longer period [3]. CE is considered by many practitioners and policy makers as a framework for the transition toward environmental, economic, and social sustainability, which are key pillars of sustainable development [1, 3, 4]. Environmentally, CE seeks to reduce impacts such as global warming potential, resource depletion, and ecosystem degradation by encouraging reuse, recycling, and waste reduction [5]. Economically, CE can drive cost savings by enhancing resource efficiency and reducing waste. This leads to lower material and utility costs, as well as decreased disposal expenses. Socially, CE focuses on promoting social welfare and community well-being with actions like creating jobs and ensuring fair wages, education, and access to public infrastructure like waste management facilities. Overall, CE is regarded as being essential for achieving sustainable development by integrating economic growth, environmental preservation, and social well-being simultaneously [6].

To effectively implement CE practices globally, it is crucial to systematically assess alternative pathways and scenarios and develop comprehensive metrics to evaluate various aspects of CE [1, 3, 7]. First, a clear definition of CE is required, based on a set of goals that ensures minimized environmental impact, including reduced harmful emissions, decreased material losses, increased use of renewable resources, and enhanced durability. Then, understanding the elements to be measured, determining the appropriate metrics, setting benchmarks for evaluation, and establishing targets are crucial [3, 8]. Various approaches have been proposed in the literature to assess different CE aspects; however, they differ in goals, scopes, and applications, leading to a fragmented approach [9]. Frameworks and metrics for CE assessment have also been introduced, although they frequently focus only on a subset of environmental evaluation metrics [10]. Life Cycle Assessment (LCA) is a widely used methodology for evaluating the environmental impacts of a product or process throughout its entire life cycle. Given its environmental focus, some studies combine LCA with methodologies that assess economic aspects, such as Techno-Economic Analysis (TEA), to provide a more comprehensive assessment. However, while this integrated approach offers valuable insights, it does not fully address all the CE aspects. Avraamidou et al. [3] highlight contributions from the process systems engineering area to various CE goals, yet some circularity aspects like durability or use of renewable energy sometimes remain unaddressed. Consequently, there still remains a lack of standardized and comprehensive methods in CE assessment restricting its effective implementation [4].

There are several studies that analyze some CE metrics suggested in the literature. Moraga et al. [11] propose a classification framework that categorizes indicators based on six CE strategies and three measurement scope according to their Life Cycle Thinking (LCT) approach. This offers a novel perspective, because that it categorizes indicators without being limited by specific definitions. Among the limitations of the study, the framework is unable to distinguish between indicators measuring inputs and outputs, and the connections between CE and also its impact on sustainable development are not completely established. Additionally, there is a lack of consideration of nonmaterial flows like land use or renewable energy. Sassanelli et al. [12] conduct a systematic literature review and, based on the findings, propose a positioning framework that includes three key dimensions: product life cycle stages, variables, and a circularity degree. The goal of this framework is to identify gaps that can be addressed with a new circular performance assessment method. However, the analysis shows a lack of overall CE evaluation and a strong emphasis on the environmental and material aspect.

Recent work on CE assessment exhibits that current approaches still inadequately consider CE strategies and fail to address all the sustainability dimensions. Elia et al. [13] perform an umbrella review and classify several assessment approaches, arriving at this conclusion and additionally highlighting that the social dimension, in particular, remains significantly neglected. Muoz et al. [14] propose the 9R circularity index that uses a material flow analysis perspective for the *micro-level*. The work suggests a set of indicators to evaluate the effectiveness of CE strategies, aligned with the 9R framework. It utilizes the Analytic Hierarchy Process (AHP) to integrate these indicators into a single metric called the "9R circularity index." However, the index does not consider other important environmental impacts like greenhouse gas emissions and water consumption or the other economic and social dimensions. Miatto et al. [15] study measuring circularity at the national economy and sector levels using a material flow analysis of the Australian economy and propose calculations for the circularity rate and circularity gap. These encompass factors related to domestic resource consumption and the use of recycled materials, therefore focusing on material circularity. A material footprint assessment is also conducted; however, both assessments are not integrated into a single metric.

Incorporating CE practices in large critical industries is essential due to their scale, as it can lead to substantial impacts and drive significant improvements. Embracing CE in key sectors such as energy systems, food supply chains (FSC), and the chemical industry is essential for advancing the sustainability goals and promoting a more sustainable and balanced future. The Food Supply Chain (FSC) is fundamental to meet societal needs, the chemical industry supplies critical commodities that support global functioning, and the energy sector supports the functioning of both these industries and many others. However, developing metrics that can be applied across different scales, economic sectors, and geographic areas remains an ongoing challenge. Due to the distinct operations of each sector, a single CE metric that comprehensively assesses all sectors is very challenging to be developed. Instead, different frameworks and sector-specific indicators tailored to each sector's unique characteristics are being developed for effective evaluation.

Sections 9.2–9.4 of this chapter will explore the advances reported in the literature in CE assessment concerning the Food, Chemical, and Energy sectors.

9.2 Circular Economy Assessment in the Food Sector

Zero Hunger and Sustainable Production and Consumption are key United Nations Sustainable Development Goals that highlight the critical need and the global commitment to fostering sustainable FSC. FSC demand increased attention due to their substantial contributions to global emissions, watershed pollution, and resource consumption. Approximately, one-third of the food produced worldwide is lost or wasted, leading to substantial financial losses, resource waste, and unnecessary environmental emissions. According to the Food and Agriculture Organization, this accounts for up to 10% of global greenhouse gas emissions [16]. Moreover, food waste exacerbates food insecurity and undermines social equity by wasting resources that could otherwise feed those in need. CE practices offer a solution by minimizing food loss and waste (FLW), thus improving environmental impact and resource efficiency throughout the supply chain. Proposed CE strategies include increasing the use of renewable resources and energy, reducing water use and waste generation at various stages of the supply chain, and enhancing product durability.

Systematic and quantitative methods are essential to identify and assess alternative production, distribution, and recycling pathways in the FSC and to ensure that CE principles can be effectively implemented in practical settings [1, 17–19]. Assessing the sustainability impacts of CE practices prior to implementation is crucial because CE solutions don't necessarily imply sustainability [17]. For example, while food packaging recycling is a key element of CE, some recycling technologies for complex products can be more resource intensive than using virgin raw materials [20]. Hence, it is essential to develop comprehensive metrics and frameworks that assess both specific circular strategies and their overall impact on closing material loops and leveraging ecological material cycles [21].

The state-of-the-art in FSC CE assessment methods includes various metrics and approaches applied to different areas of the supply chain, such as food production, packaging, and distribution. Table 9.1 provides a literature review of studies in these areas, outlining the methodologies and indicators used. Some studies offer green alternatives focusing on renewable resources and achieving zero waste; however, the assessment methods often emphasize solely on some environmental impacts, usually greenhouse gas emissions, overlooking other critical factors.

LCA is a widely used methodology for evaluating and comparing the environmental impacts of CE pathways across agricultural systems, food processing, production activities, food waste management, and food packaging [22–26]. Some LCA studies, such as those examining FLW, focus on specific stages of the supply chain, while others consider the entire system [27]. Krishnan et al. [28] use LCA to identify operational and resource inefficiencies in the mango FSC and propose a framework for redesigning the FSC but focusing on improving environmental sustainability. The study uses eight impact categories developed by the Centre of Environmental Science, including global warming potential, human toxicity, acidification, and eutrophication, which are commonly used in LCA studies of FSC [29]. Examples of other LCA-based studies can be found in Table 9.1. However, LCA does not address social and economic dimensions or directly tackle all the CE goals, such as increasing the durability of products. Nevertheless, as previously mentioned combining LCA with methodologies as TEA is an alternative for evaluating both environmental and economic impacts, as demonstrated in studies by Neha et al. [30] and del Carmen Munguıa-López et al. [31] on food waste and food packaging recycling technologies.

Recent work attempts to expand the assessment scope in the FSC, by defining circularity indicators. The Material Circularity Indicator (MCI), pioneered by the MacArthur Foundation, serves as a key reference point in many studies. It measures product circularity solely based on material flows, but recommends the use of complementary indicators [32]. Gallo et al. [33] propose enhancing circularity assessments by integrating the MCI with LCA in the agri-food sector, correlating circularity with various environmental impacts. This approach considers life cycle impact categories such as global warming potential (GWP), acidification potential (AP), eutrophication potential (EP), abiotic depletion potential (ADP), and water scarcity footprint (WSF). These indicators complement the MCI, and the comparison between both metrics was illustrated using a Cartesian graph. Despite its innovative approach, more case studies are required to prove that the integrated model is robust and effective. This is required to validate the assumptions of the model, determine adjustments in the weighing system of technical and biological cycles, and determine whether the findings can serve as general guidelines for standardized CE assessments. Pauer et al. [34] present an extended LCA and a circularity assessment based on material input-related and material output-related indicators and energy indicators, applying this broader framework to food packaging. The environmental impact of packaging involves assessing 16 impact categories like climate change, ozone depletion, and human toxicity. The study also incorporates indicators related to packaged food, such as climate change and food-to-packaging ratio, therefore, addressing key aspects of packaging sustainability. However, this research is limited to a specific type of packaging, and further studies are needed to determine how improving circularity indicators can effectively reduce environmental impact across the entire packaging life cycle. Additionally, more data is necessary to accurately estimate recycling output rates for different packaging types. Baratsas et al. [4] develop a *micro-level* framework designed to measure circularity across various economic sectors using metrics and indicators in categories derived from the CE goals. This framework was applied to companies in the food manufacturing sector, allowing the monitoring of circularity over time by tracking the overall circularity index. This study aims to cover indicators across all CE goals. However, to ensure equitable weighting, standardization is required. Additionally, the calculator should be adjusted to encompass all stages of the supply chain, not just the manufacturing phase. Despite the advances described, some CE goals like increasing the durability of products and reducing the use of natural resources are still less explored in the literature.

To achieve a comprehensive perspective on sustainability aligned with CE principles, Silvestri et al. [35] suggest the use of additional LCT tools to assess economic and social dimensions. These include Life Cycle Costing (LCC) to evaluate economic impacts and Social Life Cycle Assessment (S-LCA) to assess social impacts [36]. LCC differs from other tools like TEA, given that it considers all potential expenses associated with the project, offering a complete overview of the financial commitments over its entire life cycle, while TEA evaluates the economic viability of the process. Some economic impacts that can be found in LCC studies are related to initial, operational, and end-of-life costs, for instance, input, labor, and machinery costs [37]. Within S-LCA, social indicators related to subcategories like fair wages, health and safety, local employment, and water and land rights can be found in food-related studies like those in the sugar industry [38]. Furthermore, Falcone et al. [17] offer a comparative analysis of linear versus circular scenarios in olive oil production, incorporating LCA with Environmental Life Cycle Costing (ELCC), an LCT tool, along with the previous mentioned metric MCI. However, integrating the appropriate social and economic dimension, which are less extensively studied, into CE assessment remains an ongoing challenge.

Additionally, optimization techniques have been used as supplementary tools to evaluate trade-offs, identify sustainable value chains and CE strategies. These methods have been applied to areas such as reverse logistics, closed-loop systems, and green supply chains [17, 39]. Various optimization models have been proposed for efficient resource utilization in different phases of the FSC, including network design, waste reduction, production planning, and facility location, as shown in Table 9.1. Optimization based models can effectively integrate methodologies like LCA and TEA by considering different objective functions like minimizing CO_2 emissions and costs. Multi-objective optimization is required, since that it allows the optimization of various objective functions simultaneously. This approach can integrate all these indicators and, by utilizing techniques like the weighted sum method, the ϵ-constraint method, and various advanced algorithms, solve the model. Some studies extend their focus including circularity metrics like MCI and objectives like water, energy, and raw material consumption, waste generation, and most specific environmental impacts like marine ecotoxicity [1, 20]. Few studies address impacts from the perspective of all sustainability dimensions.

Table 9.1 Literature review of CE assessment in food supply chains.

Source	Methodology	Indicators	Application
Pagotto and Halog [42]	Material flow analysis (MFA)	Volume-based indicators, impact based-indicators, monetary indicators	Agri-food industry
Genovese et al. [43]	LCA, multi-regional input–output model	Emissions, waste recovered, resources use, and carbon maps	Waste cooking oil supply chain
Noya et al. [24]	LCA	LCA impact categories	Pork value chain
Pauer et al. [34]	Framework based on LCA and circularity	LCA impact categories and input and output circularity indicators	Food packaging
Baratsas et al. [4]	Framework based on Global Reporting Initiative standards	Metrics and indicators for waste, water, procurement, energy, emissions, and durability	Food production companies
Falcone et al. [17]	Framework based on LC methodologies and circularity indicators	LCA, Environmental Life Cycle Costing, and Material Circularity Indicator (MCI)	Closep-loop agri-food pathways: Olive Oil
Sazdovski et al. [44]	Comparison of circular metrics	MCI and Product Circular Chain	Food packaging
Oglethorpe [40]	Goal programming	Emissions, water consumption, sales, number of jobs, and fat content	Food supply chain: Sausage
Validi et al. [45]	Multi-objective optimization (MOO)	Emissions and costs	Milk distribution
Azadnia et al. [41]	MOO	Total cost and social and environmental scores	Packaging plastics
Baratsas et al. [1]	MOO	Material and water consumption, energy, waste, and emissions	Food (coffee) supply chain optimization
Sakamoto et al. [39]	MOO, LCA	Emissions and costs	Coffee supply chain
Karayılan et al. [20]	Optimization	Climate change, marine ecotoxicity, and MCI	Plastic packaging

Oglethorpe [40] evaluates key goals like emissions, water consumption, and number of jobs using goal programming. This approach sets specific targets and potential deviations, with the objective function aimed at minimizing the total deviations from these goals. Meanwhile, other studies define their own economic, environmental, and social scores [41]. Nevertheless, there is still no consensus on the most suitable CE indicators for evaluating circularity in FSC.

A major challenge for the transition of the food industry to circularity is the lack of standardized CE evaluation approach. Each study often develops its own methodology and metrics, complicating the comparison of findings across different research efforts. While existing literature addresses sustainability from various perspectives, there is still a significant gap in developing a comprehensive approach for benchmarking circularity within the food industry [46]. Consequently, there remains a critical need for a comprehensive and standardized assessment method that integrates all dimensions and spans the entire FSC. This would enable more effective evaluation, comparison, and improvement of circular practices across various FSC.

9.2.1 Example: CE Assessment for Food Packaging Waste Management Technologies

FSC studies primarily concentrate on production and distribution pathways, but there are also significant challenges related to food packaging and waste management. Plastic packaging plays a fundamental role in the food industry, avoiding food waste and facilitating food access [47]. For instance, Chialdikas et al. [48] adapt the CE assessment framework developed by Baratsas et al. [4] called "MICRON" (MIcro CirculaR ecOnomy iNdex) to measure the circularity of companies of different economic sectors, applying it to a case study on mechanical recycling, a key technology for food packaging recycling. Following the five CE goals defined by [49], the calculator is structured around five dimensions: waste, water and procurement, energy, emissions, and substitutability (Figure 9.1). Each dimension includes specific indicators and metrics that enable the measurement of circularity, ranging from 0 (linear) to 1 (circular). Then, an overall index is calculated to reflect the circularity of the technology.

The analysis covers four end cases: film to film, bottle to film, bottle to bottle, and film to bottle (Figure 9.2). The category with the highest circularity is emissions and spillages, while the least circular categories are water and procurement and energy, highlighting that mechanical recycling is a water and energy-intensive process. This assessment helps identify the areas that need improvement to create a more circular process. For each end case, the overall circularity scores are calculated, and the bottle to bottle and film to film are identified as the most circular cases based on their higher substitutability score.

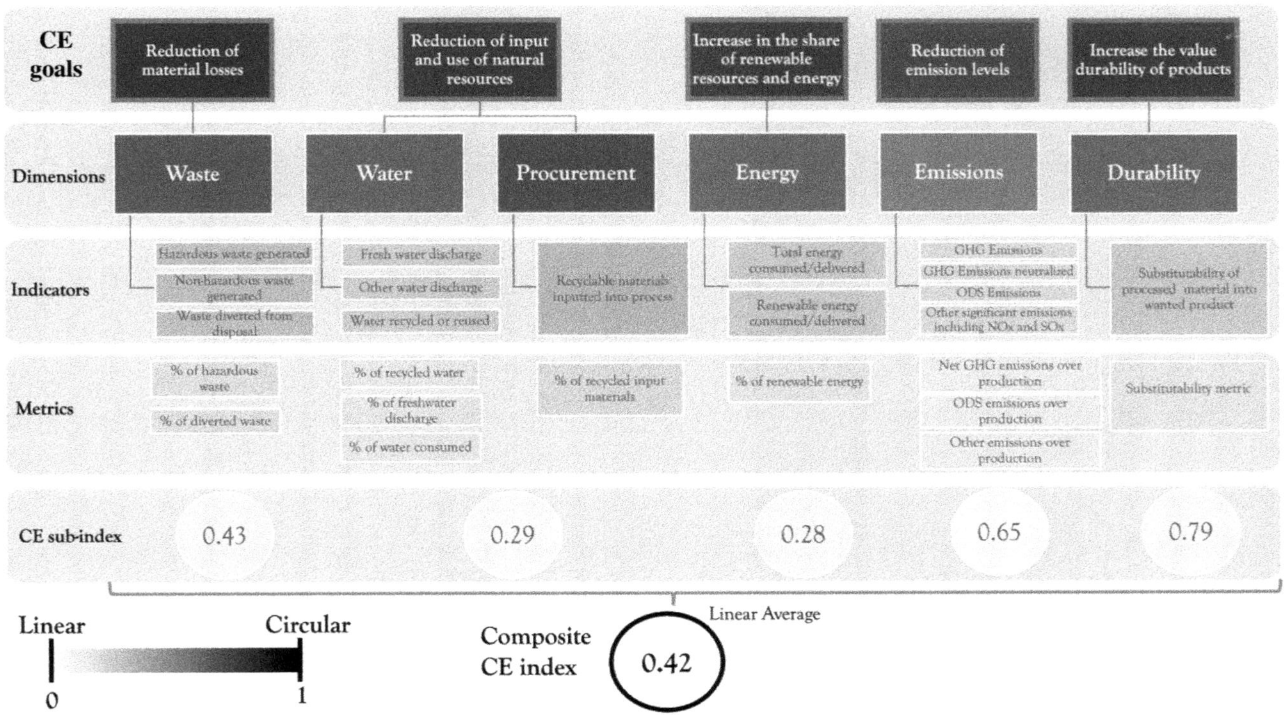

Figure 9.1 A representation of the CE goals, dimensions, indicators, and metrics defined as part of the MICRON framework used to assess circular economy of mechanical recycling of packaging.

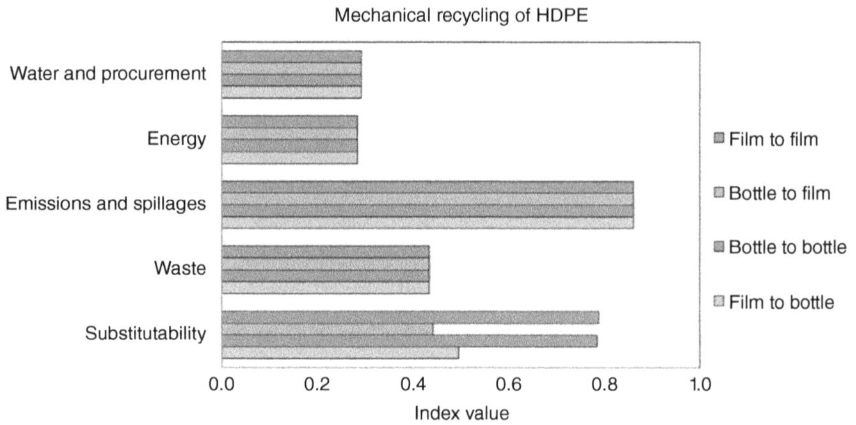

Figure 9.2 Circularity subindices calculated for the mechanical recycling of HDPE food packaging using MICRON.

9.3 Circular Economy Assessment in the Chemical Industry

Driven by rapid industrialization and population growth, the chemical sector is becoming the largest industrial energy consumer and the third-largest direct carbon dioxide emitter. It remains heavily reliant on oil and gas feedstocks, with over 35% of its process energy derived from coal [50]. Transitioning to a CE is essential not only to reduce fossil fuel dependence and stabilize markets amid geopolitical and economic fluctuations but also to address the sector's substantial environmental impact. By prioritizing the reuse, recycling, and regeneration of materials, CE reduces the demand for raw resources, lowers greenhouse gas emission (GHG), and diminishes the sector's carbon footprint. This shift enhances energy efficiency, as circular processes often require less energy than extracting and processing virgin materials, leading to significant reductions in energy consumption. Economically, CE fosters resilience by creating sustainable business opportunities like waste recovery and valorization, and drives innovation in product design and manufacturing, enhancing competitiveness and reducing costs. Socially, CE improves community well-being by creating jobs and promoting healthier environments by reducing pollution. However, despite these benefits, the chemical sector's reliance on fossil fuels and linear economic models continues to impede progress toward minimizing environmental impacts.

Chemical industries face challenges in quantifying sustainability metrics due to the diverse nature of chemical operations that include basic chemicals, specialty chemicals, polymers, agrochemicals, and pharmaceuticals, each requiring tailored indicators. To address this, the chemical sector must first define indicators and build CE frameworks that effectively measure the circularity of their processes. Keijer et al. [51] propose the 12 principles of circular chemistry as a guideline for chemical industries towards CE. This framework was developed in response to the growing awareness of the finite nature of resources and the need for a sustainable chemical industry that can optimize resource efficiency across the entire life cycle of chemical products. These principles focus on designing processes that reuse and recycle resources, transform waste into marketable products, and maximize resource conservation. Circular chemistry emphasizes energy efficiency, non-toxicity, and innovations that enable on-site reuse while also considering optimal end-of-life designs. It promotes the 11R framework (reject, reduce, reuse, redistribute, repair, refurbish, repurpose, remanufacture, recycle, recover, return) to achieve the highest end-of-life outcomes. It also encourages service-based business models like chemical leasing and foster collaboration between industry and policymakers [51]. Therefore, building CE assessments that align with these guidelines is key to transition to CE.

There are consistent challenges in achieving CE across the chemical sector. The lack of sector-specific indicators that address all the goals of CE hinders accurate circularity evaluation, as tailored metrics are essential for meaningful comparisons among alternative chemical processes and systems. Velasco-Muoz et al. [52] classify 41 different indicators that are used to measure the circularity performance of agricultural systems in CE strategies and sustainability dimensions. However, the authors observe that most indicators are heavily weighted toward technical and environmental aspects, with fewer indicators focusing on economic and social dimensions. The authors conclude that there is an urgent need to develop new sets of indicators that can better reflect the variety of activities and processes within the agricultural production systems. These new indicators should be able to measure circularity across different scales (e.g. farm, regional, and national) and consider the full spectrum of sustainability dimensions. Meanwhile, Liang et al. [53] center their assessment on indicators specific to the phosphorus production processes within chemical firms, incorporating the economic and environmental dimensions of CE. For the economic aspects, they propose indicators such as resource consumption per unit of industrial added value, energy consumption per unit of industrial added value, and comprehensive utilization of associated resources. On the environmental side, they include indicators like the loss rate of phosphate ore, dilution rate of phosphate ore, solid waste recycling rate, wastewater recycling rate, exhaust cycling rate, and investment in environmental governance. The assessment did not include social indicators on its framework. The selected indicators were chosen because phosphorus production is resource-intensive and generates substantial waste and emissions, making efficient management crucial for this non-renewable resource that is vital for agriculture.

Another challenge that is rarely addressed is the development of comprehensive CE assessments in the chemical sector. Few studies explore this, often concentrating on particular chemical industries and overlooking the full range of CE dimensions. Liang et al. [53] develop a tailored evaluation framework for phosphorus chemical firms, focusing on specific aspects of CE performance. The framework uses a ranking method (VIKOR-QUALIFLEX) to assess sustainability benefits, including resource conservation, energy efficiency, waste recycling rates, emission reduction, and the implementation of CE regulations within firms. However, it does not provide a comprehensive CE assessment, as it does not cover the social aspects that would be necessary for a broader evaluation of CE. In the specialty chemical industry, Fadel and Tarabieh [54] develop a composite index for solvent-based processes in large-scale chemical and pharmaceutical plants to assess sustainability, integrating LCA with the 12 principles of Green Chemistry. This index provides metrics aligned with the CE goals, guiding industries in making decisions that promote circular chemistry. While the index includes environmental metrics, the economic and social dimensions of CE are not explored. Baratsas et al. [8] introduce the MICRON framework, a comprehensive methodology designed to assess circularity through sector-specific indicators, as described in Section 9.2. In their case study on an industrial gas company, they apply this framework to evaluate the company's circularity over five years, using a circularity index that resulted from a diverse set of indicators tailored to the industrial sector. This case study demonstrates the practical utility of this framework in providing a holistic assessment of a company's circularity, identifying both improvements and ongoing challenges. Other industries should consider implementing a similar circular framework to effectively track and enhance their sustainability performance.

Another prevalent problem is the lack of robust resource efficiency and recycling techniques that align with circular chemistry principles across the chemical sector. In the basic chemicals industry, Aurisano et al. [55] highlight the significant environmental and health challenges posed by plastics, particularly due to over 1518 potential chemical additives of concern, and only a fraction has been adequately assessed for toxicity and exposure pathways. To address these challenges, the authors propose developing a global open-access database for chemicals in plastics to provide essential data for LCA, risk assessment, chemical management, and the promotion of safer alternatives. In the pharmaceutical manufacturing (PM), the traditional waste management techniques that focus on reducing waste after it is generated lack the proactive strategy to prevent resource overconsumption and energy inefficiencies by minimizing initial resource use. Recognizing these gaps, Liang et al. [53] design a sector-specific 9R framework, the PM9R, which adapts the generic 9R model to address the unique challenges of PM, such as the reclamation of solvents and efficient resource management, thereby prioritizing CE.

To design closed-loop chemical industries that effectively minimize resource input, waste, emissions, and energy consumption and increase the value durability and the share of renewable resources within regenerative systems, it is crucial to optimize multiple dimensions of the CE. In the hydrogen production industry, Kabir et al. [56] investigate the use of machine learning (ML)-based optimization models to enhance resource efficiency in green hydrogen production from water industries. Their study applies ML techniques, such as Random Forest and K-Nearest Neighbors, to predict and optimize the efficiency of hydrogen production processes like dark fermentation and proton exchange membrane electrolysis. By identifying and fine-tuning key process parameters such as chemical oxygen demand, temperature, and catalyst loadings-the models significantly improve resource utilization and reduce waste generation. In the phosphorus chemical industry, challenges related to resource efficiency and environmental impacts have been identified, particularly concerning the low utilization rates and high waste production associated with phosphorus processing. Therefore, Ma et al. [57] propose an optimization model for this case study focusing on maximizing economic benefit while imposing constraints on CE driven metrics such as resource efficiency ratios and recycling rates.

Transitioning the chemical sector to a CE is impeded by inconsistent regulations, a gap between academic research and industry practices, and a lack of standardized, global frameworks. The sector often prioritizes waste management over other circular strategies, highlighting the need for stronger collaboration and standardized metrics. Developing robust frameworks that integrate circular chemistry principles with the triple bottom line is crucial for driving sustainable growth and ensuring the future of the chemical industry. Moreover, integrating metrics into optimization models to design closed-loop processes in the chemical sector is essential for maximizing resource efficiency, reducing environmental impact, and achieving sustainable economic growth.

9.3.1 Example: Circular Economy Assessment of Viable for Fuels for Mobility

The increasing GHG emissions and the potential depletion of lithium due to the growing adoption of battery electric vehicles highlight the urgent need for alternative sustainable transportation fuels. Addressing this challenge, Martinez et al. [58] apply the MICRON framework proposed by Baratsas et al. [4] to evaluate the sustainability of various fuel production pathways, as shown in Figure 9.3. This graph shows the circularity score of various fuel production

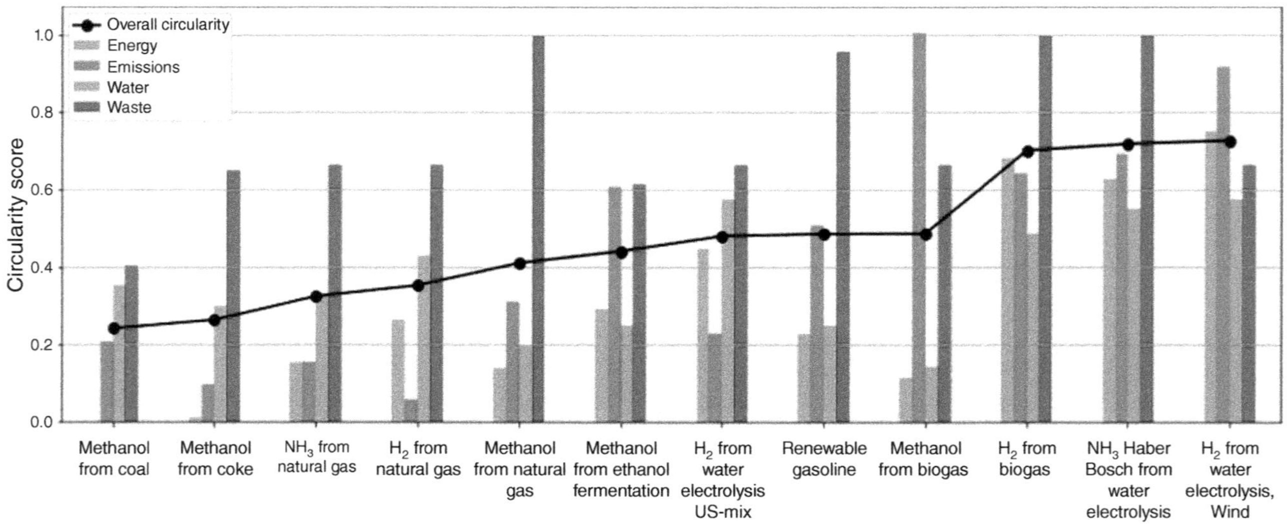

Figure 9.3 Circularity framework for fuel production technologies [4, 59–79].

pathways, where each pathway is evaluated based on category-based indexes for the energy consumption, emissions, water usage, and waste generation categories. The overall circularity trend (black line) indicates an increase from fossil-based methods like methanol from coal to more sustainable options like hydrogen from water electrolysis with wind energy, suggesting that renewable energy-based processes generally achieve higher circularity scores compared to traditional fossil fuel-based methods. The study primarily examines fuel production, with plans to extend the framework to include fuel transportation, storage, and usage, providing a more holistic assessment of life cycle impacts.

9.4 Circular Economy Metrics for Energy Systems

CE plays a key role in the energy transition by promoting resource efficiency and reducing environmental impact throughout the lifecycle of energy systems [80]. CE measures offer multifaceted benefits beyond decarbonization and achieving net zero emission goals. They also stimulate economic growth, promote job creation, promote environmental justice, reduce pollution, and improve supply chain security [81]. Current manufacturing processes for renewable technologies, such as photovoltaic modules, are highly dependent on carbon-intensive methods [82, 83]. Decarbonization and implementing circular practices within these technologies are imperative to facilitate a sustainable energy transition and enhance the resilience of urban energy systems. This involves minimizing carbon emissions during production, optimizing resource efficiency through innovative product design, extending the useful life of the product, and improving end-of-life recycling and reuse strategies [84]. A CE approach will also help determine whether we are indeed moving toward a CE by reducing the consumption of fossil fuels and increasing renewable energy.

One of the significant challenges in promoting sustainability within energy systems is accurately measuring circularity, a key metric for tracking progress, but one that is difficult to implement effectively [85]. Potting et al. [86] emphasize the importance of developing appropriate indicators for evaluating environmental conditions and industrial performance. Energy-specific indicators are crucial for decision-makers and policymakers to support the adoption of CE practices in energy systems and to monitor their impact [9, 87]. Various approaches and metrics have been proposed to measure different aspects of CE in the context of energy systems. Several studies have conducted economic and environmental evaluations, focusing on their planning, design, management, and operation, as seen in Table 9.2. However, these studies often focus on cost and carbon emissions while neglecting other effects, assuming that increased circularity always equals reduced environmental impacts, which is not always accurate [88, 89].

Table 9.2 Literature review for objective functions in energy systems.

Source/developer	Objective function	Application
Meza et al. [93]	Costs, CO_2 emissions, and fuel price risks	Power generation expansion planning of electric systems
Bernal-Agustín and Dufo-López [94]	Total cost, CO_2 emissions, and unmet load	Design of isolated hybrid systems (PV–wind–diesel hydrogen battery)
Zheng et al. [95]	Cost, CO_2 emissions, and the annual energy bill	Operational planning of a CCHP system
Eriksson and Gray [96]	Costs, loss of power supply probability, carbon footprint, and sociopolitical index	Simulate and optimize a renewable energy system
Al Shidhani et al. [97]	Costs, carbon emissions, land use, and social opposition	Power system expansion planning
Mayer et al., 2020	Net present costs and environmental footprint	Hybrid renewable energy systems
Serrano-Arévalo et al. [98]	Total cost, harmful emissions (CO_2, NO_x, SO_2, and particle emissions), and water consumption	Planning of an energy system (conventional and renewable technologies)
Chen et al. [99]	Costs, CO_2 emissions, and water consumption	Optimal design of integrated urban energy systems
Cook et al. [100]	Costs, water consumption, and land use	Optimal design of renewable energy systems
Horasan and Kilic [101]	Costs, CO_2 emissions, technical score, and job creation	Renewable energy planning
Serrano-Arévalo et al. [102]	Total cost, CO_2 emissions, and CE indicators	Design and operation of an electrical supply system

The extraction of raw materials and waste generated in energy systems have enormous environmental and socioeconomic impacts, including climate change, loss of biodiversity, depletion of natural resources, and pollution [87]. For example, implementing and managing a 100 MW wind power plant involves the consumption of 67,850 t of materials, including metals, composites, and polymers, which pose challenges for recycling [90]. As a result, most wind turbine waste ends up in landfills or incinerators, leading to resource loss and environmental impacts [91]. Current research is exploring alternative end-of-life management solutions that incorporate CE principles, although comprehensive studies on the circularity and environmental sustainability of wind turbines are lacking [92].

Recent work attempts to address these gaps. Baratsas et al. [4] develop a *micro-level* CE evaluation framework, "MICRON," resulting in a comprehensive circularity index that is applied to the utilities and energy sector, among others, and can be adapted to different applications. Cook et al. [100] develop a multi-scale mathematical modeling and optimization framework capable of holistically addressing energy–water–land nexus interactions in renewable energy systems in the context of infrastructure planning. While this work effectively integrates these interactions, a limitation of the framework is its lack of a comprehensive sustainability index that incorporates broader CE principles, such as material circularity and waste reduction, which are essential for thoroughly assessing the sustainability of renewable energy systems. Serrano-Arévalo et al. [102] propose three new indices to measure CE: relative material impact, absolute material impact, and energy self-sufficiency. These indicators account for the material requirements of the infrastructure participating in the energy system and analyze the relationship between producer and consumer systems, specifying energy self-sufficiency. In addition, the quantification of materials is normalized through the ADP to compare the depletion of *non-renewable* resources. However, the gap in this work lies in the potential need to integrate broader CE goals, such as waste reduction, product life extension, and reuse/recycling of materials. Čuček et al. [103] propose a dimensionality reduction method for a multi-objective optimization problem that considers multiple environmental footprints, such as carbon, energy, water, water pollution, and land footprint, but misses certain CE goals, such as the comprehensive integration of material circularity and product life extension. The method is based on similarities among the total footprints, resulting in a number of representative footprints. Although this avoids subjective weighting of footprints, a disadvantage is that it can oversimplify the complexity of interactions between different environmental impacts. Addressing these gaps is crucial for understanding how CE principles can be effectively implemented and measured.

CE is indispensable for achieving a sustainable energy transition, providing a comprehensive framework to enhance resource efficiency and mitigate environmental impacts throughout the life cycle of energy systems. Recent advances in CE evaluation frameworks demonstrate a growing emphasis on quantifying and optimizing the circularity of energy infrastructures. These efforts are essential for reducing carbon emissions, meeting net-zero emissions targets, and ensuring broader economic, environmental, and social benefits. However, as the implementation of CE in energy systems continues to evolve, it is vital to develop more refined and comprehensive metrics to address the complexities of sustainability beyond just decarbonization. This holistic approach will be crucial for steering the energy sector toward a more resilient and sustainable future.

References

1 Baratsas, S.G., Pistikopoulos, E.N., and Avraamidou, S. (2021c). A systems engineering framework for the optimization of food supply chains under circular economy considerations. *Science of the Total Environment* 794: 148726. https://doi.org/10.1016/j.scitotenv.2021.148726.

2 Lloret, A. (2016). Modeling corporate sustainability strategy. *Journal of Business Research* 69 (2): 418–425. https://doi.org/10.1016/j.jbusres.2015.06.047.

3 Avraamidou, S., Baratsas, S.G., Tian, Y., and Pistikopoulos, E.N. (2020). Circular economy –a challenge and an opportunity for process systems engineering. *Computers & Chemical Engineering* 133: 106629. https://doi.org/10.1016/j.compchemeng.2019.106629.

4 Baratsas, S.G., Pistikopoulos, E.N., and Avraamidou, S. (2022). A quantitative and holistic circular economy assessment framework at the micro level. *Computers & Chemical Engineering* 160: 107697. https://doi.org/10.1016/j.compchemeng.2022.107697.

5 Hassan, M., Saraç, M., and Alam, A. (2020). *Circular Economy, Sustainable Development and the Role of Islamic Finance*, 1–26. Istanbul University Press https://doi.org/10.26650/b/ss10.2020.017.01.

6 Kirchherr, J., Reike, D., and Hekkert, M. (2017). Conceptualizing the circular economy: an analysis of 114 definitions. *Resources, Conservation and Recycling* 127: 221–232. https://doi.org/10.1016/j.resconrec.2017.09.005.

7 Silk, D., Mazzali, B., Gargalo, C.L. et al. (2020). A decision-support framework for techno-economic-sustainability assessment of resource recovery alternatives. *Journal of Cleaner Production* 266: 121854. https://doi.org/10.1016/j.jclepro.2020.121854.

8 Baratsas, S.G., Masoud, N., Pappa, V.A. et al. (2021). Towards a circular economy calculator for measuring the "circularity" of companies. In: *Computer Aided Chemical Engineering* (ed. M. Türkay and R. Gani), 1547–1552. Elsevier https://doi.org/10.1016/B978-0-323-88506-5.50239-4.

9 Saidani, M., Yannou, B., Leroy, Y. et al. (2019). A taxonomy of circular economy indicators. *Journal of Cleaner Production* 207: 542–559. https://doi.org/10.1016/j.jclepro.2018.10.014.

10 Elia, V., Gnoni, M.G., and Tornese, F. (2017). Measuring circular economy strategies through index methods: a critical analysis. *Journal of Cleaner Production* 142: 2741–2751. https://doi.org/10.1016/j.jclepro.2016.10.196.

11 Moraga, G., Huysveld, S., Mathieux, F. et al. (2019). Circular economy indicators: what do they measure? *Resources, Conservation and Recycling* 146: 452–461. https://doi.org/10.1016/j.resconrec.2019.03.045.

12 Sassanelli, C., Rosa, P., Rocca, R., and Terzi, S. (2019). Circular economy performance assessment methods: a systematic literature review. *Journal of Cleaner Production* 229: 440–453. https://doi.org/10.1016/j.jclepro.2019.05.019.

13 Elia, V., Gnoni, M.G., and Tornese, F. (2024). Integrating circular economy and sustainability assessment on the micro-level: an umbrella review. *Sustainable Production and Consumption* 50: 405–415. doi: 10.1016/j.spc.2024.08.012.

14 Muoz, S., Hosseini, M.R., and Crawford, R.H. (2024). Towards a holistic assessment of circular economy strategies: the 9R circularity index. *Sustainable Production and Consumption* 47: 400–412. https://doi.org/10.1016/j.spc.2024.04.015.

15 Miatto, A., Emami, N., Goodwin, K. et al. (2024). Australia's circular economy metrics and indicators. *Journal of Industrial Ecology* 28: 216–231. https://doi.org/10.1111/jiec.13458.

16 FAO (2022). Technical platform on the measurement and reduction of food loss and waste, 9.

17 Falcone, G., Stillitano, T., Iofrida, N. et al. (2022). Life cycle and circularity metrics to measure the sustainability of closed-loop agri-food pathways. *Frontiers in Sustainable Food Systems* 6: 12. https://doi.org/10.3389/fsufs.2022.1014228.

18 Feng, B., Zheng, C., Zhang, W. et al. (2020). Analyzing the role of spatial features when cooperating hyperspectral and lidar data for the tree species classification in a subtropical plantation forest area. *Journal of Applied Remote Sensing* 14 (1): 4. https://doi.org/10.1117/1.JRS.14.022213.

19 Saif, Y., Rizwan, M., Almansoori, A., and Elkamel, A. (2017). A circular economy solid waste supply chain management based approach under uncertainty. *Energy Procedia* 142: 2971–2976. https://doi.org/10.1016/j.egypro.2017.12.316.

20 Karayılan, S., Yılmaz, Ö., Uysal, Ç., and Naneci, S. (2021). Prospective evaluation of circular economy practices within plastic packaging value chain through optimization of life cycle impacts and circularity. *Resources, Conservation and Recycling* 173 (10): 105691. https://doi.org/10.1016/j.resconrec.2021.105691.

21 Haas, W., Krausmann, F., Wiedenhofer, D., and Heinz, M. (2015). How circular is the global economy? An assessment of material flows, waste production, and recycling in the European Union and the world in 2005. *Journal of Industrial Ecology* 19: 765–777. https://doi.org/10.1111/jiec.12244.

22 Nazir, N. (2017). Understanding life cycle thinking and its practical application to agri-food system. *International Journal on Advanced Science, Engineering and Information Technology* 7 (5): 1861–1870.

23 Notarnicola, B., Tassielli, G., Renzulli, P.A., and Monforti, F. (2017). Energy flows and greenhouses gases of EU (European Union) national breads using an LCA (life cycle assessment) approach. *Journal of Cleaner Production* 140: 455–469.

24 Noya, I., Aldea, X., González-Garcıa, S. et al. (2017). Environmental assessment of the entire pork value chain in Catalonia –a strategy to work towards circular economy. *Science of the Total Environment* 589: 122–129. https://doi.org/10.1016/j.scitotenv.2017.02.186.

25 Siracusa, V., Ingrao, C., Giudice, A.L. et al. (2014). Environmental assessment of a multilayer polymer bag for food packaging and preservation: an LCA approach. *Food Research International* 62: 151–161. https://doi.org/10.1016/j.foodres.2014.02.010.

26 Toniolo, S., Mazzi, A., Niero, M. et al. (2013). Comparative LCA to evaluate how much recycling is environmentally favourable for food packaging. *Resources, Conservation and Recycling* 77: 61–68. https://doi.org/10.1016/j.resconrec.2013.06.003.

27 Omolayo, Y., Feingold, B.J., Neff, R.A., and Romeiko, X.X. (2021). Life cycle assessment of food loss and waste in the food supply chain. *Resources, Conservation and Recycling* 164: 105119.

28 Krishnan, R., Agarwal, R., Bajada, C., and Arshinder, K. (2020). Redesigning a food supply chain for environmental sustainability –an analysis of resource use and recovery. *Journal of Cleaner Production* 242: 118374. https://doi.org/10.1016/j.jclepro.2019.118374.

29 Manfredi, M. and Vignali, G. (2014). Life cycle assessment of a packaged tomato puree: a comparison of environmental impacts produced by different life cycle phases. *Journal of Cleaner Production* 73: 275–284. https://doi.org/10.1016/j.jclepro.2013.10.010.

30 Neha, S., Ramesh, K.P.K., and Remya, N. (2022). Techno-economic analysis and life cycle assessment of microwave co-pyrolysis of food waste and low-density polyethylene. *Sustainable Energy Technologies and Assessments* 52: 102356. https://doi.org/10.1016/j.seta.2022.102356.

31 del Carmen Munguıa-López, A., Göreke, D., Sánchez-Rivera, K.L. et al. (2023). Quantifying the environmental benefits of a solvent-based separation process for multilayer plastic films. *Green Chemistry* 25: 1611–1625. https://doi.org/10.1039/d2gc04262b.

32 Ellen MacArthur Foundation (2024). Material circularity indicator. https://www.ellenmacarthurfoundation.org/material-circularity-indicator (accessed 30 August 2024).

33 Gallo, F., Manzardo, A., Camana, D. et al. (2023). Integration of a circular economy metric with life cycle assessment: methodological proposal of compared agri-food products. *International Journal of Life Cycle Assessment* 8: 1359–1379. https://doi.org/10.1007/s11367-022-02130-0.

34 Pauer, E., Wohner, B., Heinrich, V., and Tacker, M. (2019). Assessing the environmental sustainability of food packaging: an extended life cycle assessment including packaging-related food losses and waste and circularity assessment. *Sustainability (Switzerland)* 11 (2): 925. https://doi.org/10.3390/su11030925.

35 Silvestri, C., Silvestri, L., Piccarozzi, M., and Ruggieri, A. (2022). Toward a framework for selecting indicators of measuring sustainability and circular economy in the agri-food sector: a systematic literature review. *International Journal of Life Cycle Assessment* 29: 1446–1484. https://doi.org/10.1007/s11367-022-02032-1.

36 UNEP (2020). Guidelines for social life cycle assessment of products and organizations 2020.

37 Degieter, M., Gellynck, X., Goyal, S. et al. (2022). Life cycle cost analysis of agri-food products: a systematic review. *Science of the Total Environment* 850: 158012. https://doi.org/10.1016/j.scitotenv.2022.158012.

38 Prasara-A, J. and Gheewala, S.H. (2018). Applying social life cycle assessment in the Thai sugar industry: challenges from the field. *Journal of Cleaner Production* 172: 335–346. https://doi.org/10.1016/j.jclepro.2017.10.120.

39 Sakamoto, H., Bruschi, L.T., Kulay, L., and Yamakami, A. (2023). Using the life cycle approach for multiobjective optimization in the context of the green supply chain: a case study of Brazilian coffee. *Sustainability* 15 (9): https://doi.org/10.3390/su151813987.

40 Oglethorpe, D. (2010). Optimising economic, environmental, and social objectives: a goal-programming approach in the food sector. *Environment and Planning A* 42: 1239–1254. https://doi.org/10.1068/a42292.

41 Azadnia, A.H., Mat Saman, M.Z., and Wong, K.Y. (2015). Sustainable supplier selection and order lot-sizing: an integrated multi-objective decision-making process. *International Journal of Production Research* 53: 383–408. https://doi.org/10.1080/00207543.2014.935827.

42 Pagotto, M. and Halog, A. (2016). Towards a circular economy in Australian agri-food industry: an application of input-output oriented approaches for analyzing resource efficiency and competitiveness potential. *Journal of Industrial Ecology* 20: 1176–1186. https://doi.org/10.1111/jiec.12373.

43 Genovese, A., Acquaye, A.A., Figueroa, A., and Lenny Koh, S.C. (2017). Sustainable supply chain management and the transition towards a circular economy: evidence and some applications. *Omega (United Kingdom)* 66: 344–357. https://doi.org/10.1016/j.omega.2015.05.015.

44 Sazdovski, I., Batlle-Bayer, L., Bala, A. et al. (2024). Comparative assessment of two circularity indicators for the case of reusable versus single-use

secondary packages for fresh foods in Spain. *Heliyon* 10 (3): e27922. https://doi.org/10.1016/j.heliyon.2024.e27922.
45 Validi, S., Bhattacharya, A., and Byrne, P.J. (2014). A case analysis of a sustainable food supply chain distribution system –a multi-objective approach. *International Journal of Production Economics* 152: 71–87. https://doi.org/10.1016/j.ijpe.2014.02.003.
46 de Koning, I., Kassahun, A., and Tekinerdogan, B. (2024). Benchmarking circularity in supply chains: a systematic literature review. *Journal of Environmental Management* 366: 121676.
47 Munoz-Briones, P. A., Munguia-Lopez, A. D. C., Sanchez-Rivera, K. L., et al. (2024). Optimal design of food packaging considering waste management technologies to achieve circular economy. In *Foundations of Computer Aided Process Design*, 820–828. https://doi.org/10.69997/sct.154335.
48 Chialdikas, E., del Carmen Munguia-Lopez, A., Aguirre-Villegas, H., and Avraamidou, S. (2023). *A Framework for the Evaluation of the Circularity of Plastic Waste Management Systems: A Case Study on Mechanical Recycling of HDPE*. Foundations of Computer Aided Process Operations/Chemical Process Control.
49 Reichel, A., Schoenmaker, M. D., and Gillabel, J. (2016). Circular economy in Europe: developing the knowledge base. Report 2/2016, European Environment Agency.
50 International Energy Agency 2023. Chemicals. https://www.iea.org/energy-system/industry/chemicals (accessed 22 August 2023).
51 Keijer, T., Bakker, V., and Chris Slootweg, J. (2019). Circular chemistry to enable a circular economy. *Nature Chemistry* 11 (3): 190–195. https://doi.org/10.1038/s41557-019-0226-9.
52 Velasco-Muoz, J.F., Mendoza, J.M.F., Aznar-Sánchez, J.A., and Gallego-Schmid, A. (2021). Circular economy implementation in the agricultural sector: definition, strategies and indicators. *Resources, Conservation and Recycling* 170: 105618. https://doi.org/10.1016/j.resconrec.2021.105618.
53 Liang, W., Zhao, G., and Hong, C. (2018). Performance assessment of circular economy for phosphorus chemical firms based on vikor-qualiflex method. *Journal of Cleaner Production* 196: 1365–1378. https://doi.org/10.1016/j.jclepro.2018.06.099.
54 Fadel, C. and Tarabieh, K. (2019, 2019). Development of an Industrial Environmental Index to assess the sustainability of industrial solvent-based processes. *Resources* 8 (2): 115. https://doi.org/10.3390/resources8020115.
55 Aurisano, N., Weber, R., and Fantke, P. (2021). Enabling a circular economy for chemicals in plastics. *Current Opinion in Green and Sustainable Chemistry* 31: 100513. https://doi.org/10.1016/j.cogsc.2021.100513.

56 Kabir, M.M.R., Roy, S.K., Alam, F. et al. (2023). Machine learning-based prediction and optimization of green hydrogen production technologies from water industries for a circular economy. *Desalination* 567: 116992. https://doi.org/10.1016/j.desal.2023.116992.

57 Ma, S., Shu, H., Chen, D., and Zhu, B. (2015). A case study of a phosphorus chemical firm's application of resource efficiency and eco-efficiency in industrial metabolism under circular economy. *Journal of Cleaner Production* 87: 839–849. https://doi.org/10.1016/j.jclepro.2014.10.079.

58 Martinez, K.J., Munoz-Briones, P.A., and Avraamidou, S. (2024). Circular economy assessment of viable for fuels for mobility. In: *34th European Symposium on Computer Aided Process Engineering/15th International Symposium on Process Systems Engineering, volume 53 of Computer Aided Chemical Engineering* (ed. F. Manenti and G.V. Reklaitis), 2335–2340. Elsevier https://doi.org/10.1016/B978-0-443-28824-1.50390-2.

59 Akhtar, M.S. and Liu, J.J. (2022). Life cycle assessment of green hydrogen transportation and distribution pathways. In: *Computer Aided Chemical Engineering*, vol. 49 (ed. Y. Yamashita and M. Kano), 1897–1902. Elsevier https://doi.org/10.1016/B978-0-323-85159-6.50316-X.

60 Bicer, Y. and Dincer, I. (2018). Life cycle assessment of ammonia utilization in city transportation and power generation. *Journal of Cleaner Production* 170: 1594–1601. https://doi.org/10.1016/j.jclepro.2017.09.243.

61 Bicer, Y., Dincer, I., Vezina, G., and Raso, F. (2017). Impact assessment and environmental evaluation of various ammonia production processes. *Environmental Management* 59 (5): 842–855. https://doi.org/10.1007/s00267-017-0831-6.

62 Chatterjee, S., Parsapur, R.K., and Huang, K.-W. (2021). Limitations of ammonia as a hydrogen energy carrier for the transportation sector. *ACS Energy Letters* 6 (12): 4390–4394. https://doi.org/10.1021/acsenergylett.1c02189.

63 De Abreu, V.H.S., Da Costa, M.G., Da Costa, V.X. et al. (2022). The role of the circular economy in road transport to mitigate climate change and reduce resource depletion. *Sustainability* 14 (14): 14. https://doi.org/10.3390/su14148951.

64 Eggemann, L., Escobar, N., Peters, R. et al. (2020). Life cycle assessment of a small-scale methanol production system: a power-to-fuel strategy for biogas plants. *Journal of Cleaner Production* 271: 122476. https://doi.org/10.1016/j.jclepro.2020.122476.

65 Hajjaji, N., Martinez, S., Trably, E. et al. (2016). Life cycle assessment of hydrogen production from biogas reforming. *International Journal of Hydrogen Energy* 41 (14): 6064–6075. https://doi.org/10.1016/j.ijhydene.2016.03.006.

66 Henriksen, M.S., Matthews, H.S., White, J. et al. (2024). Tradeoffs in life cycle water use and greenhouse gas emissions of hydrogen production pathways. *International Journal of Hydrogen Energy* 49: 1221–1234. https://doi.org/10.1016/j.ijhydene.2023.08.079.

67 Ji, M. and Wang, J. (2021). Review and comparison of various hydrogen production methods based on costs and life cycle impact assessment indicators. *International Journal of Hydrogen Energy* 46 (78): 38612–38635. https://doi.org/10.1016/j.ijhydene.2021.09.142.

68 Kang, D. and Han, J. (2021). Lifecycle assessment of methanol production from blast furnace gas. *Environmental Science and Pollution Research* 28 (43): 61601–61607. https://doi.org/10.1007/s11356-021-15063-2.

69 Li, C., Jia, T., Wang, H. et al. (2023a). Assessing the prospect of deploying green methanol vehicles in china from energy, environmental and economic perspectives. *Energy* 263: 125967. https://doi.org/10.1016/j.energy.2022.125967.

70 Li, J., Ma, X., Liu, H., and Zhang, X. (2018). Life cycle assessment and economic analysis of methanol production from coke oven gas compared with coal and natural gas routes. *Journal of Cleaner Production* 185: 299–308. https://doi.org/10.1016/j.jclepro.2018.02.100.

71 Li, M., Ming, P., Huo, R., and Mu, H. (2023b). Economic assessment and comparative analysis of hydrogen transportation with various technical processes. *Journal of Renewable and Sustainable Energy* 15 (2): 025904. https://doi.org/10.1063/5.0141098.

72 Liu, X., Elgowainy, A., and Wang, M. (2020). Life cycle energy use and greenhouse gas emissions of ammonia production from renewable resources and industrial by-products. *Green Chemistry* 22 (17): 5751–5761. https://doi.org/10.1039/d0gc02301a.

73 Mann, M. and Spath, P. (2004). Life cycle assessment of renewable hydrogen production via wind/electrolysis: milestone completion report (NREL/MP-560-35404). National Renewable Energy Lab., Golden, CO (US). https://www.osti.gov/biblio/15006927. Accessed: 2024-08-28.

74 Matzen, M. and Demirel, Y. (2016). Methanol and dimethyl ether from renewable hydrogen and carbon dioxide: alternative fuels production and life-cycle assessment. *Journal of Cleaner Production* 139: 1068–1077. https://doi.org/10.1016/j.jclepro.2016.08.163.

75 Saidani, M., Yannou, B., Leroy, Y., and Cluzel, F. (2018). Heavy vehicles on the road towards the circular economy: analysis and comparison with the automotive industry. *Resources, Conservation and Recycling* 135: 108–122. https://doi.org/10.1016/j.resconrec.2017.06.017.

76 Spath, P. L. and Mann, M. K. (2000). Life cycle assessment of hydrogen production via natural gas steam reforming (NREL/TP-570-27637, 764485; p. NREL/TP-570-27637, 764485). https://doi.org/10.2172/764485.

77 Spath, P. L. and Mann, M. K. (2004). Life cycle assessment of renewable hydrogen production via wind/electrolysis: milestone completion report. National Renewable Energy Lab., Golden, CO (US).

78 Vidal-Barrero, F., Baena-Moreno, F.M., Preciado-Cárdenas, C. et al. (2022). Hydrogen production from landfill biogas: profitability analysis of a real case study. *Fuel* 324: 124438. https://doi.org/10.1016/j.fuel.2022.124438.

79 Zupko, R. (2019). Life cycle assessment of the production of gasoline and diesel from forest residues using integrated hydropyrolysis and hydroconversion. *The International Journal of Life Cycle Assessment* 24 (10): 1793–1804. https://doi.org/10.1007/s11367-019-01616-8.

80 Tiwari, S., Mohammed, K.S., Mentel, G. et al. (2024). Role of circular economy, energy transition, environmental policy stringency, and supply chain pressure on CO2 emissions in emerging economies. *Geoscience Frontiers* 15 (3): 101682. https://doi.org/10.1016/j.gsf.2023.101682.

81 Schröder, P., Anggraeni, K., and Weber, U. (2019). The relevance of circular economy practices to the sustainable development goals. *Journal of Industrial Ecology* 23 (1): 77–95. https://doi.org/10.1111/jiec.12732.

82 Mirletz, H., Ovaitt, S., Sridhar, S., and Barnes, T.M. (2022). Circular economy priorities for photovoltaics in the energy transition. *PLoS One* 17: 1–21. https://doi.org/10.1371/journal.pone.0274351.

83 Mirletz, H., Ovaitt, S., Sridhar, S., and Barnes, T.M. (2024). Prioritizing circular economy strategies for sustainable PV deployment at the TW scale. *EPJ Photovoltaics* 15: 18. https://doi.org/10.1051/epjpv/2024015.

84 Velenturf, A.P.M. and Purnell, P. (2021). Principles for a sustainable circular economy. *Sustainable Production and Consumption* 27: 1437–1457. https://doi.org/10.1016/j.spc.2021.02.018.

85 Rincón-Moreno, J., Ormazábal, M., Álvarez, M.J., and Jaca, C. (2021). Advancing circular economy performance indicators and their application in Spanish companies. *Journal of Cleaner Production* 279: 123605. https://doi.org/10.1016/j.jclepro.2020.123605.

86 Potting, J., Hekkert, M., Worrell, E., and Hanemaaijer, A. (2017). Circular economy: measuring innovation in the product chain –policy report. Technical Report January, PBL Netherlands Environmental Assessment Agency.

87 Avraamidou, S. and Torres, A.I. (2023). *Circular Economy: Definitions, Challenges, and Opportunities*. Foundations of Computer Aided Process Operations/Chemical Process Control.

88 Corona, B., Shen, L., Reike, D. et al. (2019). Towards sustainable development through the circular economy –a review and critical assessment on current circularity metrics. *Resources, Conservation and Recycling* 151: 104498. https://doi.org/10.1016/j.resconrec.2019.104498.

89 Parchomenko, A., Nelen, D., Gillabel, J., and Rechberger, H. (2019). Measuring the circular economy –a multiple correspondence analysis of 63 metrics. *Journal of Cleaner Production* 210: 200–216. https://doi.org/10.1016/j.jclepro.2018.10.357.

90 Mali, S. and Garrett, P. (2022). Life cycle assessment of electricity production from an onshore v150-4.2mw wind plant. Technical report, Vestas, Aarhus, Denmark. https://www.vestas.com/content/dam/vestas-com/global/en/sustainability/reports-and-ratings/lcas/LCA%20of%20Electricity%20Production%20from%20an%20onshore%20V150-4.2,%204.5MW%20Wind%20Plant_Final.Web.pdf.coredownload.inline.pdf (accessed 28 August 2024).

91 Nagle, A.J., Delaney, E.L., Bank, L.C., and Leahy, P.G. (2020). A comparative life cycle assessment between landfilling and co-processing of waste from decommissioned Irish wind turbine blades. *Journal of Cleaner Production* 277: 123321. https://doi.org/10.1016/j.jclepro.2020.123321.

92 Bertoldi, P., Geller, H., and Ruggiero, S. (2023). Assessing the potential of low-carbon technologies for industrial decarbonization: a case study of the Italian cement industry. *Waste Management* 164: 143–157. https://doi.org/10.1016/j.wasman.2023.03.041.

93 Meza, J.L.C., Yildirim, M.B., and Masud, A.S.M. (2007). A model for the multiperiod multiobjective power generation expansion problem. *IEEE Transactions on Power Systems* 22 (2): 871–878. https://doi.org/10.1109/TPWRS.2007.895178.

94 Bernal-Agustín, J.L. and Dufo-López, R. (2009). Simulation and optimization of stand-alone hybrid renewable energy systems. *Renewable and Sustainable Energy Reviews* 13 (8): 2111–2118. https://doi.org/10.1016/j.rser.2009.01.010.

95 Zheng, X., Guoce, W., Qiu, Y. et al. (2018). A MINLP multi-objective optimization model for operational planning of a case study CCHP system in urban China. *Applied Energy* 210: 1126–1140. https://doi.org/10.1016/j.apenergy.2017.06.038.

96 Eriksson, E.L.V. and Gray, E.M.A. (2019). Optimization of renewable hybrid energy systems –a multi-objective approach. *Renewable Energy* 133: 971–999. https://doi.org/10.1016/j.renene.2018.10.053.

97 Al Shidhani, T., Ioannou, A., and Falcone, G. (2020). Multi-objective optimisation for power system planning integrating sustainability indicators. *Energies* 13 (9): 2199. https://doi.org/10.3390/en13092199.

98 Serrano-Arévalo, T., Juárez-García, M., and Ponce-Ortega, J. (2020). Optimal planning for satisfying future electricity demands involving simultaneously economic, emissions, and water concerns. *Process Integration and Optimization for Sustainability* 4: 379–389. https://doi.org/10.1007/s41660-020-00125-8.

99 Chen, Z., Avraamidou, S., Liu, P. et al. (2021). Optimal design of integrated urban energy systems under uncertainty and sustainability requirements. *Computers and Chemical Engineering* 155: 107502. https://doi.org/10.1016/j.compchemeng.2021.107502.

100 Cook, J., Marcello Di Martino, R., Allen, C. et al. (2022). A decision-making framework for the optimal design of renewable energy systems under energy-water-land nexus considerations. *Science of the Total Environment* 827: 154185. https://doi.org/10.1016/j.scitotenv.2022.154185.

101 Horasan, M.B. and Kilic, H.S. (2022). A multi-objective decision-making model for renewable energy planning: the case of Turkey. *Renewable Energy* 193: 484–504. https://doi.org/10.1016/j.renene.2022.04.158.

102 Serrano-Arévalo, T.I., Dıaz-Alvarado, F.A., Tovar-Facio, J., and Ponce-Ortega, J.M. (2024). The impact of circular economy indicators in the optimal planning of energy systems. *Sustainable Production and Consumption* 44: 234–249. https://doi.org/10.1016/j.spc.2023.12.018.

103 Čuček, L., Klemeš, J.J., and Kravanja, Z. (2014). Objective dimensionality reduction method within multi-objective optimisation considering total footprints. *Journal of Cleaner Production* 71: 75–86. https://doi.org/10.1016/j.jclepro.2013.12.035.

10

Decarbonization of Steam Cracking for Clean Olefins Production: Optimal Microgrid Scheduling

Saba Ghasemi Naraghi[1], *Tylee Kareck*[1], *Lingyun Xiao*[2], *Richard Reed*[3], *Paritosh Ramanan*[3], *and Zheyu Jiang*[1]

[1] *School of Chemical Engineering, Oklahoma State University, Stillwater, OK, USA*
[2] *Operations Research and Industrial Engineering, University of Texas, Austin, TX, USA*
[3] *School of Industrial Engineering and Management, Oklahoma State University, Stillwater, OK, USA*

10.1 Introduction

Olefins are hydrocarbons containing one or more double bonds between two adjacent carbon atoms. Olefins are widely used as crucial precursors and essential building blocks in the manufacturing of chemical products, including plastic, detergent, adhesive, rubber, and food packaging. Ethylene is the most important olefin with global annual production exceeding 200 million metric tons [1]. Like the United States, the global ethylene market is expected to grow by more than 60% to 240 billion USD between 2024 and 2034 [2]. Currently, ethylene is almost entirely produced via steam cracking of gaseous and liquid hydrocarbon feedstocks such as ethane, propane, and naphtha. The highly endothermic conversion takes place inside cracking furnaces where preheated feedstock passes through and reacts in furnace coils at very high temperatures (e.g. around 850 °C) in a very short residence time (in the order of milliseconds) [3]. The heat is provided by burning the methane fraction byproduct and natural gas in the cracking furnaces. This makes steam cracking one of the most energy and carbon-intensive processes in the chemical industry. Depending on the specific feedstock, it is estimated that 1–1.6 tons of CO_2 are released for every ton of ethylene produced [3]. In fact, steam cracking has been identified by the US Department of Energy as one of the top five energy-intensive refining processes that account for the majority of U.S. refining CO_2 emissions and represent the most cost-effective

R&D opportunities to reduce refining emissions. Thus, there is a global search for technological advancements and process optimization and intensification to improve the energy efficiency and sustainability of the steam cracking process.

As the US energy landscape continues to transition toward clean, renewable electricity, one promising solution to decarbonize the steam cracking process is to implement electric cracking technology. In 2022, Dow Chemical and Shell started up an experimental electrified cracking unit in Amsterdam, whereas the start-up of a multi-megawatt (MW) pilot plant is expected to take place in 2025 [4]. Meanwhile, BASF, SABIC, and Linde in partnership have built a demonstration plant in Ludwigshafen, Germany in early 2024 to validate two electric heating concepts of direct (resistive heating directly applied on coils) or indirect (resistive heating elements installed at furnace walls) heating [5]. While only 40–45% of the total firing duty can be transferred through the coils to the process fluids in conventional fired cracking furnaces [6], this number is at least 95% for electrified cracking furnaces [5, 7]. Nevertheless, due to (i) the sheer size of most ethylene plants in the United States (whose ethylene production capacities are greater than 1 million tons per year), (ii) the need to run these plants around the clock, and (iii) the intermittent nature of variable renewable electricity (VRE) from solar and wind whose proportion in the US electricity generation will continue to increase (this number was 14% in 2023 [8] and is projected to be 36% by 2050 [9]), it would be economically unrealistic and practically impossible to install massive energy storage systems (e.g. batteries) or perform complete plant reconfiguration to accommodate such a large power demand from electrified crackers [10].

Accounting for these complications and practical limitations, our vision for using electricity to provide process heat for steam cracking is shown in Figure 10.1. We envision that the electrification of steam cracking will take place gradually due to the large capital investment associated with the decommissioning of existing crackers and the installation of new cracker units. Thus, in Figure 10.1, both electrified and conventional crackers are present in the superstructure. Battery storage, electrolyzer, and hydrogen storage are used in conjunction with VRE generated onsite to support round-the-clock ethylene plant operation. Electrified crackers can be powered by electricity from the main grid, electricity generated in-house from dispatchable generators and fuel cell units, as well as from batteries. On the other hand, conventional crackers can be powered by fresh natural gas feedstock, as well as the methane fraction byproduct (containing CH_4 and H_2) from both conventional and electrified crackers. Essentially, the future ethylene plant becomes a microgrid, a local electric grid that acts as a single controllable entity with respect to the grid [11]. A microgrid can operate in either grid-connected mode or islanded mode [12–16] to account for potential power outages, price fluctuations, and other instabilities associated with the

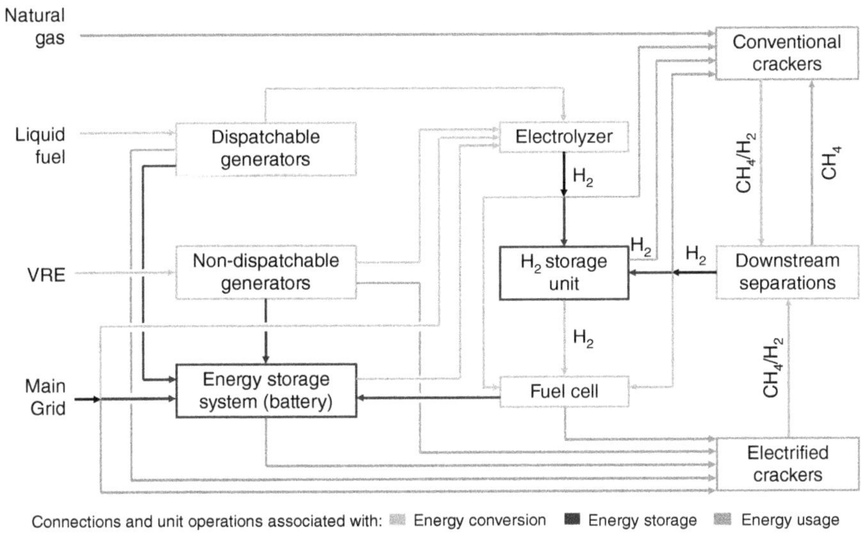

Figure 10.1 Our envisioned framework for using electricity to supply process heat for steam cracking. Diverse energy sources supply heat for both electrified and conventional crackers that are present in the superstructure. Depending on the nature of the energy carriers, the connections shown in the superstructure can represent either energy or mass flows.

main grid [14, 17–21]. Therefore, compared with directly plugging into the power system in a centralized manner, microgrid-like steam cracking provides benefits such as improved resilience, economic operation, and flexibility [22].

This chapter presents the first study of the optimal microgrid scheduling problem for clean olefins production based on the proposed superstructure, as shown in Figure 10.1. We consider a hypothetical ethylene plant located on Texas Gulf Coast. Ethane is chosen to be the major feedstock, which is consistent with most US crackers [5]. We consider a deterministic, steady-state operation of the ethylene plant with a fixed total energy demand, ignoring coke formation due to its slow kinetics [3]. On the other hand, the uncertainties associated with VRE generation and market price predictions are considered [23]. Note that for this problem, we have a choice of whether to formulate it as a centralized or distributed optimization problem, just like other microgrid scheduling problems studied in the literature [20, 24–27]. It turns out that the centralized formulation solves the problem in much less time compared to the distributed formulation based on Benders decomposition. In future works when considering more complex superstructure (e.g. involving downstream processes and plant-wide energy integration pathways), problem specifications (e.g. accounting for coking and decoking), and model formulation (e.g. joint optimization of microgrid and cracker operations), distributed or decentralized formulations may be needed to achieve desired computational performance.

The rest of this chapter is organized as follows. In Section 10.2, we determine the energy requirement of conventional and electrified cracking by developing a differential-algebraic equation (DAE) numerical model and solving the resulting dynamic optimization problem in `pyomo.dae` [28]. The energy demand obtained from the mechanistic model is then used to formulate the optimal scheduling problem for the microgrid in Section 10.3. This scenario-based mixed-integer linear programming (MILP) model is solved in Section 10.4, where several insights are drawn from the case study. Finally, we summarize the results in the concluding remarks in Section 10.5.

10.2 Dynamic Optimization of Steam Cracking Process

The modeling, simulation, and optimization of steam cracking reactors processing different feedstocks have been active research areas for decades following the emergence and growth of petrochemical industries across the world, the recent shale gas boom, and more recently, the industrial decarbonization trend. While process simulator (e.g. Aspen Plus and SPYRO)-based modeling and simulation studies are prevalent in the literature [29–33], mathematical modeling of steam cracking process has also gained considerable research interests [34, 35], made possible by the pioneering work of Sundaram and Froment [36], who developed the first rigorous reaction model. Note that the complete reaction model is based on a free radical mechanism, which is comprehensive and accurate but often results in a stiff nonlinear DAE system due to the large differences in concentration gradient between molecular and free radical species [35]. Instead, various simplified molecular kinetic models have been proposed [37–40] to overcome this computational challenge. In this study, we implement an eight-reaction system (see Tables 10.1 and 10.2), previously adopted by [37, 38] for modeling major ethane cracking reactions.

Following most mathematical models developed in the literature, we model the cracking furnace coils as a one-dimensional plug flow reactor (PFR). This is a reasonable approximation as fluid flow inside the coil is highly turbulent due to the short residence time required to suppress undesired side reactions and coke formation. The configuration specifications of the PFR, which operates in a steady state, are summarized in Table 10.3. While they are tunable parameters, the default specifications represent common industrial furnaces for ethane cracking. Furthermore, we assume that the key difference between conventional and electrified furnaces lies in the thermal efficiency value η [5]. Thus, we can

Table 10.1 Major molecular reactions taking place in ethane cracking.

Reaction	Standard enthalpy of reaction $\left(\frac{\text{kJ}}{\text{mol}}\right)$	Rate expression
$C_2H_6 \longleftrightarrow C_2H_4 + H_2$	$\Delta H_1^0 = 136.33$	$r_1 = k_1\left[\frac{F_{C_2H_6}}{F_{tot}}\left(\frac{P}{RT}\right)\right] - K_{e,1}\left[\frac{F_{C_2H_4}F_{H_2}}{F_{tot}^2}\left(\frac{P}{RT}\right)^2\right]$
$2\,C_2H_6 \longrightarrow C_3H_8 + CH_4$	$\Delta H_2^0 = -11.56$	$r_2 = k_2\left[\frac{F_{C_2H_6}}{F_{tot}}\left(\frac{P}{RT}\right)\right]$
$C_3H_8 \longrightarrow C_3H_6 + H_2$	$\Delta H_3^0 = 124.91$	$r_3 = k_3\left[\frac{F_{C_3H_8}}{F_{tot}}\left(\frac{P}{RT}\right)\right]$
$C_3H_8 \longrightarrow C_2H_4 + CH_4$	$\Delta H_4^0 = 82.67$	$r_4 = k_4\left[\frac{F_{C_3H_8}}{F_{tot}}\left(\frac{P}{RT}\right)\right]$
$C_3H_6 \longleftrightarrow C_2H_2 + CH_4$	$\Delta H_5^0 = 133.45$	$r_5 = k_5\left[\frac{F_{C_3H_6}}{F_{tot}}\left(\frac{P}{RT}\right)\right] - K_{e,5}\left[\frac{F_{C_2H_2}F_{CH_4}}{F_{tot}^2}\left(\frac{P}{RT}\right)^2\right]$
$C_2H_2 + C_2H_4 \longrightarrow C_4H_6$	$\Delta H_6^0 = -171.47$	$r_6 = k_6\left[\frac{F_{C_2H_2}F_{C_2H_4}}{F_{tot}^2}\left(\frac{P}{RT}\right)^2\right]$
$2\,C_2H_6 \longrightarrow C_2H_4 + 2\,CH_4$	$\Delta H_7^0 = 71.10$	$r_7 = k_7\left[\frac{F_{C_2H_6}}{F_{tot}}\left(\frac{P}{RT}\right)\right]$
$C_2H_6 + C_2H_4 \longrightarrow C_3H_6 + CH_4$	$\Delta H_8^0 = -22.98$	$r_8 = k_8\left[\frac{F_{C_2H_6}F_{C_2H_4}}{F_{tot}^2}\left(\frac{P}{RT}\right)^2\right]$

Table 10.2 The rate coefficient expressions in the Arrhenius form.

$k_i = A_i \exp\left(\frac{-E_i}{RT}\right)$	Arrhenius constant A_i	Activation energy E_i $\left(\frac{\text{kJ}}{\text{mol}}\right)$
k_1	$4.65 \times 10^{13}\left(\frac{m^3}{mol\cdot s}\right)$	273.0
k_2	$3.85 \times 10^{11}\left(\frac{1}{s}\right)$	273.0
k_3	$5.89 \times 10^{10}\left(\frac{1}{s}\right)$	215.0
k_4	$4.69 \times 10^{10}\left(\frac{1}{s}\right)$	212.0
k_5	$7.28 \times 10^{12}\left(\frac{m^3}{mol\cdot s}\right)$	154.0
k_6	$1.03 \times 10^{9}\left(\frac{1}{s}\right)$	173.0
k_7	$6.37 \times 10^{23}\left(\frac{1}{s}\right)$	530.0
k_8	$7.08 \times 10^{10}\left(\frac{1}{s}\right)$	253.0
$K_{e,1}$	$8.49 \times 10^{8}\left(\frac{m^3}{mol\cdot s}\right)$	136.5
$K_{e,5}$	$3.81 \times 10^{8}\left(\frac{m^3}{mol\cdot s}\right)$	147.2

Table 10.3 Design and operating specifications of ethane cracking furnace.

Parameter	Default value
Residence time (s)	0.2
Coil length L (m)	100
Coil inner diameter d_t (m)	0.1
Inlet steam–ethane mass ratio	0.3
η of conventional furnace	0.4
η of electrified furnace	0.971
Coil inlet temperature T_{in} (°C)	650
Coil outlet temperature T_{out} (°C)	850
Coil inlet pressure P_{in} (bar)	3.03
Coil outlet pressure P_{out} (bar)	1.95
Ethylene yield (kg/kg)	0.5

The ethylene yield, which is the main product specification, is defined as the mass flow ratio between ethane feed and ethylene product.

estimate the total energy demand of electrified crackers by directly scaling the heat duty results from the dynamic optimization.

The objective of dynamic optimization is to minimize the total energy demand to achieve an ethylene yield of at least 50% (which is typical for ethane cracking), subject to mass and energy balances. This is done by adjusting the temperature profile inside the coil, which in turn changes the heat flux profile. While the rigorous momentum balance equation can be incorporated in the model to accurately calculate the pressure drop profile within the coil by accounting for friction losses, P–V–T relationships, and additional losses in the coil bends [35], we have identified that incorporating this highly nonlinear ODE does not affect the optimal objective function value, but significantly increases model complexity and computational time. Therefore, we employ a simple analytical expression of Eq. (10.2e) to approximate the pressure drop profile. This equation is derived from the well-known pressure drop expression for flow in pipes [41], namely

$$\frac{P(x)}{P_{in}} = \sqrt{1 - \alpha_p V(x)}, \qquad (10.1)$$

where α_p is a coil-specific parameter, and $V(x) = A_t x$ is the total pipe volume already passed by the fluid.

With this, the dynamic optimization problem is formulated as follows:

$$\text{minimize} \ \frac{1}{\eta} \int_0^L q(x)dx \qquad (10.2a)$$

$$\text{subject to: } \frac{dF_j}{dx} = \frac{\pi d_t^2}{4} \sum_{i \in I} s_{i,j} r_i, \qquad \forall j \in J \tag{10.2b}$$

$$\sum_{j \in J} F_j C_{p,j} \frac{dT}{dx} = q(x) + \frac{\pi d_t^2}{4} \sum_{i \in I} r_i (-\Delta H_i) \tag{10.2c}$$

$$\frac{dT}{dx} \geq 0 \tag{10.2d}$$

$$P(x) = P_{in} \sqrt{1 - \frac{x}{L} \left[1 - \left(\frac{P_{out}}{P_{in}} \right)^2 \right]} \tag{10.2e}$$

$$\Delta H_i(x) = \Delta H_i^0 + \Delta C_{p,i}(T - 298), \qquad \forall i \in I \tag{10.2f}$$

$$\Delta C_{p,i} = \sum_{j \in J} C_{p,j} s_{i,j} \tag{10.2g}$$

$$F_{C_2H_4}(L) \geq y_{C_2H_4} \frac{MW_{C_2H_6}}{MW_{C_2H_4}} F_{C_2H_6,in} \tag{10.2h}$$

$$0 \leq x \leq L; \quad T_{in} \leq T \leq T_{out}. \tag{10.2i}$$

Here, I and J are sets consisting of all reactions and species included in Table 10.1, respectively. $q(x)$ represents the heat flux (in, e.g. MW), $s_{i,j}$ is the stoichiometry of component j in reaction i ($s_{i,j} > 0$ for products and < 0 for reactants), molar heat capacity $C_{p,j}$ of pure component j is approximated to be independent of temperature (whose value is obtained at $\frac{T_{in}+T_{out}}{2} = 750\,°C$ from Aspen Properties), and $y_{C_2H_4}$ is the ethylene yield specification. Note that by implementing Eq. (10.2d), we ensure that the direction of heat transfer matches with the temperature gradient in the furnace (keep in mind that the flame temperature is as high as 1150 °C, which is much higher than the fluid temperature inside the coil).

This DAE-constrained optimization problem is discretized using an orthogonal collocation method with 13 finite elements and 3 Radau collocation points, resulting in a discretized nonlinear program with 1863 variables, 1824 equality constraints, and 39 degrees of freedom. We model the problem in `pyomo.dae` [28] and find an optimal solution using IPOPT [42] (version 3.14.9) in 34.154 seconds on a Dell Precision 7920 Tower workstation (Intel Xeon Gold 6226R 2.9 GHz, 12 × 8 GB RAM). The outlet product composition (excluding steam) is shown in Table 10.4. The minimum energy requirements for the conventional and electrified crackers are given by 4.27 and 1.75 MWh per ton of ethylene produced. These numbers are used in Sections 10.3 and 10.4 to study the optimal microgrid scheduling problem.

Table 10.4 Outlet product composition of major species (excluding steam) at the optimal solution obtained by dynamic optimization.

Component	Molar fraction	Mass fraction
Ethane (C_2H_6)	0.2384	0.3753
Ethylene (C_2H_4)	0.3406	0.5
Propane (C_3H_8)	0.0016	0.0038
Propylene (C_3H_6)	0.0208	0.0458
Acetylene (C_2H_2)	0.0005	0.0006
Butadiene (C_4H_6)	0.0011	0.0032
Methane (CH_4)	0.0397	0.0333
Hydrogen (H_2)	0.3573	0.0377

10.3 Scenario-Based Optimal Microgrid Scheduling Problem

Before formulating the microgrid scheduling problem, we perform a quick calculation [10] using the results obtained in Section 10.2 to motivate the need for a diverse portfolio of energy sources. For a plant with 1 million tons/year of ethylene production capacity via electrified cracking, assuming that VRE is on average available for 30% of a day, then at least 70% of daily energy (i.e. 3356 MWh of electricity) needs to be stored for around-the-clock operation of electrified furnaces. Battery storage, based on 100 kWh of the battery pack in a Tesla Model S electric car, would require the battery capacity of 33,560 Tesla Model S cars! Moreover, the actual amount of energy that would need to be stored is likely to be one to two orders of magnitude larger because of daily and seasonal weather variations. This justifies the need to consider a hybrid process heating landscape similar to the superstructure drawn in Figure 10.1. Note that in the superstructure, we have both dispatchable and non-dispatchable generators. Dispatchable generators are typically controlled by the microgrid master controllers [14] and are subject to constraints associated with the generation capacity, ramping rates, and minimum on/off time. On the other hand, non-dispatchable generators cannot be controlled by the master controller as their operation depends solely on the availability and capacity of renewable sources.

In this study, we develop a single-stage scenario-based mixed-integer linear programming (MILP) model to identify the optimal schedule for the microgrid and the optimal degree of electrification for ethane cracking. The sets, parameters, and parameters used in this MILP model are defined below:

Abbreviation

NG	Natural gas
CC, EC	Conventional cracker, electrified cracker
ESS	Energy storage system (battery)
FC	Fuel cell
EL	Electrolyzer
C, DC	Charging, discharging
SU, SD	Startup, shutdown
HS	Hydrogen storage
PV, WT	photovoltaic panel, wind turbine

Indices

g	Index of dispatchable generator groups $\mathcal{G} = \{1, \ldots, G\}$
i	Index of dispatchable generators belonging to group $g \in \mathcal{G}$, $i \in \mathcal{I}_g$
t, τ	Indices of hourly time periods $\mathcal{T} = \{1, \ldots, 24\}$
ω	Index of set of scenarios Ω

Parameters

Δt	Time interval (1 hour)
$\dot{Q}_{NG}$	Lower heating value of natural gas (13.826 MWh/ton)
$\dot{Q}_{H_2}$	Lower heating value of hydrogen (33.320 MW/ton)
$\dot{Q}_{CH_4}$	Lower heating value of methane (13.896 MW/ton)
$\dot{Q}_g$	Lower heating value of fuel group g (12.222 MWh/ton)
$F_{CC}^{CH_4/H_2}$	Total outlet flow of CH_4 and H_2 from conventional cracker (ton/h)
$F_{EC}^{CH_4/H_2}$	Total outlet flow of CH_4 and H_2 from electrified cracker (ton/h)
f_{CH_4}	Mass fraction of CH_4 in outlet CH_4/H_2 mixture (0.4692 from Table 10.4)
$r_{CH_4,sep}$	CH_4 recovery in downstream separation units (0.997)
$r_{H_2,sep}$	H_2 recovery in downstream separation units (0.99)
HSC	Hydrogen storage capacity (ton)
$M_{HS,start}^{H_2}$	Hydrogen storage level at $t = 0$

ELC	PEM electrolyzer hydrogen production capacity (ton/h)
η_{EL}	PEM electrolyzer efficiency (0.736) [43]
η_{FC}	Fuel cell efficiency (0.65)
η_g	Generators efficiency (0.60)
P^{H_2}	Electrolysis energy requirement with no efficiency loss (39.4 MWh/ton)
P_{CC}	Power requirement for conventional cracker (MW)
P_{EC}	Power requirement for electrified cracker (MW)
$\underline{P}^{FC}, \overline{P}^{FC}$	Minimum and maximum generation capacity of the fuel cell (MW)
$\overline{P}^{G}$	Maximum power that can be purchased from the grid (MW)
$P^{WT}_{t,\omega}$	Wind power generation (MW)
$P^{PV}_{t,\omega}$	Solar power generation (MW)
$\underline{P}^{ESS}_{C}, \overline{P}^{ESS}_{C}$	Minimum and maximum ESS (battery) charging power (MW)
$\underline{P}^{ESS}_{DC}, \overline{P}^{ESS}_{DC}$	Minimum and maximum ESS (battery) discharging power (MW)
ESC	Energy storage system capacity (MWh)
E^{ESS}_{start}	Starting ESS charged level at $t = 0$
MC^{ESS}, MDC^{ESS}	Minimum battery charging and discharging time (hour)
$\underline{P}^{D}, \overline{P}^{D}$	Minimum and maximum generator capacity of dispatchable units (MW)
RU^{D}_{g}, RD^{D}_{g}	Ramp-up and ramp-down rate for dispatchable group g (MW/h)
UT^{D}_{i}, DT^{D}_{i}	Minimum up and down time for dispatchable unit i (h)
c^{Fuel}_{g}	Fuel group g price ($/ton)
$c^{G}_{t,\omega}$	Electricity market price ($/MWh)
c^{D}_{g}	Dispatchable units generation cost ($/MWh)
$c^{D}_{SU,g}$	Dispatchable units startup cost ($/MWh)
$c^{D}_{SD,g}$	Dispatchable units shutdown cost ($/MWh)
c^{HS}	Hydrogen storage cost ($/ton)

c^{EL}	PEM electrolyzer operating cost (\$/ton)
c^{FC}	Fuel cell generation cost (\$/MWh)
ρ_w	Probability of scenario ω

Binary variables

$x_{t,\omega}^{FC}$	Fuel cell commitment status at time t and scenario ω
$x_{C,t,\omega}^{ESS}$	ESS (battery) charging status at time t and scenario ω
$x_{DC,t,\omega}^{ESS}$	ESS (battery) discharging status at time t and scenario ω
$x_{i,t,\omega}^{D}$	Dispatchable unit i commitment status at time t and scenario ω

Continuous variables (all under time t and scenario ω)

$F_{CC,t,\omega}^{NG}$	Natural gas usage by conventional crackers (ton/h)
$F_{i,t,\omega}^{D}$	Fuel usage by generator $i \in \mathcal{I}_g$ (ton/h)
$F_{CC,sep,t,\omega}^{CH_4/H_2}$	Amount of CH_4 and H_2 from CCs to separation unit for energy generation purposes (ton/h)
$F_{CC,sep,t,\omega}^{CH_4/H_2}$	Amount of CH_4 and H_2 from ECs to separation unit for energy generation purposes (ton/h)
$F_{sep,CC,t,\omega}^{CH_4}$	Amount of CH_4 from the separation unit to CCs (ton/h)
$F_{sep,CC,t,\omega}^{H_2}$	Amount of H_2 from the separation unit to CCs (ton/h)
$F_{sep,HS,t,\omega}^{H_2}$	Amount of H_2 from the separation unit to HS (ton/h)
$F_{sep,FC,t,\omega}^{H_2}$	Amount of H_2 from the separation unit to FC (ton/h)
$F_{HS,FC,t,\omega}^{H_2}$	Amount of H_2 from the HS to FC (ton/h)
$F_{HS,CC,t,\omega}^{H_2}$	Amount of H_2 from the HS to CCs (ton/h)
$M_{HS,t,\omega}^{H_2}$	Amount of H_2 stored in HS (ton)
$F_{EL,FC,t,\omega}^{H_2}$	Amount of H_2 produced from the PEM electrolyzer to FC (ton/h)

$F^{H_2}_{EL,CC,t,\omega}$	Amount of H_2 produced from the PEM electrolyzer to CCs (ton/h)
$F^{H_2}_{EL,HS,t,\omega}$	Amount of H_2 produced from the PEM electrolyzer to HS (ton/h)
$F^{H_2}_{EL,t,\omega}$	Amount of H_2 produced from PEM electrolyzer (ton/h)
$P^{D}_{EC,g,t,\omega}$	Amount of power from dispatchable units to ECs (MW)
$P^{G}_{EC,t,\omega}$	Amount of power from the grid to the ECs (MW)
$P^{ND}_{EC,t,\omega}$	Amount of power from non-dispatchable units to the ECs (MW)
$P^{ESS}_{EC,t,\omega}$	Amount of power from the ESS to ECs (MW)
$P^{FC}_{EC,t,\omega}$	Amount of power from the fuel cell to ECs (MW)
$P^{D}_{EL,g,t,\omega}$	Amount of power from dispatchable units to PEM electrolyzer (MW)
$P^{G}_{EL,t,\omega}$	Amount of power from the grid to PEM electrolyzer (MW)
$P^{ND}_{EL,t,\omega}$	Amount of power from non-dispatchable units to PEM electrolyzer (MW)
$P^{ESS}_{EL,t,\omega}$	Amount of power from ESS to PEM electrolyzer (MW)
$P^{FC}_{ESS,t,\omega}$	Amount of power from the fuel cell to ESS (MW)
$P^{FC}_{t,\omega}$	Amount of fuel cell generated power (MW)
$P^{G}_{ESS,t,\omega}$	Amount of power from the grid to ESS (MW)
$P^{G}_{t,\omega}$	Amount of power from the grid (MW)
$P^{ND}_{ESS,t,\omega}$	Amount of power from non-dispatchable units to ESS (MW)
$P^{D}_{ESS,g,t,\omega}$	Amount of power from dispatchable units to ESS (MW)
$P^{ESS}_{C,t,\omega}$	Amount of power used to charge the ESS (MW)
$P^{ESS}_{DC,t,\omega}$	Amount of power discharged from ESS (MW)
$E^{ESS}_{t,\omega}$	Amount of stored power in the ESS (MWh)
$P^{D}_{i,t,\omega}$	Amount of power generated by the dispatchable unit i (MW)
$P^{D}_{g,t,\omega}$	Total amount of power generated by dispatchable group g (MW)

The objective function (10.3) minimizes the expected total costs, which consist of fuel costs, electricity costs from the grid, energy generation costs, startup and shutdown costs, and hydrogen storage costs. The startup and shutdown costs in the objective function can be easily linearized:

$$\min \sum_{\omega \in \Omega} \sum_{t \in \mathcal{T}} \rho_\omega \left[c_{NG}^{Fuel} F_{CC,t,\omega}^{NG} \Delta t + \sum_{g \in \mathcal{G}} \sum_{i \in \mathcal{I}_g} \left(c_g^{Fuel} F_{i,t,\omega}^{D} \Delta t + c_g^{D} P_{i,t,\omega}^{D} \right.\right.$$
$$+ c_{SU,g}^{D} \max\{0, x_{i,t,\omega}^{D} - x_{i,t-1,\omega}^{D}\} + c_{SD,g}^{D} \max\{0, x_{i,t-1,\omega}^{D} - x_{i,t,\omega}^{D}\} \Big) \tag{10.3}$$
$$\left. + c_{t,\omega}^{G} P_{t,\omega}^{G} + c_t^{FC} P_{t,\omega}^{FC} + c_t^{EL} F_{EL,t,\omega}^{H_2} \Delta t + c_t^{HS} M_{HS,t,\omega}^{H_2} \right].$$

In terms of the constraints, Eq. (10.4) says that conventional crackers can be powered by natural gas fuel, CH_4 and H_2 produced from cracker units, as well as H_2 coming from the electrolyzer and H_2 storage unit:

$$\begin{aligned} P_{CC} = {} & \dot{Q}_{NG} F_{CC,t,\omega}^{NG} + \dot{Q}_{CH_4} F_{sep,CC,t,\omega}^{CH_4} \\ & + \dot{Q}_{H_2} \left(F_{sep,CC,t,\omega}^{H_2} + F_{EL,CC,t,\omega}^{H_2} + F_{HS,CC,t,\omega}^{H_2} \right) \end{aligned} \qquad \forall t \in \mathcal{T}, \omega \in \Omega. \tag{10.4}$$

Equation (10.5) says that electric crackers can be powered by the main grid, as well as other local fuel-based generators, fuel cells, wind turbines, photovoltaic panels, and energy storage systems:

$$P_{EC} = \sum_{g \in \mathcal{G}} P_{EC,g,t,\omega}^{D} + P_{EC,t,\omega}^{ND} + P_{EC,t,\omega}^{G} + P_{EC,t,\omega}^{ESS} + P_{EC,t,\omega}^{FC} \qquad \forall t \in \mathcal{T}, \omega \in \Omega. \tag{10.5}$$

After steam cracking, the product stream undergoes a series of cooling, drying, compression, and separation operations downstream to obtain individual components from the cracked gas mixture. These steps are required to produce ethylene, propylene, hydrogen, and other value-added hydrocarbons that meet certain specifications. In particular, we are interested in tracking the flows of CH_4 and H_2, which can be used to power conventional crackers, produce electricity by the fuel cell, or be stored [44], as described in Eq. (10.6). Note that Eq. (10.6) is expressed as inequalities because not all CH_4 and H_2 will be reused.

$$\begin{aligned} & F_{CC,sep,t,\omega}^{CH_4/H_2} \le F_{CC}^{CH_4/H_2}, \\ & F_{EC,sep,t,\omega}^{CH_4/H_2} \le F_{EC}^{CH_4/H_2}, \\ & F_{sep,CC,t,\omega}^{CH_4} = r_{CH_4,sep} f_{CH_4} \left(F_{CC,sep,t,\omega}^{CH_4/H_2} + F_{EC,sep,t,\omega}^{CH_4/H_2} \right), \qquad \forall t \in \mathcal{T}, \omega \in \Omega. \\ & F_{sep,CC,t,\omega}^{H_2} + F_{sep,HS,t,\omega}^{H_2} + F_{sep,FC,t,\omega}^{H_2} \\ & \quad = r_{H_2,sep} (1 - f_{CH_4}) \left(F_{CC,sep,t,\omega}^{CH_4/H_2} + F_{EC,sep,t,\omega}^{CH_4/H_2} \right), \end{aligned} \tag{10.6}$$

Equation (10.7) describes the H_2 mass balances around hydrogen storage unit and electrolyzer:

$$M^{H_2}_{HS,t-1,\omega} + \left(F^{H_2}_{sep,HS,t,\omega} + F^{H_2}_{EL,HS,t,\omega} - F^{H_2}_{HS,FC,t,\omega} - F^{H_2}_{HS,CC,t,\omega}\right)\Delta t = M^{H_2}_{HS,t,\omega},$$
$$F^{H_2}_{EL,FC,t,\omega} + F^{H_2}_{EL,HS,t,\omega} + F^{H_2}_{EL,CC,t,\omega} = F^{H_2}_{EL,t,\omega},$$
$$0 \le M^{H_2}_{HS,t,\omega} \le \text{HSC},\ M^{H_2}_{HS,0,\omega} = M^{H_2}_{HS,start},$$
$$\forall t \in \mathcal{T},\ \omega \in \Omega. \tag{10.7}$$

Equation (10.8) models hydrogen production in a PEM electrolyzer using electricity from various sources:

$$F^{H_2}_{EL,t,\omega} = \frac{\eta_{EL}}{p^{H_2}}\left(\sum_{g\in\mathcal{G}} P^{D}_{EL,g,t,\omega} + P^{ND}_{EL,t,\omega} + P^{G}_{EL,t,\omega} + P^{ESS}_{EL,t,\omega}\right),$$
$$0 \le F^{H_2}_{EL,t,\omega} \le \text{ELC}, \tag{10.8}$$
$$\forall t \in \mathcal{T},\ \omega \in \Omega.$$

Equation (10.9) describes constraints related to the fuel cell:

$$P^{FC}_{t,\omega} = \eta_{FC}\dot{Q}_{H_2}\left(F^{H_2}_{HS,FC,t,\omega} + F^{H_2}_{EL,FC,t,\omega} + F^{H_2}_{sep,FC,t,\omega}\right),$$
$$\overline{P}^{FC} x^{FC}_{t,\omega} \le P^{FC}_{t,\omega} \le \overline{P}^{FC} x^{FC}_{t,\omega}, \qquad \forall t \in \mathcal{T},\ \omega \in \Omega.$$
$$P^{FC}_{t,\omega} = P^{FC}_{EC,t,\omega} + P^{FC}_{ESS,t,\omega}, \tag{10.9}$$

The power distribution constraints associated with the main grid are given in Eq. (10.10):

$$P^{G}_{t,\omega} = P^{G}_{ESS,t,\omega} + P^{G}_{EC,t,\omega} + P^{G}_{EL,t,\omega},$$
$$0 \le P^{G}_{t,\omega} \le \overline{P}^{G}, \qquad \forall t \in \mathcal{T},\ \omega \in \Omega. \tag{10.10}$$

Similarly, the power distribution constraints associated with the local non-dispatchable units are shown in Eq. (10.11):

$$P^{ND}_{ESS,t,\omega} + P^{ND}_{EC,t,\omega} + P^{ND}_{EL,t,\omega} = P^{WT}_{t,\omega} + P^{PV}_{t,\omega}, \qquad \forall t \in \mathcal{T},\ \omega \in \Omega. \tag{10.11}$$

For fuel-based local generators, Eq. (10.12) considers the power distribution, generation limits, ramping, and up/downtime constraints:

$$P^{D}_{i,t,\omega} = \eta_g \dot{Q}_g F^{D}_{i,t,\omega}, \quad \forall i \in \mathcal{I}_g,$$
$$\underline{P}^{D}_{g} x^{D}_{i,t,\omega} \le P^{D}_{i,t,\omega} \le \overline{P}^{D}_{g} x^{D}_{i,t,\omega}, \quad \forall i \in \mathcal{I}_g,$$

$$
\begin{aligned}
& P^{\mathrm{D}}_{g,t,\omega} = \sum_{i\in\mathcal{I}_g} P^{\mathrm{D}}_{i,t,\omega},\\
& P^{\mathrm{D}}_{g,t,\omega} = P^{\mathrm{D}}_{\mathrm{ESS},g,t,\omega} + P^{\mathrm{D}}_{\mathrm{EC},g,t,\omega} + P^{\mathrm{D}}_{\mathrm{EL},g,t,\omega},\\
& P^{\mathrm{D}}_{i,t,\omega} - P^{\mathrm{D}}_{i,t-1,\omega} \le \mathrm{RU}^{\mathrm{D}}_g, \quad \forall i \in \mathcal{I}_g,\\
& P^{\mathrm{D}}_{i,t-1,\omega} - P^{\mathrm{D}}_{i,t,\omega} \le \mathrm{RD}^{\mathrm{D}}_g, \quad \forall i \in \mathcal{I}_g,\\
& x^{\mathrm{D}}_{i,\tau,\omega} \ge x^{\mathrm{D}}_{i,t,\omega} - x^{\mathrm{D}}_{i,t-1,\omega}, \quad \forall i \in \mathcal{I}_g,\ \tau \in \{t+1,\dots,\min\{|\mathcal{T}|, t+\mathrm{UT}_i-1\}\},\\
& 1 - x^{\mathrm{D}}_{i,\tau,\omega} \ge x^{\mathrm{D}}_{i,t-1,\omega} - x^{\mathrm{D}}_{i,t,\omega}, \quad \forall i \in \mathcal{I}_g,\ \tau \in \{t+1,\dots,\min\{|\mathcal{T}|, t+\mathrm{DT}_i-1\}\},\\
& \hspace{20em} \forall g \in \mathcal{G},\ t \in \mathcal{T},\ \omega \in \Omega.
\end{aligned}
\tag{10.12}
$$

The last set of constraints, Eq. (10.13), considers the power balance and charging/discharging mechanisms of the energy storage system made of batteries. For simplicity, we assume that every battery cell behaves the same.

$$
\begin{aligned}
& x^{\mathrm{ESS}}_{\mathrm{C},t,\omega} + x^{\mathrm{ESS}}_{\mathrm{DC},t,\omega} \le 1,\\
& x^{\mathrm{ESS}}_{\mathrm{C},\tau,\omega} \ge x^{\mathrm{ESS}}_{\mathrm{C},t,\omega} - x^{\mathrm{ESS}}_{\mathrm{C},t-1,\omega}, \quad \forall \tau \in \{t+1,\dots,\min\{|\mathcal{T}|, t+\mathrm{MC}^{\mathrm{ESS}}-1\}\},\\
& x^{\mathrm{ESS}}_{\mathrm{DC},\tau,\omega} \ge x^{\mathrm{ESS}}_{\mathrm{DC},t,\omega} - x^{\mathrm{ESS}}_{\mathrm{DC},t-1,\omega}, \quad \forall \tau \in \{t+1,\dots,\min\{|\mathcal{T}|, t+\mathrm{MDC}^{\mathrm{ESS}}-1\}\},\\
& \underline{P}^{\mathrm{ESS}}_{\mathrm{C}} x^{\mathrm{ESS}}_{\mathrm{C},t,\omega} \le P^{\mathrm{ESS}}_{\mathrm{C},t,\omega} \le \overline{P}^{\mathrm{ESS}}_{\mathrm{C}} x^{\mathrm{ESS}}_{\mathrm{C},t,\omega},\\
& \underline{P}^{\mathrm{ESS}}_{\mathrm{DC}} x^{\mathrm{ESS}}_{\mathrm{DC},t,\omega} \le P^{\mathrm{ESS}}_{\mathrm{DC},t,\omega} \le \overline{P}^{\mathrm{ESS}}_{\mathrm{DC}} x^{\mathrm{ESS}}_{\mathrm{DC},t,\omega},\\
& P^{\mathrm{ESS}}_{\mathrm{C},t,\omega} = P^{\mathrm{G}}_{\mathrm{ESS},t,\omega} + \sum_{g\in\mathcal{G}} P^{\mathrm{D}}_{\mathrm{ESS},g,t,\omega} + P^{\mathrm{ND}}_{\mathrm{ESS},t,\omega} + P^{\mathrm{FC}}_{\mathrm{ESS},t,\omega},\\
& P^{\mathrm{ESS}}_{\mathrm{DC},t,\omega} = P^{\mathrm{ESS}}_{\mathrm{EC},t,\omega} + P^{\mathrm{ESS}}_{\mathrm{EL},t,\omega},\\
& E^{\mathrm{ESS}}_{t,\omega} = E^{\mathrm{ESS}}_{t-1,\omega} + \left(P^{\mathrm{ESS}}_{\mathrm{C},t,\omega} - P^{\mathrm{ESS}}_{\mathrm{DC},t,\omega}\right)\Delta t,\\
& 0 \le E^{\mathrm{ESS}}_{t,\omega} \le \mathrm{ESC},\ E^{\mathrm{ESS}}_{0,\omega} = E^{\mathrm{ESS}}_{\mathrm{start}},\\
& \hspace{20em} \forall t \in \mathcal{T},\ \omega \in \Omega.
\end{aligned}
\tag{10.13}
$$

10.4 Illustrative Case Studies

10.4.1 Problem Setting

In this section, the proposed scenario-based MILP model is applied to a hypothetical ethylene plant for illustrative case studies. The plant has an ethylene production capacity of 1 million tons/year. Ethane cracking furnaces in the US have capacities ranging from 100,000 to 250,000 tons of ethylene/year. Thus, we

Table 10.5 Characteristics of fossil-based dispatchable units, ESS, and hydrogen storage.

Fuel group	# of units	Operating cost (US$/MWh)	Min–max capacity (MW)	Min up/down time (h)	Ramp-up/down rate (MW/h)
Natural gas	20	33.4	1–5	3	2.5
Hydrogen	1	30	10^{-5}–1	N/A	N/A
Storage type	**# of units**	**Storage cost (US$/ton)**	**Capacity**	**Min–max charging/discharging power (MW)**	**Min charging/discharging time (h)**
ESS	1	N/A	20 (MWh)	0.8–4	5
HSC	1	10,000	10 (ton)	N/A	N/A

Source: Adapted from [14].

assume that the plant has five ethane crackers with 200,000 tons of ethylene/year capacity. By solving this optimal scheduling problem, we hope to provide the first quantitative insights about how steam cracking electrification shall be conducted economically and sustainably. The features of dispatchable generators, energy storage systems, fuel cells, and hydrogen storage are summarized in Table 10.5.

The electricity market price as well as solar and wind energy generation are intermittent in nature. To account for these uncertainties, we synthesize five scenarios for each uncertain parameter using Monte Carlo simulations on a log-normal distribution to prevent getting any negative parameter values. These scenarios are generated based on publicly available Texas grid data from ERCOT in August 2024. For locational marginal pricing (LMP), we randomly selected Bus TC-KO without loss of generality. Table 10.6 summarizes the probability of scenarios for each uncertain parameter. Note that there still will be $5^3 = 125$ combinations of these scenarios. To mitigate this, we further perform a scenario reduction to obtain five representative combinations using a probability distance algorithm based on the Kantorovich distance [23] (see Table 10.7). Figures 10.2–10.4 illustrate the mean and average electricity market price, wind power, and solar power for the five combined scenarios.

In this chapter, we will study the optimal scheduling problem under both the grid-connected and islanded modes. We will also closely examine two different scenarios in the grid-connected mode to draw insights regarding how the microgrid should be scheduled and operated.

Table 10.6 Probability of generated scenarios for LMP, wind power, and PV.

Scenario number used in Table 10.7	LMP (%)	Probability of uncertain parameter WP (%)	PV (%)
1	64.50	13.48	4.86
2	2.92	1.52	30.31
3	3.26	64.91	15.81
4	1.29	6.18	32.96
5	28.03	13.91	16.06

Table 10.7 Probability of generated scenarios (after scenario reduction) for the microgrid problem after scenario reduction.

Combined scenario	Selected LMP scenario	Selected WP scenario	Selected PV scenario	Probability (%)
1	5	3	2	18.21
2	1	3	4	45.56
3	1	1	3	4.54
4	5	3	3	9.50
5	1	3	5	22.19

10.4.2 Grid-Connected Mode

In this case, we consider the (partially) electrified ethylene plant to be operated in the grid-connected mode for a 24-hour horizon. The ESS is considered to be half-charged at $t = 0$. Specifically, since we envision a gradual electrification progress in chemical plants, we minimize the total cost of Eq. (10.3), while considering 0%, 10%, 20%, 30%, 40%, and 50% of the ethylene is produced in electrified crackers. In Figure 10.5, we show the amount of natural gas consumed by conventional crackers and fossil-based generators, as well as power usage from the grid and ESS charging status. In Figure 10.6, we show the hourly CO_2 equivalent emissions and the objective function values corresponding to different electrification levels.

The results show that, as the plant gets more electrified, the dependency of the microgrid on the main grid increases, and the expected total cost increases as well. At 50% electrification, the expected total cost is 4.6 times that with 0%

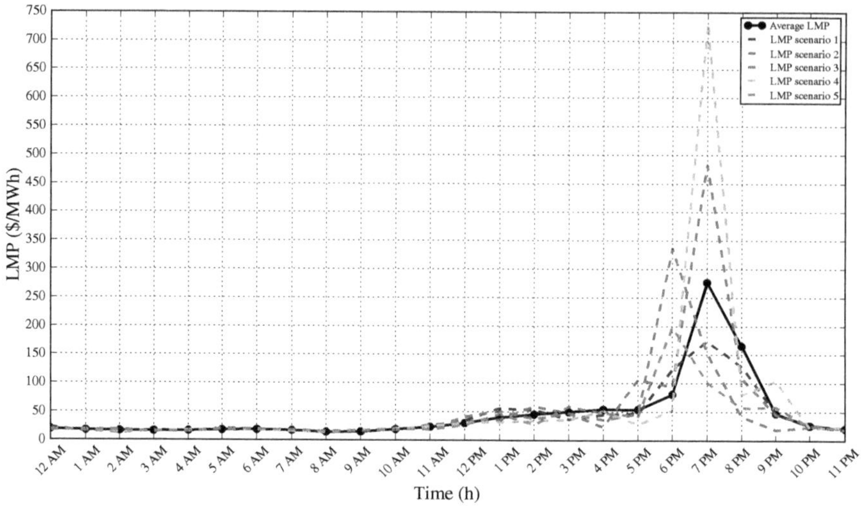

Figure 10.2 The output electricity market prices of five scenarios considered.

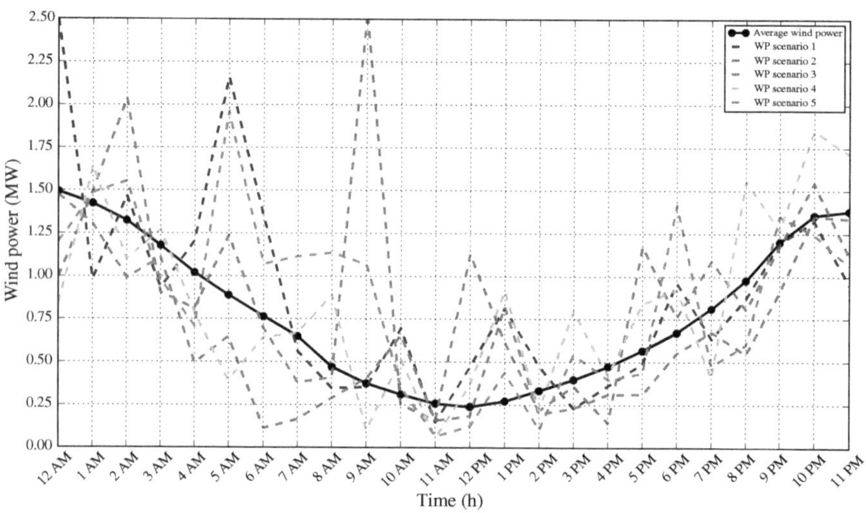

Figure 10.3 The output power of wind turbines of five scenarios considered.

electrification. On the other hand, the average hourly CO_2 equivalent emissions corresponding to 0% through 50% electrification are 38.116, 34.303, 30.951, 34.093, 37.234, and 39.219 tons, respectively. This is an interesting observation because increasing the electrification level does not always lead to a reduction in carbon intensity. Rather, the optimal degree of electrification lies somewhere 10% to 30%

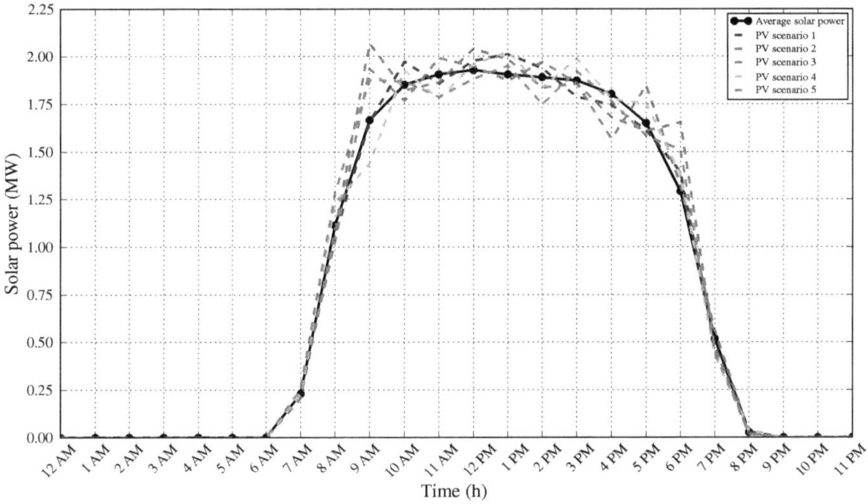

Figure 10.4 The output power of photovoltaic panels of five scenarios considered.

from a sustainability point of view. For example, at 20% electrification, the average hourly CO_2 equivalent emissions at the optimal solution (from a total expected cost perspective) is 18.8% lower than the conventional case with 0% electrification. Meanwhile, the optimal expected total cost is only 68% higher than the conventional case and is significantly lower than the 50% electrification case.

By cross-referencing Figures 10.5 and 10.6, we believe the reason behind this is the following. Between 0% and 50% electrification, the average natural gas consumption by conventional crackers declines as expected. Nevertheless, the rate of decrease is the greatest between 0% and 20% electrification and then diminishes beyond 20% electrification. On the other hand, the natural gas consumed by dispatchable generators shows a steady increase as the level of electrification increases. In fact, the average hourly natural gas consumption by conventional crackers as well as by fossil-based generators for 0%, 10%, 20%, 30%, 40%, and 50% electrification are 14.493, 11.541, 8.814, 8.549, 8.284, and 8.429 ton, respectively. As a result, the overall natural gas consumption flattens out at and after 20% of electrification. Meanwhile, when renewable electricity generation technologies (e.g. fuel cells) are not yet mature enough to be cost-competitive compared to conventional fossil-based generators, as the power demand from electrified cracker units increases, the microgrid must produce more electricity using its local generators, especially during peak hours when the LMP is high. However, due to efficiency losses, this will incur more energy demand than directly using natural gas as fuel for conventional crackers, thereby incurring more costs and higher carbon intensity.

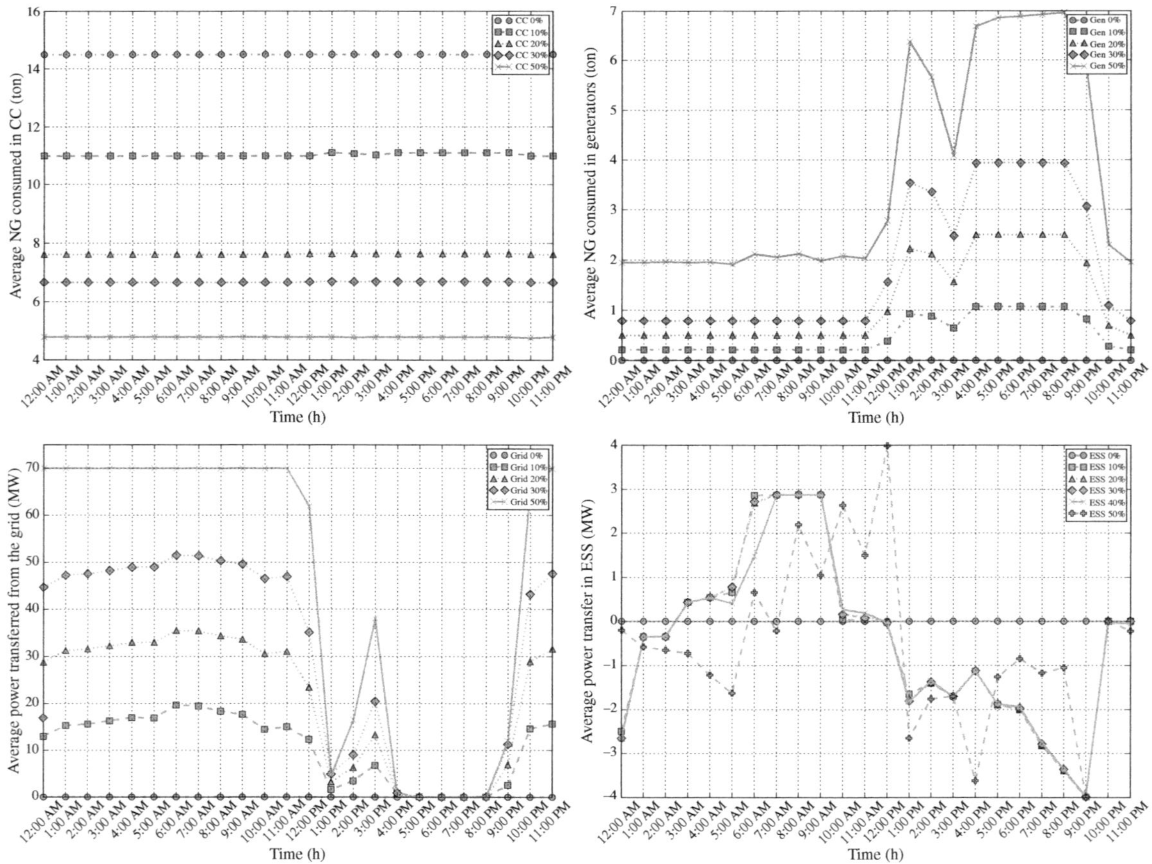

Figure 10.5 Average natural gas (NG) consumption for conventional crackers, natural gas consumption for dispatchable generators, power usage from the main grid, and ESS charging/discharging status for grid-connected mode under different degrees of electrification.

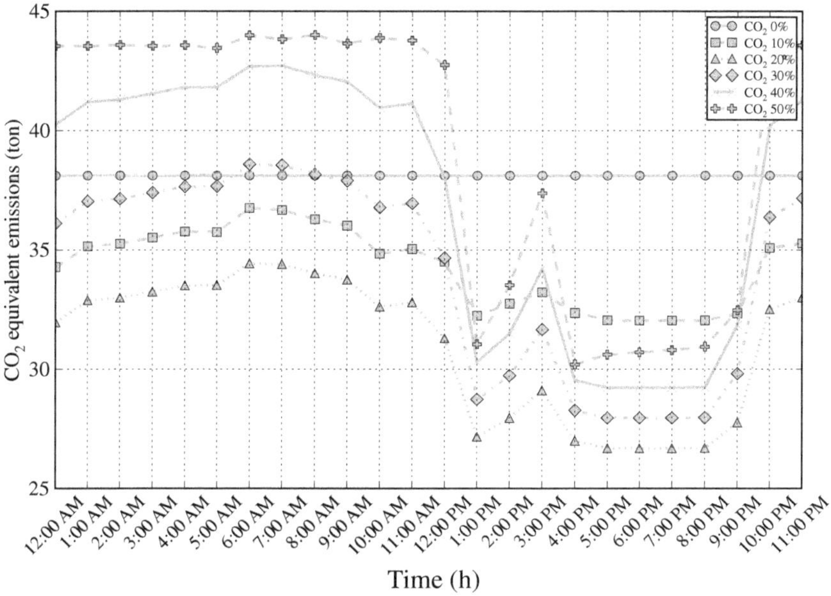

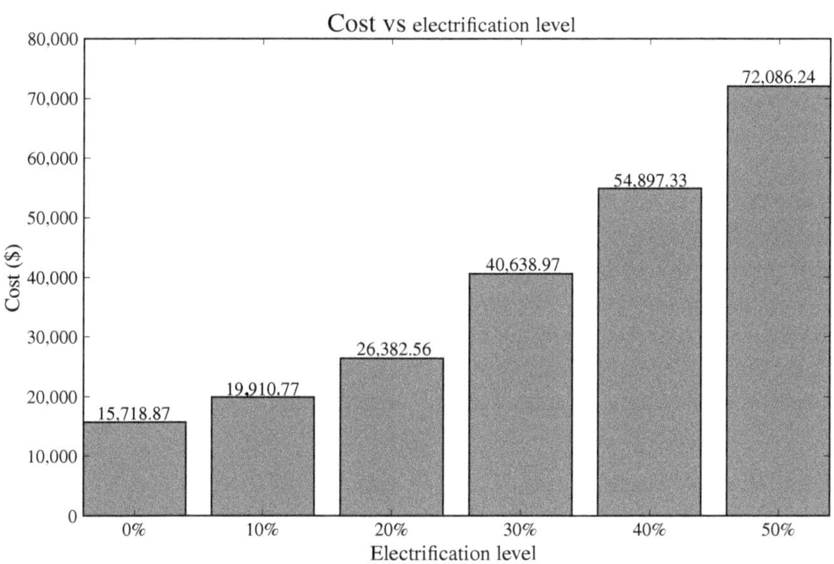

Figure 10.6 Grid-connected mode: average CO_2 equivalent emissions and expected total cost at different electrification levels.

Another observation we can draw is that, when the electricity price is high, the power withdrawn from the main grid decreases drastically to 0. In turn, the local generators will pick up the power demand, and the ESS will start discharging to compensate for the missing demand. During non-peak hours, the ESS will take advantage of the low electricity price to charge itself and prepare for the anticipated demand during peak hours. When the electricity price is high, the CO_2 equivalent emissions decrease significantly, which is consistent with the observation that local renewable electricity generation technologies are participating more heavily in operating the crackers. Meanwhile, we remark that in order to achieve holistic decarbonization via electrification, it is important for both power systems and chemical plant stakeholders to decarbonize at a similar pace. For example, if clean energy sources contribute a higher proportion to ERCOT's energy profile, then a higher degree of electrification in the ethylene plant will translate to lower carbon emissions (which is not what we currently observe in this case study).

Next, we investigate the impact of electricity prices from the grid on electrification progress. As shown in Figure 10.7, we compare the CO_2 equivalent emissions and total expected cost of the optimal solution when using the lowest hourly electricity market price among the five scenarios as well as when using the highest. As we can see, with higher electricity prices, the overall carbon intensity is reduced as the plant prefers to generate its power locally. This is because the microgrid, which harnesses the full decarbonization potential (see Figure 10.1) by considering various renewable options, is generally more renewable compared to the main grid. On the other hand, when using the lowest hourly electricity price, the plant prefers to be more grid-dependent. And the average hourly CO_2 equivalent emissions are 38.116, 32.524, 33.928, 38.798, 41.792, and 42.454 tons for electrification levels of 0% to 50%, respectively, which are higher than the carbon intensity values shown in Figure 10.6. Clearly, there is a trade-off between process economics (total cost) and sustainability (carbon intensity) attributed to the market price and carbon intensity of power from the main grid. From this illustrative example, we conclude that the decarbonization of chemical manufacturing via electrification must be accompanied by the simultaneous decarbonization of power systems.

10.4.3 Islanded Mode

Lastly, we consider the case where the ethylene plant is responsible for satisfying all of its energy demand locally, which could occur during a power system outage. We assume that the energy storage system is initially half-charged. In Figure 10.8, we show the total hourly natural gas consumption (accounting for natural gas used for conventional crackers and dispatchable generators), energy storage charging/discharging status, CO_2 equivalent emissions, and total expected cost for 0%, 10%, 20%, 30%, 40%, and 50% electrification. In terms of CO_2 equivalent

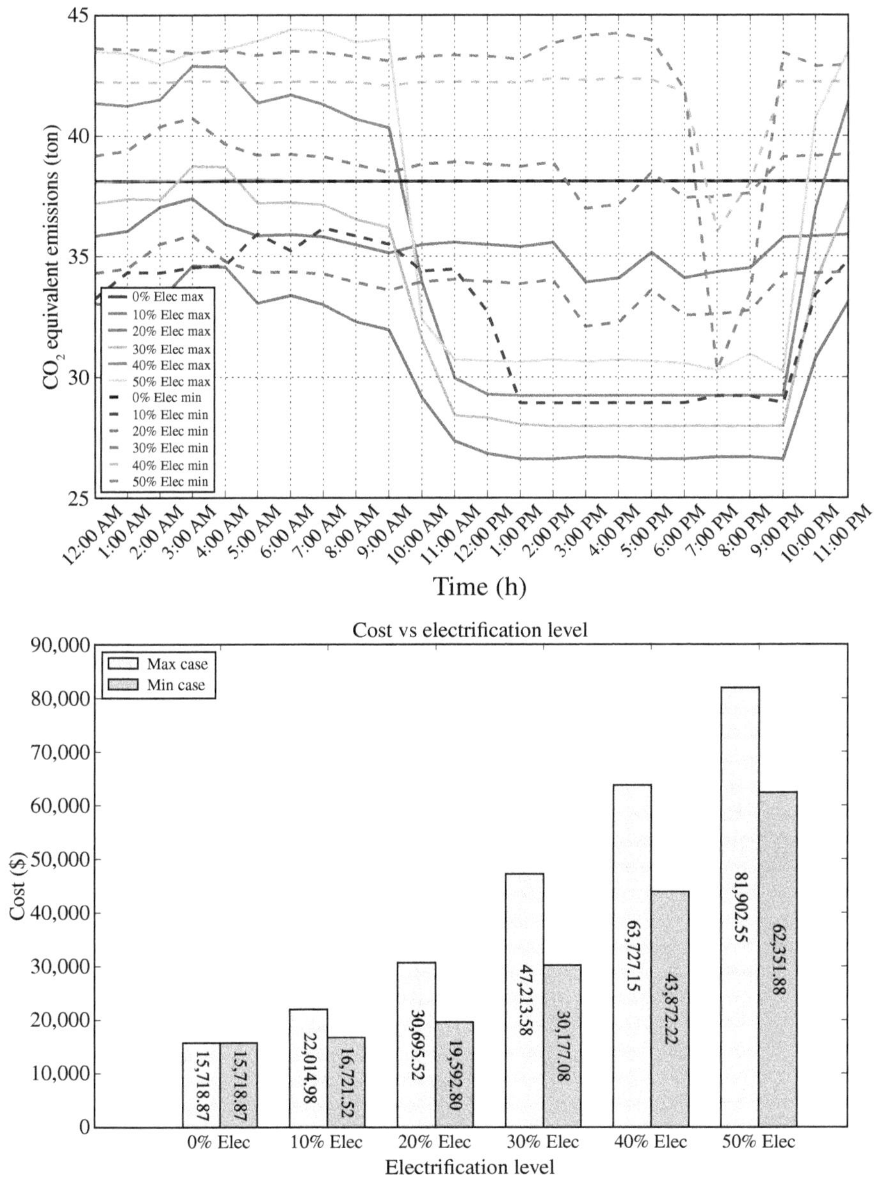

Figure 10.7 Comparing CO_2 equivalent emissions and expected total cost under different electricity prices.

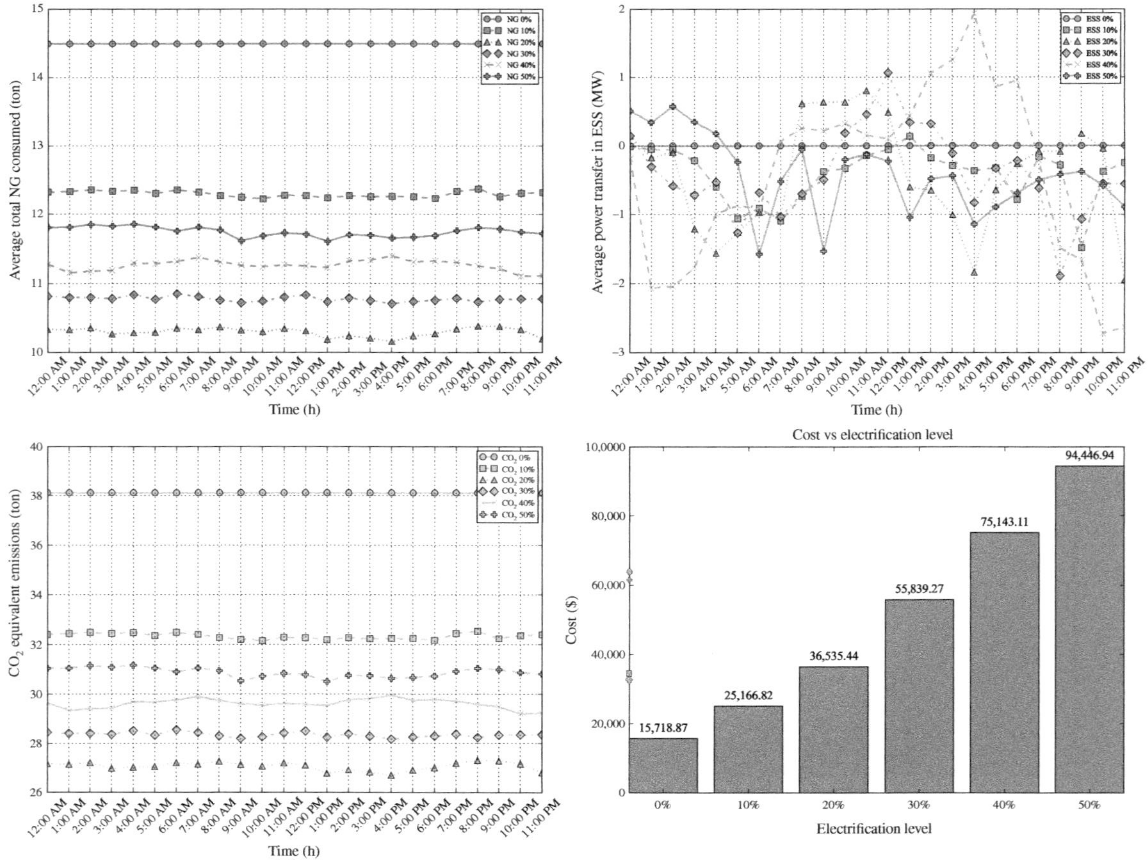

Figure 10.8 Islanded mode: average hourly total natural gas consumption, energy storage system charging/discharging status, average CO_2 equivalent emissions, and expected total cost.

emissions, we find that, even though the carbon intensity is much lower when the plant is operated in islanded mode, the optimal degree of electrification under the current technology status is still around 20% (i.e. roughly one electrified cracker and four conventional crackers). By increasing ESS and H_2 storage capacities, we may debottleneck the decarbonization barrier and achieve lower emissions with a higher degree of electrification.

10.5 Conclusion

Steam cracking is one of the most important and energy-intensive chemical processes that exhibit great opportunities for decarbonization. In this book chapter, we develop a microgrid superstructure for the steam cracking process considering diverse energy sources and storage systems. To obtain the minimum energy requirement for conventional and electrified steam cracking, we formulate a DAE-based constrained optimization problem and solve it as an NLP by discretization using the orthogonal collocation method in `pyomo.dae`. Furthermore, we present a single-stage scenario-based MILP formulation for optimal microgrid scheduling. Using actual VRE profiles and electricity market prices, we study an illustrative case study of microgrid scheduling under grid-connected and islanded modes. The carbon intensity and expected total cost under different decarbonization levels are determined along with the corresponding electricity and natural gas usage. The results show that, given the current status of the power grid and renewable energy generation technologies, the process economics and sustainability of electrified steam cracking do not always favor higher decarbonization levels. To achieve true decarbonization of olefins production, electricity from the main grid must be cleaner and cheaper, and energy storage (ESS and H_2 storage) costs per unit stored must go down. Furthermore, it is important for both chemical and power systems stakeholders must seamlessly coordinate with each other to pursue joint optimization of operation and maintenance.

Acknowledgments

This work is supported by the U.S. National Science Foundation under Award 2343072 and the start-up fund of the College of Engineering, Architecture, and Technology at Oklahoma State University.

References

1 Statista.com (2022). Production capacity of ethylene worldwide from 2018 to 2022 (accessed 16 September 2024).
2 Precedence Research (2024). Ethylene market size, share, and trends 2024 to 2033 (accessed 16 September 2024.
3 Zimmermann, H. and Walzl, R. (2009). *Ethylene*. Wiley.
4 Shell (2022). Shell and Dow start up e-cracking furnace experimental unit (accessed 16 September 2024).
5 Sinn, T., Hofstaetter, M., Kemper, R., and Kracker, G. (2024). Powering the transition to net zero with electric cracking technology. *Chemical Engineering Progress* 120 (8): 42–49.
6 Agrawal, R., Chen, Z., and Oladipupo, P. (June 2021). Electrically heated dehydrogenation process. (US Patent 11,578,019 B2). U.S. Patent and Trademark Office.
7 Maporti, D., Galli, F., Mocellin, P., and Pauletto, G. (2023). Flexible ethylene production: electrified ethane cracking coupled with oxidative dehydrogenation. *Energy Conversion and Management* 298: 117761.
8 U.S. EIA (February 2024). Electric power monthly. https://www.eia.gov/electricity/monthly/ (accessed 16 September 2024).
9 U.S. EIA (2022). Annual energy outlook. https://www.eia.gov/outlooks/aeo/pdf/AEO2022_ChartLibrary_Electricity.pdf (accessed 16 September 2024).
10 Agrawal, R. and Siirola, J. (2023). Decarbonization of chemical process industries via electrification. In: *The Bridge on Engineering the Energy Transition* (ed. R. Latanision), 32–40. The U.S. National Academy of Engineering.
11 Ton, D.T. and Smith, M.A. (2012). The U.S. department of energy's microgrid initiative. *The Electricity Journal* 25 (8): 84–94.
12 Shahidehpour, M. and Clair, J.F. (2012). A functional microgrid for enhancing reliability, sustainability, and energy efficiency. *The Electricity Journal* 25 (8): 21–28.
13 Shahidehpour, M. (2010). Role of smart microgrid in a perfect power system. In: *IEEE PES General Meeting*, 1. IEEE.
14 Khodaei, A. (2013). Microgrid optimal scheduling with multi-period islanding constraints. *IEEE Transactions on Power Systems* 29 (3): 1383–1392.
15 Bahramirad, S., Reder, W., and Khodaei, A. (2012). Reliability-constrained optimal sizing of energy storage system in a microgrid. *IEEE Transactions on Smart Grid* 3 (4): 2056–2062.
16 Flueck, A. and Li, Z. (2008). Destination: perfection. *IEEE Power and Energy Magazine* 6 (6): 36–47.
17 Kroposki, B., Lasseter, R., Ise, T. et al. (2008). Making microgrids work. *IEEE Power and Energy Magazine* 6 (3): 40–53.

18 Kennedy, S. and Marden, M.M. (2009). Reliability of islanded microgrids with stochastic generation and prioritized load. In: *2009 IEEE Bucharest PowerTech*, 1–7. IEEE.

19 Lopez-Prado, J.L., Vélez, J.I., and Garcia-Llinas, G.A. (2020). Reliability evaluation in distribution networks with microgrids: review and classification of the literature. *Energies* 13 (23): 6189.

20 Tsikalakis, A.G. and Hatziargyriou, N.D. (2011). Centralized control for optimizing microgrids operation. In: *2011 IEEE Power and Energy Society General Meeting*, 1–8. IEEE.

21 Al-Ismail, F.S. (2021). Dc microgrid planning, operation, and control: a comprehensive review. *IEEE Access* 9: 36154–36172.

22 Parhizi, S., Lotfi, H., Khodaei, A., and Bahramirad, S. (2015). State of the art in research on microgrids: a review. *IEEE Access* 3: 890–925.

23 Gholami, A., Shekari, T., Aminifar, F., and Shahidehpour, M. (2016). Microgrid scheduling with uncertainty: the quest for resilience. *IEEE Transactions on Smart Grid* 7 (6): 2849–2858.

24 Elmouatamid, A., Ouladsine, R., Bakhouya, M. et al. (2020). Review of control and energy management approaches in micro-grid systems. *Energies* 14 (1): 168.

25 Hatziargyriou, N., Contaxis, G., Matos, M. et al. (2002). Energy management and control of island power systems with increased penetration from renewable sources. In: *2002 IEEE Power Engineering Society Winter Meeting. Conference Proceedings (Cat. No. 02CH37309)*, vol. 1, 335–339. IEEE.

26 Korpas, M. and Holen, A.T. (2006). Operation planning of hydrogen storage connected to wind power operating in a power market. *IEEE Transactions on Energy Conversion* 21 (3): 742–749.

27 Palma-Behnke, R., Benavides, C., Lanas, F. et al. (2013). A microgrid energy management system based on the rolling horizon strategy. *IEEE Transactions on Smart Grid* 4 (2): 996–1006.

28 Nicholson, B., Siirola, J., Watson, J.-P. et al. (2018). Pyomo.dae: a modeling and automatic discretization framework for optimization with differential and algebraic equations. *Mathematical Programming Computation* 10: 187–223.

29 Nazarudin Rosli, M. and Aziz, N. (2018). Steady state modelling of steam cracking furnace radiant section using Aspen Plus. *Materials Today: Proceedings* 5 (10, Part 2): 21780–21789.

30 Haribal, V.P., Chen, Y., Neal, L., and Li, F. (2018). Intensification of ethylene production from naphtha via a redox oxy-cracking scheme: process simulations and analysis. *Engineering* 4 (5): 714–721.

31 Zebert, T.L., Lokhat, D., Kurella, S., and Meikap, B. (2023). Modeling and simulation of ethane cracker reactor using Aspen Plus. *South African Journal of Chemical Engineering* 43: 204–214.

32 Ranjan, P., Kannan, P., Al Shoaibi, A., and Srinivasakannan, C. (2012). Modeling of ethane thermal cracking kinetics in a pyrocracker. *Chemical Engineering & Technology* 35 (6): 1093–1097.

33 Chen, Y., Kuo, M.J., Lobo, R., and Ierapetritou, M. (2024). Ethylene production: process design, techno-economic and life-cycle assessments. *Green Chemistry* 26: 2903–2911.

34 Caballero, D.Y., Biegler, L.T., and Guirardello, R. (2015). Simulation and optimization of the ethane cracking process to produce ethylene. In: *12th International Symposium on Process Systems Engineering and 25th European Symposium on Computer Aided Process Engineering, Computer Aided Chemical Engineering*, vol. 37 (ed. K.V. Gernaey, J.K. Huusom, and R. Gani), 917–922. Elsevier.

35 Froment, G.F., Bischoff, K.B., and Wilde, J.D. (2010). *Chemical Reactor Analysis and Design*, 3e. Wiley.

36 Sundaram, K. and Froment, G. (1977). Modeling of thermal cracking kinetics—I: thermal cracking of ethane, propane and their mixtures. *Chemical Engineering Science* 32 (6): 601–608.

37 Yancheshmeh, M.S., Haghighi, S.S., Gholipour, M. et al. (2013). Modeling of ethane pyrolysis process: a study on effects of steam and carbon dioxide on ethylene and hydrogen productions. *Chemical Engineering Journal* 215: 550–560.

38 Feli, Z., Darvishi, A., Bakhtyari, A. et al. (2017). Investigation of propane addition to the feed stream of a commercial ethane thermal cracker as supplementary feedstock. *Journal of the Taiwan Institute of Chemical Engineers* 81: 1–13.

39 Kumar, P. and Kunzru, D. (1985). Modeling of naphtha pyrolysis. *Industrial & Engineering Chemistry Process Design and Development* 24 (3): 774–782.

40 Onel, O. (2017). Advances in modeling, synthesis, and global optimization of hybrid energy systems toward the production of liquid fuels and olefins. Doctoral dissertation. Princeton University.

41 Fogler, H.S. (2015). *Elements of Chemical Reaction Engineering*, 5e. Prentice Hall.

42 Wächter, A. and Biegler, L.T. (2006). On the implementation of an interior-point filter line-search algorithm for large-scale nonlinear programming. *Mathematical Programming* 106: 25–57.

43 Gusain, D., Cvetkovic, M., Bentvelsen, R., and Palensky, P. (2020). Technical assessment of large scale pem electrolyzers as flexibility service providers. *2020 IEEE 29th International Symposium on Industrial Electronics (ISIE)*, 1074–1078.

44 Lee, D.-Y. and Elgowainy, A. (2018). By-product hydrogen from steam cracking of natural gas liquids (ngls): potential for large-scale hydrogen fuel production, life-cycle air emissions reduction, and economic benefit. *International Journal of Hydrogen Energy* 43 (43): 20143–20160.

11

Multiscale Strategies for the Use of Chemicals as Energy Storage Systems

Diego Santamaría, Antonio Sánchez, and Mariano Martín

Department of Chemical Engineering, University of Salamanca, Salamanca, Spain

11.1 Introduction

Renewable energy is plentiful; however, its availability is localized and shows temporal variability, and it is difficult to predict ahead [1]. While the amount can be enough to power humankind, the management requires storage to meet the hourly demand. While batteries have been the first approach to store renewable power, their limitations in terms of weight, costs, and capacity call for alternative or complementary alternatives. Chemicals can play a major role in demand management. E-fuels and chemicals can be produced using renewable energy and later used to produce power following thermochemical or electrical technologies. The production and use of chemicals within the power industry require a multiscale strategy [2] that considers all the way from the chemical to the facility layout and also multiscale in time, from the operation to the strategic decisions involved in the design and facility allocation. In this chapter, we first present this methodology, the challenges, and process system opportunities and, in a subsequent section different, major chemicals are discussed including hydrogen, methane, methanol, and ammonia.

11.2 Methodology

The design of renewable-based energy storage systems is to be considered as a multiscale multiperiod problem. Multiscale is referred both to the size and the time scales (Figure 11.1). While typical chemical industry has benefited from the well-known economies of scale [3] that dictate that a larger process is not proportionally more expensive, the features of renewable resources, their

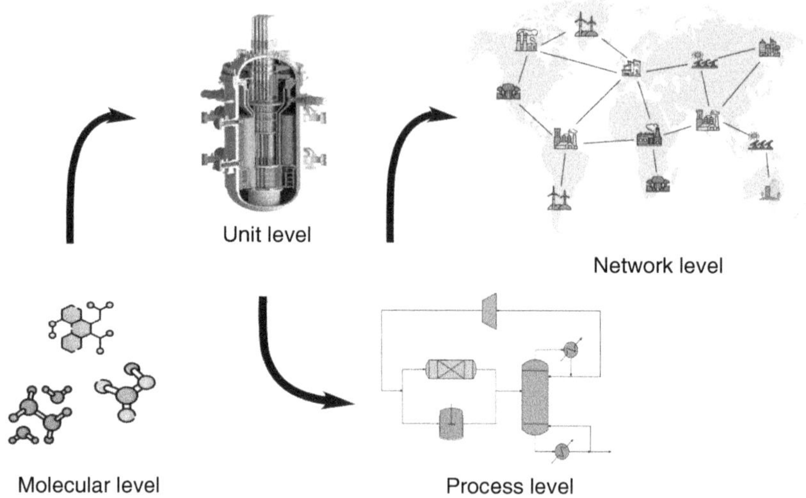

Figure 11.1 Multiscale approach in Power-to-X implementation.

distributed availability, and variability call for novel design approaches such as modularization to make the most of local resources. As a result, process scale up, or more precisely, scale down is a novel concept. Apart from the size, time scale is also important. While traditional processes are designed considering steady-state operation, the use of volatile or non-programable resources results in the need for a more comprehensive approach that integrates planning and scheduling with the units sizing and process synthesis [4]. Beyond the process, the strong link between the location of the resources, the production, or transformation facilities as well as the final consumer requires an enterprise-wide optimization approach [5]. This section comments on the approach for each of the design problems described above.

11.2.1 Process Design

Before the design of the process, different units or technologies are to be evaluated. A hybrid approach is recommended so that experimental data at laboratory and/or pilot plant validate the models used for process synthesis. Laboratory scale aims at providing the technical feasibility of the transformation. The scale up of a technology to an industrial scale is an art that requires the detail analysis of the phenomena involved such as mass, energy, and heat transfer. Reactor design and intensified separation schemes are key for the analysis of the transformation. However, the more complex the unit becomes, the more difficult the modeling. From the conventional mechanistic model such as computational fluid dynamics

(CFD) or discrete element methods (DEM) to the current addition of machine learning to the pool of tools, the design is incorporating data-driven design and optimization approaches to cope with the need for faster solutions [6].

Within process design, two dimensions are to be considered, time and size. The chemical industry has created methods and procedures for process design based on steady state that have been the base for the teaching over the years [3, 7]. However, the use of non-programable resources brings up a new paradigm. Time is to be included in the design problem to size units and decide technologies. As a result, process design must include planning and scheduling toward a simultaneous approach [4]. From a mathematical perspective, the type of problem can be described as a multiperiod (multiobjective) optimization problem under uncertainty. To address such a problem, technics such as stochastic optimization [8], robust optimization [9], or time clustering [10, 11] have been presented in the literature. Robust optimization provides a conservative solution, solving a relatively reasonable size design problem. Stochastic optimization typically results in very large mathematical problems to solve due to the number of scenarios required to capture the variability of the resources. It is here where scenario reduction techniques [12] and clustering [11] play an important role.

From a mathematical standpoint, scenario-based methodologies result in larger mathematical problems and mixed integer liner programming (MILP) formulations are typically favored instead of mixed integer non linear programing (MINLP) [8]. However, chemical processes are known to be highly nonlinear. To capture the thermodynamics, kinetics, phase, and chemical equilibria, it is possible to include a previous stage in the problem formulation where the process is optimized under a base case [13, 14]. Next, surrogate models can be developed for the entire process [14] or for sections of it [15]. The surrogate models must include operating yields but also economics. These input–output models are included in the MILP formulation that considered the time discretization. Perez-Uresti et al. [11] presented a design procedure that allowed capturing the variability of solar and wind for the design of renewable-based utilities plants. The procedure includes two steps, in sample and out of sample procedure to validate the design. Representative days are determined based on clustering techniques. It is possible to select the centroid as representative of a cluster or the closest real day. Hourly discretization level will be used, but a reduced number of days are expected to be enough to represent the design horizon. The in-sample stage presents a suggested design regarding selection of technologies and size that is validated with the out-sample. Robust optimization on the other hand benefits from the fact that the design problem is somehow similar to the one using steady state, once the design conditions are established, the process models can include nonlinear relationships.

The key result behind this paradigm is that the design of the facility is location dependent. There are two main issues, the design of the facility may change depending on the approach used, as a result of how the variability is captured by the different techniques, and the fact that the facility configuration is location dependent. For the same average production capacity, the size of the units and the selection of technologies highly depend on the availability of resources and the largest and lowest production capacity over time.

11.2.2 Process Scale Up/Down

The second dimension of the design problem is the size. Typical chemical industry has been developed on the basis of the economies of scale. The well-known six-tenths rule typically applies for scale up. However, economies of scale are based on the easy transport of both the raw material but also the products that allows locating facilities as large as the economy and market allows. However, process scale down has barely been considered. It is directly linked to new concepts such as distributed manufacturing [16] and modular design [17] where the limited availability of a resource in a region and the complex or its expensive transportation result in the need for developing smaller processing facilities or mobile ones. Scaling up a technology from an already industrial size typically maintains the basic design until the size is too big so that either structural issues or mixing issues take place recommending splitting the operation in a number of them. On the other hand, scaling down an operation may result in the change of technologies. For instance, nitrogen production is typically carried out at a large scale using the double Linde column. Its operation is highly energy demanding for gas compression, and it operates at very low temperatures. However, the production of nitrogen at lower scales can be carried out using technologies such as membranes or pressure swing adsorption (PSA) systems. In both cases, the operating pressures and temperatures are not that extreme. That affects both the safety issues of the operation and the cost. The risk associated to operating at 200 atm and 78 K is completely different compared with the 298 K and 5–30 bar that are used for membranes and PSA. With regard to the cost, while distillation columns benefit from economies of scale, membranes or PSA systems are modular ones that scale by numbers. The cost scales by the number of membrane modules or adsorption beds [18].

In terms of cost estimations, scaling up-scaling down is a complex procedure. Typically, the cost of a unit is related to the production capacity, characterized by a mass or energy flow. This value can be obtained from the analysis of the facility, its mass, and energy balances. It is more complicated with the units limited by heat or mass transfer, where the area is the main costing variable. Examples such as heat exchangers or dryers are typical. In that case, at least preliminary unit design and sizing is required [19]. Units have standard maximum and minimum sizes. When a maximum standard size is reached, an additional unit is to be set up in parallel. In this case, an optimization problem rises, the optimum size of the units depends on how the cost of each on scales. Units are more efficient close to their design point [18]. There is no incentive to heavily over design a unit beyond safety considerations or expected range of operation since it becomes an economic burden. Minimum sizes are handled differently since the technology may change. Typically, multibed reactors are used for ammonia production at large production capacities while multitubular reactors are the selected configuration for smaller production capacities [13]. The selection of technologies and the duplication of units introduces discontinuities to the problem, representing a challenge for the modeling and optimization. Disjunctive programming approaches have been developed to address such formulation [20].

The investment cost, CAPEX, of the facility can be estimated using the factorial method Sinnot & Towler [3, 21] once all the major units of the process are sized and their cost is estimated. This method is well known in the chemical industry, and a set of coefficients are used to estimate the pipping, isolation, and other cost out of the units' costs. Therefore, the scale up and down of the CAPEX can be performed by a scaling factor in the yield of the facility, as long as the particular issues related to technology change and number of units described above are taken into account. Similarly, the OPEX is also a function of the processing capacity. Most of the items are directly proportional to the production capacity, such as raw materials and utilities. However, the amortization and depreciation depend on the CAPEX, and there are other items that represent fixed costs as well such as labor, insurance, or taxes. This is important for the multiscale analysis of a system since it is possible to develop correlations or models in general for the CAPEX of a transformation as a function of the production capacity. Examples for renewable-based facilities can be found in the literature such as concentrated solar power (CSP) [22], geothermal [23], biomass-based facilities [24], renewable ammonia production [25], and X to power facilities [26]. While we can consider the entire transformation facility, it may also be interesting to prepare models by process sections so that a network is synthesized [24]. CAPEX and OPEX show a nonlinear relationship with the production capacity as well as certain discontinuities. Capturing the economies of scale or the modularity of the design results in the formulation of MINLP problems that are hard to solve [27].

11.2.3 Enterprise-Wide Level

The exploitation of the resources of any region integrates the spatial–temporal analysis of the availability, the detailed study of the transformation technologies, and the location of each of the agents. There are two approaches or lines of work to the analysis, the development of generic models and solution procedures or the solution of particular case examples. The first one targets the method, and the second the solution to a particular problem. The first one provides the tools to solve the mathematical problem, and the second is expected to provide guidelines to address the physical one. It is in this last one that the formulation of the problem with real resource data, yields, and process features plays a more important role. The integration of the process transformation through the proper scale up and down is paramount for the solution to be representative and useful for decision-makers. Again, time and space are to be considered and determine the size of the problem. There are a number of challenges. Some of them are similar for other industries or sectors such as the biorefineries [28]

1) The level of detail of the analysis is given by the discretization of the territory and the time discretization used. Note that decision-making can only be performed at a strategic level. From the spatial point of view, strategic studies can only be performed at large-scale, such as entire continents or countries [24], while smaller regions such as provinces [29, 30] are more appropriate for operation of the systems. Ideally both should be solved simultaneously, but there is still a way to go in terms of algorithms and strategies as well as computer capacity to do so. The spatial discretization can be based on administrative divisions such as shires, counties, provinces, regions, or states as well as the use of geographical information systems (GIS). The large number of divisions can lead to large mathematical problems, even intractable. GIS is the step forward providing further resolution and would result in a challenge to address such a mathematical problem as presented in the next subsection. Clustering techniques can be applied. From the temporal dimension point of view, the operation of such systems requires at least hourly definition, due to the variability of solar and wind. However, a fine time discretization together with detailed spatial discretization result is an intractable problem. Efforts toward aggregation via clustering of time and space are so being used to address the problem. Different decomposition-based optimization algorithms have been proposed with particular attention at energy systems where planning and scheduling problems can yield highly complex problems [31]. Some of these decomposition techniques are based on the graph models such as the those proposed by Mitrai et al. [32] or Jalving et al. [33]. Lara et al. [34] introduced a decomposition algorithm based on Nested Benders Decomposition for MILP problems that involved planning and scheduling decisions.
2) **Data Availability:** Typically, governmental databases collect data regarding solar and wind availability, especially NASA and EU. The main issue is the combination of the information on the resources with the discretization level. The model formulation can be independent of the discretization level, but the granulation selected affects the problem size as well as the design decisions.

If data available is given only at a certain discretization, or taken at different points, estimations/extrapolations based on them are to be performed to populate the model with a finer or broader discretization. These estimations are typically based on area or population fraction. GIS are being included for the analysis of the availability and supply chain design to evaluate the actual biomass available, its location, and cost depending on distances and transportation routes [35]. GIS allows accounting for the details of the ground such as roads, rivers, and other natural accidents, and mathematical optimization is included in the framework for decision-making including multiple criteria [36]. This approach can complement the databases available toward a more precise evaluation of the resources available [37] and provide details on the actual allocation of the resources available [38].

Conceptually, the model for a supply chain stem from a facility location problem [39, 40], where the aim is to select the resource, the location of the technology to collect, process, and use that resource as well as the deployment of transformation technologies to obtain a product. The connections between process and resources rely on the process level analysis and the scale up/down procedure as described before. Surrogate models for the facilities must include not only the yield and the feedstock but also the OPEX and CAPEX [41].

In addition, the deployment of a system or a set of technologies represents an opportunity to enhance the economy of a region by generating jobs and wealth. Heras and Martín [41] and Sánchez et al. [29] proposed a number of indexes to identify the relative status of the regions so that the deployment of the production facilities helps generate jobs and wealth in the regions where it is most needed. Unemployment ratio, population density, gross domestic product (GDP) [41] and also aging and depopulation rate among others [29] can be used as indexes for a social objective function that will consider the social aspect into the decision-making for the deployment of the new renewable-based production system. Social is to be added to environmental as objectives for the deployment of the future production system since sustainability must consider all three aspects.

11.3 Cases of Study

In this section, we present the production of hydrogen, methane, methanol, and ammonia, describing the different technologies and studies behind this multiscale approach (Figure 11.2).

11.3.1 Hydrogen

Hydrogen has garnered significant attention in recent years for its potential energy applications. The idea of using hydrogen as an energy source is not new; the

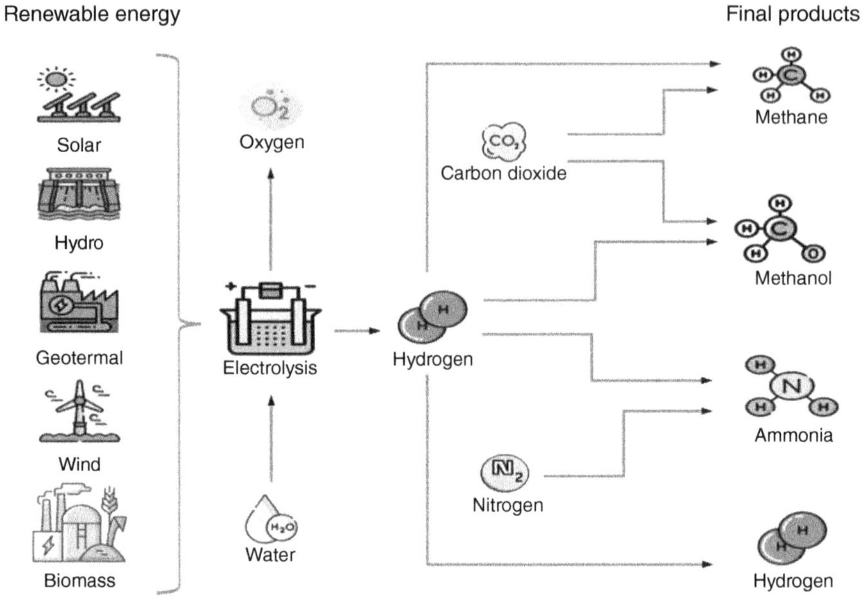

Figure 11.2 Pathways to produce the main Power-to-X products.

concept of a "hydrogen economy" was introduced by John Bockris as early as 1972 [42]. He emphasized the potential to produce green energy with reasonable efficiency, easy transportation conditions, and versatility for use in various sectors, including trucking, cars, shipping, or the chemical industry. According to BloombergNEF [43], hydrogen and its derivatives have the potential to cut roughly one-third of the current global emissions generated by fossil fuels and industrial processes. Furthermore, by 2050, hydrogen is expected to supply about 24% of the global final energy demand. Hydrogen will be essential to decarbonize some important sectors in order to achieve the final goal of climate neutrality around 2050. Oliveira et al. [44] proposed three main areas for green hydrogen: hydrogen as raw material for the chemical industry, hydrogen for transport and heating, and hydrogen as a seasonal storage system. Replacing gray hydrogen with green hydrogen in industrial applications offers a direct substitution opportunity, leading to a surge in related projects [45].

In recent years, extensive research has been conducted on green hydrogen, spanning its production and potential applications across various levels. The main production alternative to produce renewable hydrogen is water electrolysis [46]. Three main alternative technologies have been proposed: alkaline electrolysis, proton exchange membrane (PEM) electrolysis, and solid oxide (SOE) electrolysis. Alkaline electrolysis is the most mature technology nowadays, although the other alternatives are emerging with significant performance improvements. PEM electrolysis offers the advantage of operational flexibility, which is beneficial in renewable energy systems with variable electricity generation. In terms of SOE electrolysis, a higher temperature is required, between 700 and 1000 °C; however, the highest efficiency values in electrolysis can be reached. Coupled with the use of water electrolysis to produce hydrogen, it is necessary to analyze the second essential part of the use of hydrogen, its transformation into electricity. Hydrogen fuel cells are the central units in this area. As in the case of electrolyzers, different technologies have been suggested and studied. Among them, it is possible to find phosphoric acid fuel cells, alkaline fuel cells, molten carbonate fuel cells, solid oxide fuel cells, or polymer electrolyte membrane fuel cells [47].

The first level of study is the modeling of the isolated system, with special attention to electrolyzers or fuel cells. Falcão and Pinto [48] reviewed the modeling issues for PEM electrolyzer with special attention to the equations to predict the cell voltage including reversible voltage, activation losses, ohmic losses, and mass transport losses. Hernández-Gómez et al. [49] also analyzed the one-dimensional (1D) modeling of PEM electrolyzer with a particular focus on dynamic problems. A comparison of a modeling approach using MATLAB/Simulink and real units is provided by Brezak et al. [50] with an error percentage of less than 10% in all cases for an electrolyzer with a nominal capacity of 1 kW. For solid oxide electrolyzers, an electrochemical and thermodynamic modeling approach is presented by AlZahrani and Dincer [51] including also the optimization of the system. An exergy analysis is also proposed in this work yielding, in the optimized case, an exergy efficiency of around 87% working at 1168 K, 8 MPa with a current density of 5000 A/m^2. Mueller et al. [52] review the different modeling approaches proposed for solid oxide electrolysis from zero-dimensional (0D) to two-dimensional (2D) models. Hydrogen fuel cells have also been the subject of different modeling approaches [53]. Ziogou et al. [54] proposed a semiempirical dynamic modeling for PEM fuel cells implemented in gPROMS. The results obtained have been compared showing good experimental agreement. Ota et al. [55] modeled a tubular SOFC based on a 1D approach. Steady-state and transient simulations are presented showing a quick response to a load change. A comprehensive review of modeling issues for SOFCs was provided by Bao et al. [56]. For alkaline fuel cells, Sommer et al. [57] presented a fast dynamic model based on electrochemical principles, and mass, momentum, energy, and species conservation. Based on this, surrogate models for the exchange current density in the electrodes have been generated. CFD tools have been applied to model electrolysis units as well. Ni [58] proposed a 2D model for a planar solid oxide electrolyzer analyzing the influence of different parameters on the performance of the cell. The model is developed in Fortran but a validation procedure using Fluent is performed. The modeling of PEM using these computational tools is also investigated, for example, by Ma et al. [59]. They developed a rigorous model including the electrochemical package to simulate a laboratory-scale PEMEC with three serpentine flow fields. A numerical algorithm is proposed to solve the problem using ANSYS/Fluent showing good agreement with the experimental results. A unified approach is proposed by Hu et al. [60] to model the different hydrogen electrolyzers (alkaline, PEM, and SOE) using multiphysics modeling. This evaluation is performed using COMSOL Multiphysics as software tool. The results show the differences in the electrical efficiency of the different technologies ranging from around 50% to 81%. This rigorous technique has also been applied to fuel cell systems. Xu et al. [61] developed a 2D model for all porous solid oxide fuel cells implemented in COMSOL Multiphysics. The authors analyzed the effects of operating conditions and the electrolyte microstructure properties. Using the same approach, Sivertsen and Djilali [62] analyzed the PEM fuel cells under this three-dimensional (3D) formulation. Research has also explored the modeling of these units using machine learning techniques. For example, Mohamed et al. [63] introduced five different machine learning techniques to predict the behavior of a PEM electrolyzer using different design parameters. Ming et al. [64] summarized the uses of these novel modelling techniques into fuel cells in three main fields: performance evaluation, durability prediction, and application monitoring. Wilberforce and Olabi [65] evaluated the use of neural networks to predict voltage and current from a proton exchange membrane fuel cell. They suggested that the group method of data handling (GMDH) network as the best approach for this specific application. Further research in this area is expected in the coming years in different areas related to green hydrogen production and utilization.

The entire process of producing green hydrogen using water electrolysis has also been simulated and optimized. A holistic evaluation of these alternatives is performed from a technical, economic, and environmental point of view. Terlouw et al. [66] assessed the production of hydrogen under three different scenarios: grid-connected, isolated, and a hybrid situation evaluating different power sources. Current prices of hydrogen are approximately 3.7 €/kg H_2, with potential reductions to about 2.0 €/kg H_2 by 2040. The isolated case (only based on renewables) in areas with abundant wind resources exhibits the lowest green house gas emissions (GHG) emissions with levels around 2.1 kg CO_2 eq./kg H_2. However, the best economic and environmental balance corresponds to hybrid systems. Similarly, Jang et al. [67] conducted a techno-economic analysis of four different hydrogen production methods. In this case, solid oxide electrolysis coupled with a waste heat source shows the best economic results. In this study, production costs are generally higher than those reported by Terlouw et al. [66], ranging from 7 to 10 \$/kg H_2. A Monte Carlo analysis was used to evaluate cost calculation uncertainties. Nami et al. [68] compared different processes using alkaline and solid oxide electrolysis through thermodynamic and techno-economic analysis, incorporating the impact of CO_2 taxes. Their findings indicate current costs of approximately 3–4.5 €/kg H_2, with future costs potentially falling to 1.5–3 €/kg H_2 by 2050. From an environmental point of view, Weidner et al. [69] evaluated the synthesis of green, blue, and gray hydrogen. They concluded that the forecast hydrogen production in the future would transgress the allocated planetary boundaries for climate change regardless of the production technology and power generation source. The extraction and processing of metals for renewable energy infrastructure were identified as the primary contributors to this transgression.

The operation and control of this kind of system are crucial due to the inherent fluctuations associated with the main renewable resources and the possible use of hydrogen as energy storage system to guarantee a proposed demand [70]. Marocco et al. [71] proposed an MILP formulation to optimize the operation of an off-grid insular grid powered by renewables. Batteries and hydrogen are proposed as energy storage systems. In the case of hydrogen, the combination of electrolyzer, hydrogen storage, and fuel cell is included to serve as a storage system. The new optimized system achieved a cost of electricity of 455 €/MWh, competitive with the current analyzed insular system. The use of hydrogen in microgrids is also extensively studied to enhance performance ensuring secure and network-independent operation. Tostado-Véliz et al. [72] proposed a microgrid based on renewable sources and hydrogen. An MILP formulation is developed to determine the optimal scheduling proposing a three-stage solution procedure to account for uncertainty. Ban et al. [73] proposed the combination of an integrated power facility including Power-to-H_2-to-Power with the distribution of a fraction of the hydrogen for mobility applications. In this case, exhaustive search and genetic algorithms are proposed to solve the proposed problem. Yang et al. [74] analyzed the scheduling of hydrogen production in two steps. First, they developed a CFD model to evaluate the performance of an alkaline electrolyzer. Then, an MINLP (solved using BARON [75]) is proposed to determine the optimal operation of the system using a short planning horizon.

In terms of control of hydrogen technologies, Flamm et al. [76] analyzed control strategies for a hydrogen production facility at refueling station. In a first step, they conducted several experiments to develop a dynamic nonlinear model. Then, this was used to design a real-time optimal controller, using an MILP reformulation of the problem. A linear hybrid model predictive controller (HMPC) for hydrogen electrolysis was proposed by de Andrade et al. [77] showing good robustness in the control performance. A linearization and discretization strategy were suggested to obtain an MIQP problem. Zhao et al. [78] employed a data-driven modeling technique to capture the dynamic behavior of a PEM electrolyzer, in particular, to compute power consumption, hydrogen production, and temperature. Based on this, a fuzzy control strategy and a neural network predictive control strategy were implemented. Zhu et al. [79] applied these control strategies to a microgrid using hydrogen as storage alternative. A distributed economic model predictive control (DEMPC) scheme is developed. The global optimum of the problem is reached using an MINLP algorithm. In this work, three controllers were designed for the PV, alkaline electrolyzer, and PEM fuel cell systems. The control strategies for different kinds of fuel cells have also been assessed [80]. Chatrattanawet et al. [81] proposed a model predictive control (MPC) strategy to be implemented in PEM fuel cells. The inlet molar flow rates of air and hydrogen were the manipulated variables of the system showing a good performance guaranteeing the stable operation of the proton exchange membrane fuel cell (PEMFC). An improved control strategy for vehicular fuel cells was proposed by Quan et al. [82]. These authors proposed an enhanced multiple-input multiple-output (MIMO) model predictive control (MPC). The improved MPC alternative can reduce the control root mean square error (RMSE) by around 50%. In this field, Pereira et al. [83] introduced a nonlinear model predictive control (NMPC) for the controllability of a fuel cell in a hybrid electric vehicle. A recurrent neural network (RNN) was incorporated to model the PEM fuel cell, enabling highly accurate representation of the process dynamics. The results obtained were validated in an experimental device with a real 3-kW FC. Yin and Liu [84] introduced quantum parallel model predictive control (QPMPC) for the control of solid oxide fuel cells. The quantum revolving gate of the QPMPC can obtain more information about the optimal solution, significantly enhancing the efficiency and effectiveness of the solution process. A comparison with other control strategies (such as MPC and fuzzy control) was included showing a better performance in terms of fuel utilization.

A broader perspective on hydrogen use is explored in various strategic studies including supply chain, facility location problems, implementation routes, etc. Ehrenstein et al. [85] developed a hydrogen supply chain considering the environmental impact through planetary boundaries, focusing on hydrogen distribution as a road transport fuel in the United Kingdom. The results show that while significant environmental improvements are achievable with hydrogen, operating entirely within all planetary boundaries remains a challenge. The best option, from an environmental perspective, is the production of hydrogen using water electrolysis powered by wind and nuclear energy combined with compressed storage and rail distribution. In this area, Almaraz et al. [86] incorporated social aspects in the design of a hydrogen supply chain for industrial and mobility markets. In particular, four different objective functions have been introduced in the multiobjective and multiperiod MILP problem: levelized cost of hydrogen, global warming potential, safety risk, and social cost–benefit. Hungary is selected as case study revealing a levelized cost of 4.78 \$/kg H_2 with a global warming potential of 3.14 kg $CO_{2\text{-eq}}$ per kg H_2. Robles et al. [87] also evaluated a hydrogen supply chain using three different objective functions (levelized hydrogen cost, global warming potential, and a safety risk index) using genetic algorithms to address the complexities of large MILP problems. A variant of the non-dominated sorting genetic algorithm (NSGA-II) is employed. In addition, these authors have included demand uncertainty in the design of the supply chain using fuzzy concepts. Guerra et al. [88] assessed the role of energy storage in a future energy system based on renewable energy sources. In particular, biomass and hydrogen are considered in a specific region (the United Kingdom). A significant contribution of green hydrogen is expected with hydrogen turbines meeting up to 10% of the electricity demand requiring around 8 TWh of hydrogen storage. Similar analyses have been conducted for other regional areas. For example, for Spain, Brey [89] proposed the use of the surplus of renewable energy sources to produce green hydrogen for blending with natural gas. By implementing this strategy, up to 7.27 TWh of surplus renewable energy could be reutilized, and 2.54 million tons of CO_2 equivalent could be avoided annually. The role of renewable hydrogen in heavy road transport is also analyzed by Valente et al. [90]. They analyzed the implementation of this alternative from an environmental perspective using the relative metric of planetary boundaries. Among the Power-to-hydrogen alternatives, the use of wind electricity emerges as the most promising alternative.

Finally, it should be mentioned that the aforementioned works have focused on the use of water electrolysis technology. However, other alternatives for the production of hydrogen from electricity are being studied such as photocatalysis [91], thermochemical cycles [92], or plasmolysis [93].

11.3.2 Methane

The production of methane (e-methane) through Power-to-gas processes is also receiving extensive attention [94]. The main drivers are the challenges associated with the use of green hydrogen. Among them, the low volumetric energy density and the difficult storage and transport conditions emerge as the main barriers to hydrogen deployment. The use of methane is interesting for several reasons. Its synthesis is based on the combination of green hydrogen with CO_2 aligning with the carbon capture, storage, and utilization (CCUS) strategy. The composition of the produced e-methane is similar to the current natural gas; therefore, a direct substitution could be produced in several units. This e-methane can be stored and transported using the existing extensive network for natural gas. In addition, synergies between e-methane and biomethane (mainly from anaerobic digestion) are expected.

As in the case of green hydrogen, extensive multiscale research has been conducted during the last years in order to develop and evaluate this technology to be used for different energy purposes. Related to the modelling of the units, different efforts have been performed in the simulation and optimization of the methanation reactors. Schlereth and Hinrichsen [95] evaluated the modeling of an externally cooled fixed-bed reactor for the methanation of CO_2. They concluded that intraparticle mass transfer and external heat transfer must be considered to accurately describe the temperature and concentration profiles. The use of a fixed-bed membrane reactor is also considered to improve temperature control. The use of different reactor designs has also been proposed such as fluidized bed [96] or helical fixed bed [97]. An optimal catalyst particle design approach for these methanation reactors is addressed by Zimmermann et al. [98]. They analyzed the influence of particle activity, permeability, and heat conductivity on reactor performance. A nonlinear programming problem is formulated to optimize the catalyst characteristics reaching CO_2 conversion values above 90%. CFD tools have also been applied for the rigorous modeling of methanation reactors. Di Nardo et al. [99] investigated the modeling of a cooled multi-tubular reactor for CO_2 methanation through 3D CFD simulations using ANSYS-Fluent. Two fluids have been considered as coolants: nitrogen and oil. A validation of the model with experimental results is performed showing good agreement. Dynamic modeling of these reactors is crucial due to the fluctuations inherent in renewable energy sources. Bremer et al. [100] developed a dynamic 2D reactor model for methanation optimizing the reactor start-up avoiding hot spot formation using an NLP formulation. The model proposed is also adaptable to account for fluctuations or disturbances during continuous operation. A similar approach is proposed by Kreitz et al. [101] for a micro-structured fixed-bed reactor. All these works are based on the combination of CO_2 and H_2, previously produced using green methods as discussed in the previous section. However, the direct electroreduction of CO_2 to methane is also proposed [102, 103]. However, deeper research is required to implement these alternative routes at industrial levels.

At the process scale, several works have analyzed the production of e-methane from renewable sources. Davis and Martín [104] evaluated the synthesis of methane using an integrated process combining wind/solar energies, water electrolysis, and CO_2 methanation using an equation-based optimization. The process is implemented across different regions in the United States and Europe, selecting the most appropriate power source. In Spain, methane production costs of around 0.4 €/m^3 are expected. In the same line, Peters et al. [105] investigated the synthesis of e-methane using different CO_2 sources (biogas, tail gas, direct air capture, etc.) using Aspen Plus. The methane cost ranges from 3.51 to 3.88 €/kg, which is not competitive with current natural gas costs but does represent a significant improvement in GHG emissions. Recent analysis reduces the cost of renewable methane up to values around 1.75 $/kg [106]. Different works have also introduced the environmental analysis of the Power-to-methane alternative. Sternberg and Bardow [107] performed a life cycle assessment of the Power-to-methane production comparing it with the Power-to-syngas route. The global warming threshold for electricity supply using synthetic methane is equal to 0.082 kg CO_2 eq/kWh in a steady-state operation. This value can be reduced if 100% wind (0.012 kg CO_2 eq/kWh) or solar (0.046 kg CO_2 eq/kWh) is used in an intermittent operation. Blanco et al. [108] explored the potential role of e-methane in the European Union's future decarbonized energy system. In the most favorable scenario, Power-to-methane reaches 670 GW of installed capacity covering around 75% of the current natural gas demand. A 4% reduction in climate change impacts and 9% fossil depletion is expected if this scenario is implemented. De Roeck et al. [109] compared different biological and catalytic methods to combine hydrogen and CO_2. Both processes have been modeled using Aspen Plus yielding similar levels of global warming potential, around 27 kg CO_2 eq/MWh CH_4 since more than 90% of the GHG emissions in these processes were attributed to the electrolysis step.

Some works have addressed the operation and control of these Power-to-methane processes although the fluctuations are mainly absorbed by the electrolyzer. Bruns et al. [110] explored the operation of a Power-to-methane facility in Germany, highlighting that the flexibility of the methanation reactor is primarily constrained by the availability and flexible supply of carbon dioxide. A nonlinear dynamic optimization problem is developed to determine the optimal scheduling of the integrated facility. Belderbos et al. [111] considered the integration of the electricity and gas systems to manage the fluctuations in the new renewable resources. An MILP formulation is proposed including the current infrastructure and the possible Power-to-methane facilities. The methodology is applied in Belgium, as case study, showing the possibility of integrating e-methane into the current electricity and gas systems. Between 13.5 and 43.9 GWhe surplus electrical energy would be converted to synthetic methane reducing the cost of the entire system. In terms of control, Li et al. [112] analyzed the dynamics and control of a low-temperature methanation reactor. A nonlinear dynamic simulation is performed using Aspen Dynamics with mass and heat recycle. They proposed a nonlinear control strategy using the high selector module in Aspen Dynamics allowing to maintain the reactor variables. A cooled recycle reactor with a multilayered catalyst is proposed and analyzed by Jeong et al. [113]. Load variation is a significant disturbance impacting the operation of the methanation reactor. A control strategy with dynamic cascading is proposed to address these fluctuations. Two different control schemes have been evaluated: one maintains the superficial velocity in the first layer by adjusting the recycling ratio, and the other maintains a constant recycling ratio in the reactor. In this same area, Colelli et al. [114] analyzed from a dynamic perspective two different reactor configurations: adiabatic and cooled. After a steady-state design, the authors proposed a feedback control scheme for the entire process. Three different hydrogen loads have been evaluated showing that the cooled configuration responds more effectively and quickly to the imposed variations.

The potential deployment of e-methane has also been investigated during recent years by proposing different implementation strategies for this chemical and fuel. Blanco et al. [115] analyzed the potential of Power-to-methane in a future decarbonized scenario by 2050. They identified limited CO_2 storage, high CO_2 reduction targets, and high renewable penetration as the main drivers of the use of e-methane in the new energy system. The expected capacity for e-methane is estimated around 40 GW, increasing up to about 120 GW if the use of liquefied methane gas as maritime fuel is considered. Similarly, Yilmaz et al. [116] proposed a multiperiod linear optimization problem to assess the decarbonization of the electricity market using Power-to-methane. A graph representation is included representing the energy flows of the network. This analysis also included the influence of CO_2 prices on the road to decarbonization until 2050. The authors concluded that the combination of a carbon tax of 160 €/t CO_2 plus the use of e-methane is effective in decarbonizing the electricity grid by 2050. Sánchez et al. [29] assessed the potential use of e-methane as energy storage system. In this case, an integrated system is proposed including wind, solar, and biomass energies together with three different storage technologies: batteries, hydrogen, and methane. The authors studied different locations in Spain including an economic and social perspective. An MILP problem is formulated including the detailed scheduling. Methane is mainly used for seasonal storage purposes allowing to meet the demand regardless of the wind/solar availability. The cost of electricity of the integrated system is around 70–120 €/MWh. From a supply chain perspective, Carrera and Azzaro-Pantel [117] developed an MILP problem implemented in general algebraic modelling system (GAMS) to determine the best production and transportation network of a system with both a hydrogen and a methane demand. A multiperiod approach is considered (2035–2050) considering several renewable energy sources. The region of Occitanie (France) is considered as case study, and the optimization is performed under two different objective functions: economic (total annual cost) and environmental (global warming potential). Leonzio and Zondervan [118] integrated the carbon capture, storage, and utilization supply chain together with the e-methane production. A MILP model is developed by taking as case study the 10 regions in Italy with major carbon dioxide emissions. The proposed network is not economically profitable; however, with the appropriate incentives, the cost of methane could reach 19 €/MWh. This could reduce methane imports and improve energy security.

11.3.3 Methanol

The use of e-methanol is gaining attention in the last years as a powerful alternative to decarbonize some hard-to-abate sectors toward climate neutrality. In particular, the use of green methanol as feedstock for the chemical industry and as fuel for road or marine transport is widely proposed. The Power-to-methanol alternative is based on the combination of hydrogen and CO_2; therefore, the cost of carbon capture and hydrogen production directly influences the cost of e-methanol. The current prices are around 1200–1800 USD/t; however, significant improvements are expected with future costs of about 250–630 USD/t by 2050 [119]. Methanol is liquid at ambient pressure and temperature; therefore, the storage and transportation conditions are straightforward compared to other energy carriers proposed. In addition, methanol is a versatile fuel which can be used in both internal combustion engines and fuel cells. IRENA and Methanol Institute [119] forecasted an increase in the global methanol production from levels around 100 Mt in 2020 to values about 500 Mt by 2050 in a decarbonized scenario.

The use of e-methanol has been extensively researched covering all scales from reactor studies to development roadmaps. The most extended technology to produce e-methanol today is hydrogenation of CO_2 to produce methanol. Leonzio and Foscolo [120] analyzed the CO_2 hydrogenation using a packed bed reactor. The authors developed a 2D model for the methanol reactor considering two different catalyst distributions: structured and nonstructured. The problem is solved in MATLAB showing the structure configuration a CO_2 selectivity to methanol of around 82% while this value, in the case of nonstructured reactor, drops to around 54%. Torcida et al. [121] also evaluated the methanol production reactors. They focused on a multibed catalytic configuration with a first step based on a 1D modeling of the problem and a second step where the reactor configuration is optimized. A genetic algorithm is proposed to optimize these units using MATLAB as software. For methanol production, the results suggested the use of six adiabatic beds operating at 50 atm with a H_2/CO_2 feed ratio of 3.5. The results are significantly influenced by the cost of renewable H_2. Pavlišič et al. [122] analyzed the performance of methanol reactor based on CO_2 hydrogenation using a CDF approach. Kinetic Monte Carlo simulations were performed to study the evolution of a catalytic surface. The model is implemented in OpenFoam showing good agreement with the experimental results. Cui and Kær [123] proposed and evaluated three different designs for CO_2 to methanol: adiabatic, water-cooled, and gas-cooled reactors. Three different approaches have been considered to cover the different issues associated with CO_2 hydrogenation: thermodynamic analysis, CFD methods, and pseudo-homogeneous models. The water-cooled reactor demonstrated advantages in efficient heat removal, maintaining low hot-spot temperatures, and offering a relatively wide range of inlet temperatures for control.

Several works have been devoted to the hydrogenation of CO_2 toward methanol from a process design perspective. Among these efforts, Leonzio et al. [124] combined reactor and process design for methanol production from CO_2 and hydrogen. Three different alternatives have been assessed: once-through reactor, reactor with recycle of unconverted gases after separation of methanol and water, and reactor with membrane for selective steam separation. At 493 K and 55 bar, the conversion of the reaction is 38%, 69%, and 58% for the different reactor configurations. Pérez-Fortes et al. [125] developed a technical, economic, and environmental assessment of the production of e-methanol coupling this process to a coal power plant. The proposed process is simulated in Chemcad to obtain the key performance indicators. To make the CCU plant financially attractive, the price of methanol needs to increase by nearly a factor of 2, H_2 costs should decrease by approximately 2.5 times, or CO_2 should have a value of around 222 €/t. Cordero-Lanzac et al. [126] evaluated the production of e-methanol with particular attention to the different required catalysts for the methanol synthesis. The results suggest that for an e-methanol plant to be profitable, either minimum H_2 prices (1.5 $/kg) or severe CO_2 taxation (300 $/ton) are required. The environmental analysis determined that 1.75 tons of CO_2 can be abated per ton of methanol if renewable electricity is employed. Martín [15] proposed the design of a methanol production plant under uncertainty. In particular, a two-stage multiperiod stochastic MILP problem is proposed to be solved by GAMS. Two different power generation technologies have been evaluated, wind and solar. The problem formulation is applied to two cases of study, Spain and the United Kingdom. The cost of e-methanol is 0.41 €/kg, in Spain (where solar is the preferred source), and 0.45 €/kg, for the United Kingdom (where wind energy is the dominant energy source). In the process design of Power-to-methanol processes, Zhang et al. [127] introduced the use of a deep neural network to optimize the performance of the system. In addition to the most extended pathway to produce e-methanol based on the combination of CO_2 and hydrogen in a catalytic reactor, the use of the electrochemical reduction of CO_2 is also explored as a promising alternative [128]. Adnan and Kibria [129] compared different routes to produce e-methanol including the combination of hydrogen (from water electrolysis) and CO_2, the combination of hydrogen (from water electrolysis) and CO (from CO_2 electrolysis), and the direct approach based on the electroreduction of CO_2 with water. An Aspen simulation for each of these alternatives is performed to obtain the data required to determine the economics and the environmental performance of each pathway. For the case of the electroreduction of CO_2, the current selectivity of the reaction and current density are below the requirements that guarantee a technically and economically feasible process.

In addition to the production of methanol using renewable electricity, significant research has also focused on its conversion back into energy, positioning methanol as a potential energy storage and carrier system. Two main routes have been proposed: the thermochemical conversion (based on the combustion of methanol) and the electrochemical conversion (based on a fuel cell unit). In the thermochemical alternative, Liu et al. [130] considered the integration of a methanol gas turbine together with solar thermal energy, which is used to decompose the inlet methanol into hydrogen and carbon monoxide achieving a primary energy ratio of up to 75.42%. In the electrochemical pathway, Wu et al. [131] analyzed a hybrid system combining a solid oxide fuel cell (SOFC) and a PEMFC, achieving an energy efficiency of up to 66.2%. The flue gas from the SOFC, containing unreacted fuel components such as CO and H_2, is treated and then sent to a PEMFC to enhance the overall energy performance of the process. A comprehensive evaluation is presented by Blanco et al. [132] who assessed the different transformation alternatives to convert methanol into electricity from a process design perspective. In particular, two different routes were evaluated. The first one is based on the combustion of methanol but, due to its flammability properties, a co-combustion with hydrogen is proposed. The second alternative introduces the use of a direct methanol fuel cell (DMFC). The electrochemical alternatives show a higher cost (around 0.6 €/kWh) compared to the thermochemical one (about 0.3 €/kWh).

The operation and control of these methanol production systems has also attracted attention to be able to achieve a flexible operation compared to the traditional steady-state synthesis. Mucci et al. [133] analyzed it due to the variability in the electricity price and the power availability in a renewable scenario. In order to achieve a flexible operation, batteries and hydrogen storage are included in the Power-to-methanol process. The authors proposed a design and scheduling problem formulated as an MINLP and implemented in GAMS (solved with BARON). The study emphasized the critical role of storage systems, particularly hydrogen storage. In this same research area, Zheng et al. [134] evaluated the operation of an e-methanol production plant connected to the grid. A daily scheduling horizon is considered in the optimization problem formulated as an MILP. The authors included the prediction of the electricity price using a multilayer perception neural network. The uncertainty of the electricity price is introduced in the MILP formulation through a Wasserstein metric. The results show that the prices of green methanol range from 584 to 1146 €/t for different CO_2 prices and the performance of wind turbines. From a dynamic perspective, Cui et al. [135] analyzed the synthesis of e-methanol from CO_2 and H_2 using Aspen Plus Dynamics. The load change (LC) operations of the system, along with key performance indicators (KPIs) such as energy efficiency and the dynamic characteristics, were examined. In addition, and based on the dynamic analysis, a surrogate model is developed using nonlinear autoregressive exogenous model

(NARX) techniques. Process control is still under development for the fluctuation associated with renewable resources. Chiou et al. [136] evaluated different reactor configurations and then, a control strategy is proposed. The authors focused on the methanol synthesis neglecting the CO_2 capture or hydrogen production. The control strategy is implemented in Aspen Plus Dynamics showing a good rejection of disturbances such as changes in flowrate or composition.

Finally, different works have studied the implementation strategies of Power-to-methanol processes. Galán et al. [137] evaluated the potential of e-methanol within the carbon capture and utilization strategy in Spain. In particular, two different carbon capture alternatives have been evaluated, monoethanolamine (MEA) and direct air capture (DAC). A facility location problem is proposed and formulated as an MILP. Methanol prices of around 1–2.6 €/kg have been obtained. Ong et al. [138] evaluated the potential use of methanol as an energy carrier in the electricity sector. They analyzed the synthesis of methanol and also its reforming to produce a mixture of CO_2 and H_2. CO_2 is recycled to the methanol production section (including transportation) and hydrogen is sent to a fuel cell unit. Each of these stages has been simulated in Aspen Plus. An economic and environmental analysis is performed showing that the cost of electricity ranges from 106 to 494 USD/MWh with associated CO_2 emissions of around 0.2–0.4 t/MWh in a future scenario. At the supply chain level, Sánchez et al. [139] proposed the use of e-methanol among other chemicals as energy carrier in a future decarbonized energy system. The integrated supply chain determines the different facilities required and the transportation of the chemicals together with the optimal operation of each facility. A large-scale MILP is formulated proposing a heuristic decomposition to solve this problem. The use of methanol emerges as an attractive alternative, in particular, for long-term storage due to its favorable storage and transport conditions. In addition, the use of e-methanol has also been evaluated from a social perspective using a social life cycle assessment methodology [140].

11.3.4 Ammonia

Ammonia is one of the chemicals that is attracting attention to be used as fuel, energy carriers, or energy storage systems. The main advantages of ammonia are that it is liquid at reasonable conditions and there is a great deal of industrial experience in its handling. In addition, ammonia is a carbon-free chemical; therefore, no CO_2 emissions are involved in its direct utilization. This factor is essential in a future decarbonized society. The current ammonia production is around 200 Mt/year, however, according to the predictions of IRENA and AEA [141], by 2050, if the Paris agreement is met with an increase in the global temperature below 1.5 °C, the demand of ammonia rises to about 700 Mt/year. The current cost of green ammonia for the best locations in terms of solar and wind availability is around 700 USD/t; however, significant improvements are expected in the coming years with levels of around 300 USD/t by 2050 [141]. Although different green ammonia production alternatives have been proposed [142], this chapter focuses on the synthesis of e-ammonia using Power-to-ammonia processes.

Recently, research has been developed in the area of green ammonia from a multiscale perspective including from reactor design to define a roadmap to a future deployment of this technology. Khademi and Sabbaghi [143] evaluated three different ammonia reactor configurations including internal direct cooling reactor (IDCR), adiabatic quench cooling reactor (AQCR), and adiabatic indirect cooling reactor (AICR). The authors developed a 1D model which is optimized using the differential evolution algorithm (stochastic optimization). The ammonia synthesis reactor configurations were ranked from most to least favorable as follows: IDCR, 3-bed AICR, and 3-bed AQCR. Flórez-Orrego and de Oliveira Junior [144] evaluated the design of an ammonia synthesis reactor with special attention to the exergy behavior. A 1D model is proposed including mass, energy, and momentum balances to characterize the reactor performance. The authors proposed the optimization of the design to minimize the exergy losses in the unit. To solve this, the sequential quadratic programming (SQP) tool in Aspen Hysys is used. The exergy destruction breakdown reveals that the ammonia converter and the refrigeration system together account for 71–82% of the total exergy destruction in the ammonia loop. Fahr et al. [145] evaluated the part-load behavior of ammonia synthesis reactors due to the variability associated with renewable resources. A bi-objective optimization of an ammonia reactor is performed looking for minimum cost and maximum flexibility. The optimization problem was solved in OPTISIM using the SQP solver. A trade-off between load flexibility and plant cost is observed. They found that the minimum achievable load can be reduced from 22% to 10% with an additional reactor cost of only 3% to 7%, primarily due to an increase in the heat exchanger area.

At process level, several studies have addressed the technical, economic, and environmental aspects of Power-to-ammonia processes. Sánchez and Martín [25] proposed a green ammonia production process based on the combination of water electrolysis, air separation through cryogenic distillation, and the Haber-Bosch process to produce ammonia. An equation-based optimization approach is followed by solving a multiperiod problem formulated as an MINLP. The results show that a large investment is required, around 1500 M€, if the solar/wind power generation is considered. The price of ammonia is around 1.2–1.3 €/kg. Demirhan et al. [146] compared the production of ammonia using green alternatives (biomass, wind, or solar) with the business-as-usual case where natural gas is the main raw material. A process synthesis and global optimization approach is followed by solving a large-scale nonconvex MINLP.

The best alternative in terms of cost to compete with natural gas is biomass and wind and solar alternatives require further improvements. Zhang et al. [30] also compared the two main alternatives to produce green ammonia: biomass-to-ammonia and Power-to-ammonia. The overall energy efficiency is higher in Power-to-ammonia (around 74%) compared to biomass-to-ammonia (around 44%) and even with the current process (about 61%). As other authors proposed [147], biomass-to-ammonia is closer to economic profitability with respect to current processes with values of around 450 $/t. The Power-to-ammonia alternatives require further reduction in the cost of electrolyzer and renewable electricity. Del Pozo and Cloete [148] proposed an analysis of blue and green ammonia synthesis. In particular, for blue ammonia, the authors studied the gas switching reforming (GSR) concept for H_2 and N_2 co-production from natural gas with CO_2 capture. The models were implemented in Unisim Design for the different units involved. The author demonstrated the attractive cost of blue ammonia with the new GSR technology yielding a cost of around 330 €/t. By 2050, in a fully decarbonized society, the cost of green ammonia from renewable energies will be higher with levels of 400–700 €/t based on the particular location. D'Angelo et al. [149] included the environmental perspective in the Power-to-ammonia analysis. They evaluated different ammonia production alternatives including steam methane reforming with carbon capture and storage, biomass gasification, or water electrolysis powered by various energy sources. The different processes were simulated using Aspen Hysys and the results are the basis to analyze the process using the life cycle assessment (LCA) metrics and planetary boundaries. Among the analyzed options, ammonia produced from H_2 via PEM electrolysis using hydropower is the most sustainable. Biomass scenarios with carbon capture and sequestration (CCS) perform well in terms of climate change; however, they perform poorly in other categories, such as biosphere integrity, nitrogen flow, and freshwater use. In addition to the synthesis of ammonia using the Haber-Bosch process but using renewable feedstock, other variations have been suggested. Andrés-Martínez et al. [150] analyzed a system that substitutes condensation by absorption to improve overall system performance. This allows it to operate at lower pressures. However, the main transformation in the ammonia synthesis is related to the use of the electrochemical reduction of nitrogen [151]. This method presents high potential with the possibility of leaving behind the Haber-Bosch process; however, its current development is low, and more research is required in this area. In this area, Gomez et al. [152] performed a techno-economic analysis of this alternative green ammonia production. The ammonia cost for this alternative ranges around 1000–1400 €/t of green ammonia.

The utilization of green ammonia for energy purposes also requires the transformation back into energy. Different devices have been explored for several final uses such as electricity generation, hydrogen production, or ammonia as fuel.

Extensive research covers these crucial areas toward an effective deployment of green ammonia as energy storage/carrier system. Sánchez et al. [26] developed an integrated process to transform green ammonia into electricity using a combined cycle. The authors proposed the partial decomposition of ammonia to overcome the limited flammability of this fuel. The energy efficiency of the overall process is around 40–50% with costs of electricity of around 0.2–0.6 €/kWh. Cesaro et al. [153] also evaluated different alternatives to transform green ammonia into electricity considering three different phases: 100% hydrogen (produced from ammonia), 70% NH_3–30% hydrogen (produced from ammonia), or 100% ammonia. They forecasted a cost of electricity of about 167–197 $/MWh by 2040 assuming a green ammonia cost of 380 $/t. The use of direct ammonia fuel cells is also receiving great attention; however, the current efficiencies are still low, and further requirements are required to reach competitive levels. Siddiqui and Dincer [154] developed an energy and exergy analysis of a direct ammonia fuel cell (DAFC) based on an anion exchange membrane and a molten alkaline electrolyte. The levels of energy and exergy efficiency are 12.1% and 13.8%, still far from the values needed to be competitive. The utilization of ammonia as hydrogen carrier is also widely proposed due to the possibility of storing and transporting ammonia in the easiest way. At this point, the use of ammonia decomposition is suggested as the most effective alternative to produce hydrogen from ammonia. Devkota et al. [155] analyzed hydrogen production using ammonia decomposition from a technical, economic, and environmental point of view. A multiple catalytic-packed bed reactor with intermediate heating is selected using Aspen Plus and MATLAB to simulate the process. The overall efficiency of the process is 79% with a hydrogen cost of 6.05 $/kg. The price of ammonia represents a large fraction of the total cost with around 65%; therefore, a reduction in the cost of Power-to-ammonia could reduce significantly hydrogen cost. Wind energy as energy source involves the lowest global warming potential (0.014 kg of CO_2/kg of H_2). Finally, ammonia is also explored as fuel for different transportation applications. For example, Sánchez et al. [156] evaluated the use of ammonia as maritime fuel proposing a holistic design of the propulsion process. The introduction of a selective catalytic reduction unit controls the nitrogen oxide emissions below 0.5 g/kWh. The operating cost of the system is around 230 €/MWh, higher than the current fossil fuels but with great potential toward a future implementation in a green scenario.

Extensive research in the area of operation of green ammonia production processes has also been developed during recent years. Allman et al. [157] proposed a scheduling-informed design for a wind-powered ammonia generation facility. An MILP formulation is proposed by applying a receding horizon to cover the entire year updating wind and load forecasts. The selected case study is a specific location in Minnesota with a given annual anhydrous ammonia demand.

The results showed that good scheduling decisions allow to significantly reduce the operating cost of the system. Kelley et al. [158] introduced demand response techniques into the operation of an ammonia facility (focusing only on the ammonia production section) accounting for process dynamics. The optimization problem is solved in gPROMS reaching a reduction in the peak power consumption of around 3–8%. Wu et al. [131] also evaluated the design and scheduling of an ammonia plant based on the Power-to-ammonia concept. An MILP formulation is proposed to be solved in MATLAB considering a time horizon of one year. Two different scenarios were evaluated: grid-connected and isolated. The cost in the stand-alone scenario increases by 41.1% compared to the grid-connected case. In terms of control, special attention should be paid to ammonia reactors because traditional operation is based on a steady-state system. However, further analysis in this area is expected. Rosbo et al. [159] developed a dynamic model for an ammonia reactor system, which consists of a three-bed quench-cooled adiabatic reactor and a feed-effluent heat exchanger. Dynamic simulations revealed significant temperature oscillations in the reactor system when the feed temperature approached the extinction point. In addition, open-loop simulations indicated that the uncontrolled reactor system becomes unstable when the synthesis load is changed. A regulatory control structure consisting of combined feed-forward and feedback was proposed to achieve good controllability results with the exception of an abrupt change in the load. Kong et al. [160] analyzed the introduction of a nonlinear model predictive control scheme to tackle the fluctuations in the feed flowrate in ammonia synthesis. The system considered is based on three interstage-cooled reactor beds and a flash separator. The proposed control scheme aims to regulate the reactor temperatures, separation pressure, and liquid volume in the flash tank during feed transitions. Based on the results obtained, the authors conclude that after a change in the feed it takes 10 minutes to reach a steady state in the reactor temperature and 40 minutes in the case of the separation pressure. Oliveira Cabral et al. [161] proposed a learning-based model predictive control for the regulation of temperature and outlet concentrations in an ammonia reactor. In a first step, a dynamic model is developed including thermodynamics, kinetics, and transport phenomena. To reduce the computational burden, a long short-term memory (LSTM) network approximated the system dynamics. Based on this surrogate model, a model predictive control is proposed effectively regulating the maximum temperature and the outlet concentration at the targeted steady-state values.

High-level studies related to the implementation, supply chain, multiperiod deployment, etc. problems have also been addressed in the case of green ammonia. Nayak-Luke and Bañares-Alcántara [162] evaluated the isolated production of ammonia considering 534 locations in 70 countries. In the best locations with the current technology, a cost of ammonia of 473 $/t can be achieved. This cost could fall to 310 $/t if the expected technological improvements come true. Some countries, particularly in the north of Africa, have been identified as potential areas for low green ammonia cost; however, some political and financial limitations could be a barrier to its development. Wang et al. [163] conducted a global analysis of green ammonia shipping corridors by designing a network of alternative fuel production sites, transportation links, and bunkering ports capable of supporting a significant portion of global marine fuel demand. An MILP supply chain design model, which integrates a facility location formulation and a flow-capturing location formulation, was used. The solution indicated that 115 Gt of green ammonia is required to meet the fuel demand considered, with an associated cost of $341/t. The introduction of green ammonia in the current supply chain network is analyzed by Mitrai et al. [164] considering price uncertainty from existing producers. A two-stage stochastic programming approach is followed under an economic objective function. For a specific case study in Minnesota, the introduction of price uncertainty led to an increase in the share of green ammonia in the network to reduce the volatility of the conventional ammonia market. Kim et al. [165] analyzed a global supply chain including possible synergies between hydrogen and ammonia. A multi-time scale model is proposed including uncertainties from both the supply and demand sides. A two-stage stochastic programming problem is formulated with a hierarchical solution method to tackle varying timescales. The results showed the great potential of green ammonia since its superior storage and transportation properties significantly mitigate uncertainties within the supply chain. Fasihi et al. [166] evaluated the potential of implementation of green ammonia in different locations around the world. The authors considered wind and solar technologies as possible power generation systems. The current prices of green ammonia for the best locations around the world are around 440–630 €/t; however, a significant reduction is expected in the coming years with levels of 260–290 €/t by 2050. The introduction of a carbon tax could make green ammonia competitive with current technologies. Galimova et al. [167] studied the potential of green ammonia trading via pipelines and shipping including the production and transportation costs. The idea is to evaluate whether local production is attractive or whether it is more efficient to transport it from low-cost areas. Imports from Chile were considered economically feasible for Germany and Finland, with costs being 10–33% lower. However, in the case of Spain, after 2040, local solar resources are expected to enable local e-ammonia production at the same or lower costs compared to imported e-ammonia.

11.4 Conclusions

This chapter has presented the use of chemicals as a means to store renewable energy for long periods. To achieve that, a multiscale methodology is required due to the distributed availability of the resources used to produce the chemicals and the temporal availability. This methodology is presented for several cases of study including hydrogen, methane, methanol, and ammonia presenting their production and the alternative technologies available for that, costs and how they can be used to produce power. While some of the technologies are already mature and deployed at an industrial scale, most of them are still at conceptual or pilot plant level and further research is required to bring them up to the market. However, the use of chemicals for long-term energy storage is a promising alternative that also links the chemical and power industry bringing additional challenges and opportunities for both sectors and the need to work together.

Acknowledgment

This work was supported by the Regional Government of Castilla y León (Junta de Castilla y León) and by the Ministry of Science and Innovation MICIN and the European Union NextGenerationEU/PRTR (H2MetAmo project – C17.I01.P01.S21).

References

1 IRENA (2024). Renewable energy statistics 2024. International Renewable Energy Agency, Abu Dhabi.
2 Floudas, C.A., Niziolek, A.M., Onel, O., and Matthews, L.R. (2016). Multi-scale systems engineering for energy and the environment: challenges and opportunities. *AIChE Journal* 62 (3): 602–623.
3 Sinnott, R.K. and Towler, G. (2020). *Chemical Engineering Design, Principles*, 6the. Oxford. UK: Elsevier.
4 Zhang, Q., Martín, M., and Grossmann, I.E. (2019). Integrated design and operation of renewables-based fuels and power production networks. *Computers & Chemical Engineering* 122: 80–92.
5 Grossmann, I.E. (2005). Enterprise-wide optimization: a new frontier in process systems engineering. *AIChE Journal* 51 (7): 1846–1857.
6 Bradley, W., Kim, J., Klwein, Z. et al. (2022). Perspectives on the integration between first-principles and data-driven modelling. *Computers & Chemical Engineering* 166: 107898.

7 Biegler, L., Grossmann, I.E., and Westerberg, A.W. (1997). *Systematic Methods of Chemical Process Design*. Prentice Hall.

8 Li, C. and Grossmann, I.E. (2021). A review of stochastic programming methods for optimization of process systems under uncertainty. *Frontiers in Chemical Engineering* 2: 622241.

9 Li, Z. and Ierapetritou, M. (2008). Robust optimization for process scheduling under uncertainty. *Industrial & Engineering Chemistry Research* 47 (12): 4148–4157.

10 Mallapragada, D.S., Papageorgiou, D.I., Venkatesh, A. et al. (2018). Impact of model resolution on scenario outcomes for electricity sector system expansion. *Energy* 163: 1231–1244.

11 Perez-Uresti, S.I., Lima, R.M., Martín, M., and Jimenez-Gutierrez, A. (2023). On the design of renewable-based utility plants using time series clustering. *Computers & Chemical Engineering* 170: 108124.

12 Römisch, W. (2009). Scenario reduction techniques in stochastic programming. In: *International Symposium on Stochastic Algorithms*, 1–14. Berlin, Heidelberg: Springer Berlin Heidelberg.

13 Martín, M. (2016a). *Industrial Chemical process. Analysis and Design*. Elsevier Oxford.

14 Martín, M. and Grossmann, I.E. (2017). Optimal integration of a self sustained algae based facility with solar and/or wind energy. *Journal of Cleaner Production* 145: 336–347.

15 Martín, M. (2016b). Methodology for solar and wind energy chemical storage facilities design under uncertainty: methanol production from CO_2 and hydrogen. *Computers & Chemical Engineering* 92: 43–54.

16 Lara, C.L., Bernal, D.E., Li, C., and Grossmann, I.E. (2019). Global optimization algorithm for multi-period design and planning of centralized and distributed manufacturing networks. *Computers & Chemical Engineering* 127: 295–310.

17 Allman, A., Lee, C., Martín, M., and Zhang, Q. (2021). Biomass waste-to-energy supply chain optimization with mobile production modules. *Computers & Chemical Engineering* 150: 107326.

18 Sánchez, A. and Martín, M. (2018b). Scale up and scale down issues of renewable ammonia plants: towards modular design. *Sustainable Production and Consumption* 16: 176–192.

19 Almena, A. and Martín, M. (2016). Technoeconomic analysis of the production of epichlorohydrin from glycerol. *Industrial & Engineering Chemistry Research* 55 (12): 3226–3238.

20 Perez, H.D. and Grossmann, I.E. (2024). Extensions to generalized disjunctive programming: hierarchical structures and first-order logic. *Optimization and Engineering* 25 (2): 959–998.

21 Turton, R., Bailie, R.C., Whiting, W.B., and Shaeiwitz, J.A. (2008). *Analysis, Synthesis and Design of Chemical Processes*. Pearson Education.

22 Martín, L. and Martín, M. (2013). Optimal year-round operation of a concentrated solar energy plant in the south of Europe. *Applied Thermal Engineering* 59 (1–2): 627–633.

23 García-Anteportalatina, V.M. and Martín, M. (2022). Process synthesis for the valorisation of low-grade heat: geothermal brines and industrial waste streams. *Renewable Energy* 198: 733–748.

24 Martín, M. and Grossmann, I.E. (2018). Optimal integration of renewable based processes for fuels and power production: Spain case study. *Applied Energy* 213: 595–610.

25 Sánchez, A. and Martín, M. (2018a). Optimal renewable production of ammonia from water and air. *Journal of Cleaner Production* 178: 325–342.

26 Sánchez, A., Castellano, E., Martín, M., and Vega, P. (2021b). Evaluating ammonia as green fuel for power generation: a thermo-chemical perspective. *Applied Energy* 293: 116956.

27 González-Núñez, S., Guerras, L.S., and Martín, M. (2023). A multiscale analysis approach for the valorization of sludge and MSW via co-incineration. *Energy* 263: 126081.

28 Martín, M., Taifouris, M., and Galán, G. (2023). Lignocellulosic biorefineries: a multiscale approach for resource exploitation. *Bioresource Technology* 385: 129397.

29 Sánchez, A., Zhang, Q., Martín, M., and Vega, P. (2022). Towards a new renewable power system using energy storage: an economic and social analysis. *Energy Conversion and Management* 252: 115056.

30 Zhang, H., Wang, L., Maréchal, F., and Desideri, U. (2020). Techno-economic comparison of green ammonia production processes. *Applied Energy* 259: 114135.

31 Allen, R.C., Iseri, F., Demirhan, C.D. et al. (2023). Improvements for decomposition based methods utilized in the development of multi-scale energy systems. *Computers & Chemical Engineering* 170: 108135.

32 Mitrai, I., Tang, W., and Daoutidis, P. (2022). Stochastic blockmodeling for learning the structure of optimization problems. *AIChE Journal* 68 (6): e17415.

33 Jalving, J., Shin, S., and Zavala, V.M. (2022). A graph-based modeling abstraction for optimization: concepts and implementation in plasmo. jl. *Mathematical Programming Computation* 14 (4): 699–747.

34 Lara, C.L., Mallapragada, D.S., Papageorgiou, D.J. et al. (2018). Deterministic electric power infrastructure planning: mixed-integer programming model and nested decomposition algorithm. *European Journal of Operational Research* 271 (3): 1037–1054.

35 Gonzales, D.S. and Searcy, S.W. (2017). GIS-based allocation of herbaceous biomass in biorefineries and depots. *Biomass and Bioenergy* 97 (4): 1–10. https://doi.org/10.1016/j.biombioe.2016.12.009.

36 Lozano-García, D.F., Santibañez-Aguilar, J.E., Lozano, F.J., and Flores-Tlacuahuac, A. (2020). GIS-based modeling of residual biomass availability for energy and production in Mexico. *Renewable and Sustainable Energy Reviews* 120: 109610. https://doi.org/10.1016/j.rser.2019.109610.

37 You, F., Tao, L., Graziano, D.J., and Snyder, S.W. (2012). Optimal design of sustainable cellulosic biofuel supply chains: multiobjective optimization coupled with life cycle assessment and input-output analysis. *AIChE Journal* 58: 1157–1180. https://doi.org/10.1002/aic.12637.

38 O'Neill, E.G., Martinez-Feria, R.A., Basso, B., and Maravelias, C.T. (2022). Integrated spatially explicit landscape and cellulosic biofuel supply chain optimization under biomass yield uncertainty. *Computers and Chemical Engineering* 160: 107724. https://doi.org/10.1016/j.compchemeng.2022.107724.

39 Adeleke, O.J. and Olukanni, D.O. (2020). Facility location problems: models, techniques, and applications. *Waste Manage Recycling* 5 (2): 10.

40 Babinski, M.L. (1965). Integer programming: methods, uses, computations. *Management Science* 12 (3): 253–313.

41 Heras, J. and Martín, M. (2020). Social issues in the energy transition: effect on the design of the new power system. *Applied Energy* 278: 115654.

42 Bockris, J.O.M. (2013). The hydrogen economy: its history. *International Journal of Hydrogen Energy* 38 (6): 2579–2588.

43 BloombergNEF (2020). Hydrogen economy outlook. https://data.bloomberglp.com/professional/sites/24/BNEF-Hydrogen-Economy-Outlook-_Key-Messages.pdf

44 Oliveira, A.M., Beswick, R.R., and Yan, Y. (2021). A green hydrogen economy for a renewable energy society. *Current Opinion in Chemical Engineering* 33: 100701.

45 Kazi, M.K., Eljack, F., El-Halwagi, M.M., and Haouari, M. (2021). Green hydrogen for industrial sector decarbonization: costs and impacts on hydrogen economy in Qatar. *Computers & Chemical Engineering* 145: 107144.

46 Grigoriev, S.A., Fateev, V.N., Bessarabov, D.G., and Millet, P. (2020). Current status, research trends, and challenges in water electrolysis science and technology. *International Journal of Hydrogen Energy* 45 (49): 26036–26058.

47 Singla, M.K., Nijhawan, P., and Oberoi, A.S. (2021). Hydrogen fuel and fuel cell technology for cleaner future: a review. *Environmental Science and Pollution Research* 28 (13): 15607–15626.

48 Falcão, D.S. and Pinto, A.M.F.R. (2020). A review on PEM electrolyzer modelling: guidelines for beginners. *Journal of Cleaner Production* 261: 121184.

49 Hernández-Gómez, Á., Ramirez, V., and Guilbert, D. (2020). Investigation of PEM electrolyzer modeling: electrical domain, efficiency, and specific energy consumption. *International Journal of Hydrogen Energy* 45 (29): 14625–14639.

50 Brezak, D., Kovač, A., and Firak, M. (2023). MATLAB/Simulink simulation of low-pressure PEM electrolyzer stack. *International Journal of Hydrogen Energy* 48 (16): 6158–6173.

51 AlZahrani, A.A. and Dincer, I. (2018). Modeling and performance optimization of a solid oxide electrolysis system for hydrogen production. *Applied Energy* 225: 471–485.

52 Mueller, M., Klinsmann, M., Sauter, U. et al. (2024). High temperature solid oxide electrolysis–technology and modeling. *Chemie Ingenieur Technik* 96 (1–2): 143–166.

53 Wu, H.W. (2016). A review of recent development: transport and performance modeling of PEM fuel cells. *Applied Energy* 165: 81–106.

54 Ziogou, C., Voutetakis, S., Papadopoulou, S., and Georgiadis, M.C. (2011). Modeling, simulation and experimental validation of a PEM fuel cell system. *Computers & Chemical Engineering* 35 (9): 1886–1900.

55 Ota, T., Koyama, M., Wen, C.J. et al. (2003). Object-based modeling of SOFC system: dynamic behavior of micro-tube SOFC. *Journal of Power Sources* 118 (1–2): 430–439.

56 Bao, C., Wang, Y., Feng, D. et al. (2018). Macroscopic modeling of solid oxide fuel cell (SOFC) and model-based control of SOFC and gas turbine hybrid system. *Progress in Energy and Combustion Science* 66: 83–140.

57 Sommer, E.M., Martíns, L.S., Vargas, J.V.C. et al. (2012). Alkaline membrane fuel cell (AMFC) modeling and experimental validation. *Journal of Power Sources* 213: 16–30.

58 Ni, M. (2009). Computational fluid dynamics modeling of a solid oxide electrolyzer cell for hydrogen production. *International Journal of Hydrogen Energy* 34 (18): 7795–7806.

59 Ma, Z., Witteman, L., Wrubel, J.A., and Bender, G. (2021). A comprehensive modeling method for proton exchange membrane electrolyzer development. *International Journal of Hydrogen Energy* 46 (34): 17627–17643.

60 Hu, K., Fang, J., Ai, X. et al. (2022). Comparative study of alkaline water electrolysis, proton exchange membrane water electrolysis and solid oxide electrolysis through multiphysics modeling. *Applied Energy* 312: 118788.

61 Xu, H., Chen, B., Tan, P. et al. (2018). Modeling of all porous solid oxide fuel cells. *Applied Energy* 219: 105–113.

62 Sivertsen, B.R. and Djilali, N. (2005). CFD-based modelling of proton exchange membrane fuel cells. *Journal of Power Sources* 141 (1): 65–78.

63 Mohamed, A., Ibrahem, H., and Kim, K. (2022). Machine learning-based simulation for proton exchange membrane electrolyzer cell. *Energy Reports* 8: 13425–13437.

64 Ming, W., Sun, P., Zhang, Z. et al. (2023). A systematic review of machine learning methods applied to fuel cells in performance evaluation, durability prediction, and application monitoring. *International Journal of Hydrogen Energy* 48 (13): 5197–5228.

65 Wilberforce, T. and Olabi, A.G. (2021). Proton exchange membrane fuel cell performance prediction using artificial neural network. *International Journal of Hydrogen Energy* 46 (8): 6037–6050.

66 Terlouw, T., Bauer, C., McKenna, R., and Mazzotti, M. (2022). Large-scale hydrogen production via water electrolysis: a techno-economic and environmental assessment. *Energy & Environmental Science* 15 (9): 3583–3602.

67 Jang, D., Kim, J., Kim, D. et al. (2022). Techno-economic analysis and Monte Carlo simulation of green hydrogen production technology through various water electrolysis technologies. *Energy Conversion and Management* 258: 115499.

68 Nami, H., Rizvandi, O.B., Chatzichristodoulou, C. et al. (2022). Techno-economic analysis of current and emerging electrolysis technologies for green hydrogen production. *Energy Conversion and Management* 269: 116162.

69 Weidner, T., Tulus, V., and Guillén-Gosálbez, G. (2023). Environmental sustainability assessment of large-scale hydrogen production using prospective life cycle analysis. *International Journal of Hydrogen Energy* 48 (22): 8310–8327.

70 Mucci, S., Mitsos, A., and Bongartz, D. (2023a). Power-to-X processes based on PEM water electrolyzers: a review of process integration and flexible operation. *Computers & Chemical Engineering* 175: 108260.

71 Marocco, P., Ferrero, D., Martelli, E. et al. (2021). A MILP approach for the optimal design of renewable battery-hydrogen energy systems for off-grid insular communities. *Energy Conversion and Management* 245: 114564.

72 Tostado-Véliz, M., Kamel, S., Hasanien, H.M. et al. (2022). A mixed-integer-linear-logical programming interval-based model for optimal scheduling of isolated microgrids with green hydrogen-based storage considering demand response. *Journal of Energy Storage* 48: 104028.

73 Ban, M., Bai, W., Song, W. et al. (2022). Optimal scheduling for integrated energy-mobility systems based on renewable-to-hydrogen stations and tank truck fleets. *IEEE Transactions on Industry Applications* 58 (2): 2666–2676.

74 Yang, Y., De La Torre, B., Stewart, K. et al. (2022). The scheduling of alkaline water electrolysis for hydrogen production using hybrid energy sources. *Energy Conversion and Management* 257: 115408.

75 Sahinidis, N.V. (1996). BARON: a general purpose global optimization software package. *Journal of Global Optimization* 8: 201–205.

76 Flamm, B., Peter, C., Büchi, F.N., and Lygeros, J. (2021). Electrolyzer modeling and real-time control for optimized production of hydrogen gas. *Applied Energy* 281: 116031.

77 de Andrade, G.A., Mendes, P.R., García-Clúa, J.G., and Normey-Rico, J.E. (2020). Control of a grid assisted PV-H_2 production system: a comparative study between optimal control and hybrid MPC. *Journal of Process Control* 92: 220–233.

78 Zhao, D., He, Q., Yu, J. et al. (2022). A data-driven digital-twin model and control of high temperature proton exchange membrane electrolyzer cells. *International Journal of Hydrogen Energy* 47 (14): 8687–8699.

79 Zhu, Z., Liu, X., Kong, X. et al. (2024). PV/Hydrogen DC microgrid control using distributed economic model predictive control. *Renewable Energy* 222: 119871.

80 Daud, W.R.W., Rosli, R.E., Majlan, E.H. et al. (2017). PEM fuel cell system control: a review. *Renewable Energy* 113: 620–638.

81 Chatrattanawet, N., Hakhen, T., Kheawhom, S., and Arpornwichanop, A. (2017). Control structure design and robust model predictive control for controlling a proton exchange membrane fuel cell. *Journal of Cleaner Production* 148: 934–947.

82 Quan, S., Wang, Y.X., Xiao, X. et al. (2021). Disturbance prediction-based enhanced stochastic model predictive control for hydrogen supply and circulating of vehicular fuel cells. *Energy Conversion and Management* 238: 114167.

83 Pereira, D.F., da Costa Lopes, F., and Watanabe, E.H. (2020). Nonlinear model predictive control for the energy management of fuel cell hybrid electric vehicles in real time. *IEEE Transactions on Industrial Electronics* 68 (4): 3213–3223.

84 Yin, L. and Liu, D. (2023). Quantum parallel model predictive control for grid-connected solid oxide fuel cells. *Applied Soft Computing* 137: 110157.

85 Ehrenstein, M., Galán-Martín, Á., Tulus, V., and Guillén-Gosálbez, G. (2020). Optimising fuel supply chains within planetary boundaries: a case study of hydrogen for road transport in the UK. *Applied Energy* 276: 115486.

86 Almaraz, S.D.L., Rácz, V., Azzaro-Pantel, C., and Szántó, Z.O. (2022). Multiobjective and social cost-benefit optimisation for a sustainable hydrogen supply chain: application to Hungary. *Applied Energy* 325: 119882.

87 Robles, J.O., Azzaro-Pantel, C., and Aguilar-Lasserre, A. (2020). Optimization of a hydrogen supply chain network design under demand uncertainty by multi-objective genetic algorithms. *Computers & Chemical Engineering* 140: 106853.

88 Guerra, K., Welfle, A., Gutiérrez-Alvarez, R. et al. (2024). The role of energy storage in Great Britain's future power system: focus on hydrogen and biomass. *Applied Energy* 357: 122447.

89 Brey, J.J. (2021). Use of hydrogen as a seasonal energy storage system to manage renewable power deployment in Spain by 2030. *International Journal of Hydrogen Energy* 46 (33): 17447–17457.

90 Valente, A., Tulus, V., Galán-Martín, Á. et al. (2021). The role of hydrogen in heavy transport to operate within planetary boundaries. *Sustainable Energy & Fuels* 5 (18): 4637–4649.

91 Nishiyama, H., Yamada, T., Nakabayashi, M. et al. (2021). Photocatalytic solar hydrogen production from water on a 100-m^2 scale. *Nature* 598 (7880): 304–307.

92 Safari, F. and Dincer, I. (2020). A review and comparative evaluation of thermochemical water splitting cycles for hydrogen production. *Energy Conversion and Management* 205: 112182.

93 Younas, M., Shafique, S., Faisal, A. et al. (2023). Hydrogen production through water vapors using optimized corona-DBD hybrid plasma micro-reactor. *Fuel* 331: 125838.

94 Nemmour, A., Inayat, A., Janajreh, I., and Ghenai, C. (2023). Green hydrogen-based E-fuels (E-methane, E-methanol, E-ammonia) to support clean energy transition: a literature review. *International Journal of Hydrogen Energy* 48 (75): 29011–29033.

95 Schlereth, D. and Hinrichsen, O. (2014). A fixed-bed reactor modeling study on the methanation of CO_2. *Chemical Engineering Research and Design* 92 (4): 702–712.

96 Hervy, M., Maistrello, J., Brito, L. et al. (2021). Power-to-gas: CO_2 methanation in a catalytic fluidized bed reactor at demonstration scale, experimental results and simulation. *Journal of CO_2 Utilization* 50: 101610.

97 Bolt, A., Dincer, I., and Agelin-Chaab, M. (2023). Design and assessment of a new helical fixed bed type CO_2 methanation reactor. *Fuel* 337: 127176.

98 Zimmermann, R.T., Bremer, J., and Sundmacher, K. (2020). Optimal catalyst particle design for flexible fixed-bed CO_2 methanation reactors. *Chemical Engineering Journal* 387: 123704.

99 Di Nardo, A., Calchetti, G., Bassano, C., and Deiana, P. (2021). CO_2 methanation in a shell and tube reactor CFD simulations: high temperatures mitigation analysis. *Chemical Engineering Science* 246: 116871.

100 Bremer, J., Rätze, K.H., and Sundmacher, K. (2017). CO_2 methanation: optimal start-up control of a fixed-bed reactor for power-to-gas applications. *AIChE Journal* 63 (1): 23–31.

101 Kreitz, B., Wehinger, G.D., and Turek, T. (2019). Dynamic simulation of the CO_2 methanation in a micro-structured fixed-bed reactor. *Chemical Engineering Science* 195: 541–552.

102 Merino-Garcia, I., Albo, J., and Irabien, A. (2017). Productivity and selectivity of gas-phase CO_2 electroreduction to methane at copper nanoparticle-based electrodes. *Energy Technology* 5 (6): 922–928.

103 Sedighian Rasouli, A., Wang, X., Wicks, J. et al. (2020). CO_2 electroreduction to methane at production rates exceeding 100 mA/cm^2. *ACS Sustainable Chemistry & Engineering* 8 (39): 14668–14673.

104 Davis, W. and Martín, M. (2014). Optimal year-round operation for methane production from CO_2 and water using wind and/or solar energy. *Journal of Cleaner Production* 80: 252–261.

105 Peters, R., Baltruweit, M., Grube, T. et al. (2019). A techno economic analysis of the power to gas route. *Journal of CO_2 Utilization* 34: 616–634.

106 Al-Breiki, M. and Bicer, Y. (2023). Techno-economic evaluation of a power-to-methane plant: levelized cost of methane, financial performance metrics, and sensitivity analysis. *Chemical Engineering Journal* 471: 144725.

107 Sternberg, A. and Bardow, A. (2016). Life cycle assessment of power-to-gas: syngas vs methane. *ACS Sustainable Chemistry & Engineering* 4 (8): 4156–4165.

108 Blanco, H., Codina, V., Laurent, A. et al. (2020). Life cycle assessment integration into energy system models: an application for power-to-methane in the EU. *Applied Energy* 259: 114160.

109 De Roeck, F.G., Buchmayr, A., Gripekoven, J. et al. (2022). Comparative life cycle assessment of power-to-methane pathways: process simulation of biological and catalytic biogas methanation. *Journal of Cleaner Production* 380: 135033.

110 Bruns, B., Gross, M., Grünewald, M. et al. (2024). A multi-step framework for the design of a flexible power-to-methane process. *Journal of Cleaner Production* 447: 141434.

111 Belderbos, A., Valkaert, T., Bruninx, K. et al. (2020). Facilitating renewables and power-to-gas via integrated electrical power-gas system scheduling. *Applied Energy* 275: 115082.

112 Li, X., Yang, B., and Zhang, Y. (2013). Dynamics and control study on the low temperature methanation reactor with mass and heat recycle. *Journal of Process Control* 23 (10): 1360–1370.

113 Jeong, Y., Park, J., and Han, M. (2021). Design and control of a fixed-bed recycle reactor with multicatalyst layers: methanation of carbon dioxide. *Industrial & Engineering Chemistry Research* 60 (12): 4650–4667.

114 Colelli, L., Bassano, C., Verdone, N. et al. (2024). Power-to-gas: process analysis and control strategies for dynamic catalytic methanation system. *Energy Conversion and Management* 305: 118257.

115 Blanco, H., Nijs, W., Ruf, J., and Faaij, A. (2018). Potential of power-to-methane in the EU energy transition to a low carbon system using cost optimization. *Applied Energy* 232: 323–340.

116 Yilmaz, H.Ü., Kimbrough, S.O., van Dinther, C., and Keles, D. (2022). Power-to-gas: decarbonization of the European electricity system with synthetic methane. *Applied Energy* 323: 119538.

117 Carrera, E. and Azzaro-Pantel, C. (2021). Bi-objective optimal design of hydrogen and methane supply chains based on power-to-gas systems. *Chemical Engineering Science* 246: 116861.

118 Leonzio, G. and Zondervan, E. (2020). Analysis and optimization of carbon supply chains integrated to a power to gas process in Italy. *Journal of Cleaner Production* 269: 122172.

119 IRENA and Methanol Institute (2021). Innovation outlook: renewable methanol. International Renewable Energy Agency, Abu Dhabi.

120 Leonzio, G. and Foscolo, P.U. (2020). Analysis of a 2-D model of a packed bed reactor for methanol production by means of CO_2 hydrogenation. *International Journal of Hydrogen Energy* 45 (18): 10648–10663.

121 Torcida, M.F., Curto, D., and Martín, M. (2022). Design and optimization of CO_2 hydrogenation multibed reactors. *Chemical Engineering Research and Design* 181: 89–100.

122 Pavlišič, A., Huš, M., Prašnikar, A., and Likozar, B. (2020). Multiscale modelling of CO_2 reduction to methanol over industrial $Cu/ZnO/Al_2O_3$ heterogeneous catalyst: linking ab initio surface reaction kinetics with reactor fluid dynamics. *Journal of Cleaner Production* 275: 122958.

123 Cui, X. and Kær, S.K. (2020). A comparative study on three reactor types for methanol synthesis from syngas and CO_2. *Chemical Engineering Journal* 393: 124632.

124 Leonzio, G., Zondervan, E., and Foscolo, P.U. (2019). Methanol production by CO_2 hydrogenation: analysis and simulation of reactor performance. *International Journal of Hydrogen Energy* 44 (16): 7915–7933.

125 Pérez-Fortes, M., Schöneberger, J.C., Boulamanti, A., and Tzimas, E. (2016). Methanol synthesis using captured CO_2 as raw material: techno-economic and environmental assessment. *Applied Energy* 161: 718–732.

126 Cordero-Lanzac, T., Ramirez, A., Navajas, A. et al. (2022). A techno-economic and life cycle assessment for the production of green methanol from CO_2: catalyst and process bottlenecks. *Journal of Energy Chemistry* 68: 255–266.

127 Zhang, Z., Vo, D.N., Nguyen, T.B. et al. (2024). Advanced process integration and machine learning-based optimization to enhance techno-economic-environmental performance of CO_2 capture and conversion to methanol. *Energy* 293: 130758.

128 Tountas, A.A., Peng, X., Xu, Y. et al. (2023). Direct CO_2-to-renewable methanol: outlook, performance and optimization approach. *Sustainable Materials and Technologies* 36: e00630.

129 Adnan, M.A. and Kibria, M.G. (2020). Comparative techno-economic and life-cycle assessment of power-to-methanol synthesis pathways. *Applied Energy* 278: 115614.

130 Liu, T., Liu, Q., Xu, D., and Sui, J. (2017). Performance investigation of a new distributed energy system integrated a solar thermochemical process with chemical recuperation. *Applied Thermal Engineering* 119: 387–395.

131 Wu, Y., Zhao, T., Tang, S. et al. (2024). Research on design and multi-frequency scheduling optimization method for flexible green ammonia system. *Energy Conversion and Management* 300: 117976.

132 Blanco, E.C., Sánchez, A., Martín, M., and Vega, P. (2023). Methanol and ammonia as emerging green fuels: evaluation of a new power generation paradigm. *Renewable and Sustainable Energy Reviews* 175: 113195.

133 Mucci, S., Mitsos, A., and Bongartz, D. (2023b). Cost-optimal power-to-methanol: flexible operation or intermediate storage? *Journal of Energy Storage* 72: 108614.

134 Zheng, Y., You, S., Li, X. et al. (2022). Data-driven robust optimization for optimal scheduling of power to methanol. *Energy Conversion and Management* 256: 115338.

135 Cui, X., Kær, S.K., and Nielsen, M.P. (2022). Energy analysis and surrogate modeling for the green methanol production under dynamic operating conditions. *Fuel* 307: 121924.

136 Chiou, H.H., Lee, C.J., Wen, B.S. et al. (2023). Evaluation of alternative processes of methanol production from CO_2: design, optimization, control, techno-economic, and environmental analysis. *Fuel* 343: 127856.

137 Galán, G., Martín, M., and Grossmann, I.E. (2024). Multiscale analysis for the utilization of CO_2 towards the production of chemicals at the country level: case study of Spain. *Journal of Cleaner Production* 437: 140551.

138 Ong, C.W., Lin, J.X., Tsai, M.L. et al. (2024). Techno-economic and carbon emission analyses of a methanol-based international renewable energy supply chain. *International Journal of Hydrogen Energy* 49: 1572–1585.

139 Sánchez, A., Martín, M., and Zhang, Q. (2021a). Optimal design of sustainable power-to-fuels supply chains for seasonal energy storage. *Energy* 234: 121300.

140 Iribarren, D., Calvo-Serrano, R., Martín-Gamboa, M. et al. (2022). Social life cycle assessment of green methanol and benchmarking against conventional fossil methanol. *Science of the Total Environment* 824: 153840.

141 IRENA and AEA (2022). Innovation outlook: renewable ammonia. International Renewable Energy Agency, Abu Dhabi, Ammonia Energy Association, Brooklyn.

142 Wang, L., Xia, M., Wang, H. et al. (2018). Greening ammonia toward the solar ammonia refinery. *Joule* 2 (6): 1055–1074.

143 Khademi, M.H. and Sabbaghi, R.S. (2017). Comparison between three types of ammonia synthesis reactor configurations in terms of cooling methods. *Chemical Engineering Research and Design* 128: 306–317.

144 Flórez-Orrego, D. and de Oliveira Junior, S. (2017). Modeling and optimization of an industrial ammonia synthesis unit: an exergy approach. *Energy* 137: 234–250.

145 Fahr, S., Schiedeck, M., Reinke, M. et al. (2024). Simultaneous design and part-load optimization of an industrial ammonia synthesis reactor. *Chemical Engineering Journal* 480: 148302.

146 Demirhan, C.D., Tso, W.W., Powell, J.B., and Pistikopoulos, E.N. (2019). Sustainable ammonia production through process synthesis and global optimization. *AIChE Journal* 65 (7): e16498.

147 Martín, M. and Sánchez, A. (2024). Biomass pathways to produce green ammonia and urea. *Current Opinion in Green and Sustainable Chemistry* 47: 100933.

148 Del Pozo, C.A. and Cloete, S. (2022). Techno-economic assessment of blue and green ammonia as energy carriers in a low-carbon future. *Energy Conversion and Management* 255: 115312.

149 D'Angelo, S.C., Cobo, S., Tulus, V. et al. (2021). Planetary boundaries analysis of low-carbon ammonia production routes. *ACS Sustainable Chemistry & Engineering* 9 (29): 9740–9749.

150 Andrés-Martínez, O., Zhang, Q., and Daoutidis, P. (2023). Optimal operation of a reaction-absorption process for ammonia production at low pressure. *Industrial & Engineering Chemistry Research* 62 (36): 14456–14466.

151 MacFarlane, D.R., Cherepanov, P.V., Choi, J. et al. (2020). A roadmap to the ammonia economy. *Joule* 4 (6): 1186–1205.

152 Gomez, J.R., Baca, J., and Garzon, F. (2020). Techno-economic analysis and life cycle assessment for electrochemical ammonia production using proton conducting membrane. *International Journal of Hydrogen Energy* 45 (1): 721–737.

153 Cesaro, Z., Ives, M., Nayak-Luke, R. et al. (2021). Ammonia to power: forecasting the levelized cost of electricity from green ammonia in large-scale power plants. *Applied Energy* 282: 116009.

154 Siddiqui, O. and Dincer, I. (2019). Experimental investigation and assessment of direct ammonia fuel cells utilizing alkaline molten and solid electrolytes. *Energy* 169: 914–923.

155 Devkota, S., Cha, J.Y., Shin, B.J. et al. (2024). Techno-economic and environmental assessment of hydrogen production through ammonia decomposition. *Applied Energy* 358: 122605.

156 Sánchez, A., Rengel, M.M., and Martín, M. (2023). A zero CO_2 emissions large ship fuelled by an ammonia-hydrogen blend: reaching the decarbonisation goals. *Energy Conversion and Management* 293: 117497.

157 Allman, A., Palys, M.J., and Daoutidis, P. (2019). Scheduling-informed optimal design of systems with time-varying operation: a wind-powered ammonia case study. *AIChE Journal* 65 (7): e16434.

158 Kelley, M.T., Do, T.T., and Baldea, M. (2022). Evaluating the demand response potential of ammonia plants. *AIChE Journal* 68 (3): e17552.

159 Rosbo, J.W., Ritschel, T.K., Hørsholt, S. et al. (2023). Flexible operation, optimisation and stabilising control of a quench cooled ammonia reactor for power-to-ammonia. *Computers & Chemical Engineering* 176: 108316.

160 Kong, B., Zhang, Q., and Daoutidis, P. (2024). Nonlinear model predictive control of flexible ammonia production. *Control Engineering Practice* 148: 105946.

161 Oliveira Cabral, T., Bagheri, A., and Pourkargar, D.B. (2024). Learning-based model reduction and predictive control of an ammonia synthesis process. *Industrial and Engineering Chemistry Research* 63: 10325–10342.

162 Nayak-Luke, R.M. and Bañares-Alcántara, R. (2020). Techno-economic viability of islanded green ammonia as a carbon-free energy vector and as a substitute for conventional production. *Energy & Environmental Science* 13 (9): 2957–2966.

163 Wang, H., Daoutidis, P., and Zhang, Q. (2023). Ammonia-based green corridors for sustainable maritime transportation. *Digital Chemical Engineering* 6: 100082.

164 Mitrai, I., Palys, M.J., and Daoutidis, P. (2024). A two-stage stochastic programming approach for the design of renewable ammonia supply chain networks. *Processes* 12 (2): 325.

165 Kim, S., Park, J., Heo, S., and Lee, J.H. (2024). Green hydrogen vs Green ammonia: a hierarchical optimization-based integrated temporal approach for comparative techno-economic analysis of international supply chains. *Journal of Cleaner Production* 465: 142750.

166 Fasihi, M., Weiss, R., Savolainen, J., and Breyer, C. (2021). Global potential of green ammonia based on hybrid PV-wind power plants. *Applied Energy* 294: 116170.

167 Galimova, T., Fasihi, M., Bogdanov, D., and Breyer, C. (2023). Feasibility of green ammonia trading via pipelines and shipping: cases of Europe, North Africa, and South America. *Journal of Cleaner Production* 427: 139212.

12

Repurposing a Conventional Oil Refinery for Biomass Processing to Aviation Fuel: Process Design and Techno-Environmental Evaluation for a Real Operating Plant

Valeria González[1], Alejandro Pedezert[2], Soledad Gutiérrez[1], Roberto Kreimerman[3], Lucia Pittaluga[1,4], and Ana I. Torres[5]

[1] *Instituto de Ingeniería Química, Facultad de Ingeniería, Universidad de la República, Montevideo, Uruguay*
[2] *Ancap, Montevideo, Uruguay*
[3] *Departamento de Inserción Social del Ingeniero, Facultad de Ingeniería, Universidad de la República, Montevideo, Uruguay*
[4] *Instituto de Economía, Facultad de Ciencias Económicas y Administración, Universidad de la República, Montevideo, Uruguay Instituto Tecnológico Regional Centro-Sur, Universidad Tecnológica, Durazno, Uruguay*
[5] *Department of Chemical Engineering, Carnegie Mellon University, Pittsburgh, PA, USA*

12.1 Introduction

Decarbonization appears in almost every oil refinery's goal. Lower carbon emission levels can be achieved by improving energy efficiency, electrification of operations (when renewable energy is used for its production), and carbon capture, utilization, and storage [1] and by adopting low-carbon feeds, fuels, and energy sources. While navigating the transition to decarbonization is always challenging due to uncertainties in capital investments, electricity costs, and carbon taxes, a particular challenge is the reduction of scope 3 emissions. As stated in the website of the United States Environmental Protection Agency (US-EPA) (https://19january2021snapshot.epa.gov/climateleadership/scope-3-inventory-guidance:.html, accessed September 2024), scope 3 emissions "are the result of activities from assets not owned or controlled by the reporting organization, but that the organization indirectly affects in its value chain."

Scope 3 emissions include emissions related to the use of sold products [2], a use that is mostly related to burning hydrocarbons or disposing of (non-fuel) chemicals derived from hydrocarbons. Taking into account the report in [3], emissions related to the use of products account for roughly 70% of Shell's CO_2 emissions. Clearly, this value will be different for other companies in the refining business, but as long as refineries produce carbon-based fuels and chemicals, CO_2 will be emitted as a result of their use or disposal. In addition, as process-related emissions diminish, the relative importance of the emissions related to the use of the product will increase.

Hydrocarbons that are used as fuels might be partially displaced by zero-carbon alternatives such as electric vehicles, including those that use hydrogen, or electric boilers for the production of steam for process heating. However, not all transportation means have a non-carbon-based alternative; temperatures beyond those allowed by high-pressure steam are required in many process industries, and electricity cannot replace carbons in the production of organic chemicals. Hence, carbon sources will still be required even in the most electrified future setting.

Carbon-neutral pathways that use biomass or carbons retrieved from end-of-life products, such as waste plastic, to produce fuels and chemicals already exist; some of these pathways can be integrated within refinery operations. Adapting existing refineries to process biomass or carbonaceous wastes seems a natural win–win situation, as refinery decarbonization standards can be met while preserving valuable infrastructure. Along these lines, many refineries around the globe advertise the shift toward carbon neutral hydrocarbons via biomass-based, waste-based, or captured CO_2 feeds [4–9].

This work presents the results of a recent project in collaboration with a South American oil refinery group whose aim was the "Development of the biorefinery concept. Possible utilization of the available national infrastructure as a platform for its development." The activities will be located in Uruguay, a country rich in biomass with an already decarbonized electricity sector. The country does not have any proven fossil fuel reserves and depends on hydrocarbon imports for transportation fuels. Decarbonization of the transportation sector, while at the same time valorizing native biomass resources and expanding production chains, has been on the country's political agenda for quite some time, albeit with different degrees of urgency.

Here, we include the part of the study corresponding to the production of green hydrocarbons: green diesel, green naphtha, and, in particular, sustainable aviation fuels (SAFs) from local oilseeds and tallow via hydro treatment. The main contribution of this article is the application of several process systems engineering (PSE) tools to a real case study. The study is organized as follows: an overview of available biomass feed options in the country, as well as the refinery group's current infrastructure, is given in Section 12.2. This overview provides the rationale behind the selection of the pathway and feedstocks and is based on several discussions with technical experts from the refinery. Section 12.3

presents the design of a triglycerides to green fuels process that can be integrated into the refinery group's current operations. The design is presented in detail. It includes the formulation of an optimization problem to estimate the hydrocarbon composition after the hydrotreatment step for those feedstocks for which that information was not readily available. It also includes a discussion of the changes made to the base design to suit the group's current processing practices. Section 12.5 discusses the gate-to-gate environmental assessment of the final design using the GREENSCOPE [10] methodology.[1]

12.2 Overview of Feed Options, Processing Pathways and Current Infrastructure

The Refinery Group is a group of companies that includes an oil refinery and a bio-ethanol and bio-diesel biorefinery. At the time of this project, the group had the infrastructure to process:

- starches and sugarcane into ethanol through milling, sugar extraction, fermentation, distillation, and dehydration operations.
- oilseeds, waste oils, and fats into biodiesel and glycerin through transesterification and separation/purification operations. Oil extraction from the seeds is carried out by a private partner that specializes in food-grade oil processing.
- crude oil into propane, gasoline, jet fuel, diesel, fuel oil, and asphalt through atmospheric and vacuum distillation, hydrotreatment, isomerization, catalytic reforming, and catalytic cracking operations.

The biomass availability survey reported in a previous publication [11, 12] was the basis for assessing possible processing pathways to utilize current assets in a biorefinery setting in Uruguay. A summary of biomass relevant to the project is as follows:

- **Waste from Agricultural Activity:** Residues from soybean and wheat crops were initially considered as they are produced in the largest quantities. However, intermittent availability of the crops, soybean is a winter crop while wheat is a spring crop, was mentioned as a potential barrier. Barley (spring), sorghum, and corn (winter) can be used as a complement. All the residues from harvesting these crops are lignocellulosic (LG), and the processes for their fractionation and conversion to hydrocarbons are technologically similar.
- **Waste from Wood Industrialization:** Sawmill residues were initially considered as these have the advantage of being geographically concentrated. Industrialization residues from the pulp and paper industry were not considered

1 To preserve confidentiality, the economic analysis of the process is not reported.

as pulp mills already use them. As before, these residues are lignocellulosic material.

- **Specialty Oil-Based Crops and Lipid Based Material:** Brassica carinata, a nonedible oilseed due to its erucic acid content, is harvested in Uruguay and exported for green-diesel production. This crop has been promoted for the production of SAF [13] and was initially mentioned as a priority for this study. A related crop, canola, is also widely harvested in the country but has the disadvantage of competing with the food/feed chain. Canola is already used in the refinery group for biodiesel production. Used cooking oil is also collected for biodiesel production, but the quantities handled are not significant and was thus not further considered. Other oil crops already being produced in the country (soybean, sunflower), as well as tallow from the very large meat processing industry, could also be used for the production of green hydrocarbons, albeit with differences in reaction yields.

Figure 12.1 summarizes the current processing capabilities of the group (shaded in grey) and potential options for the conversion of the previous biomass-based options into hydrocarbons around these capabilities. Route 1 depicts the conversion of biomass-derived carbohydrates into ethanol; Route 2 illustrates the production of hydrocarbons from either biomass-derived carbohydrates or biomass-derived triacylglycerides; Route 3 considers the conversion of ethanol into hydrocarbons.

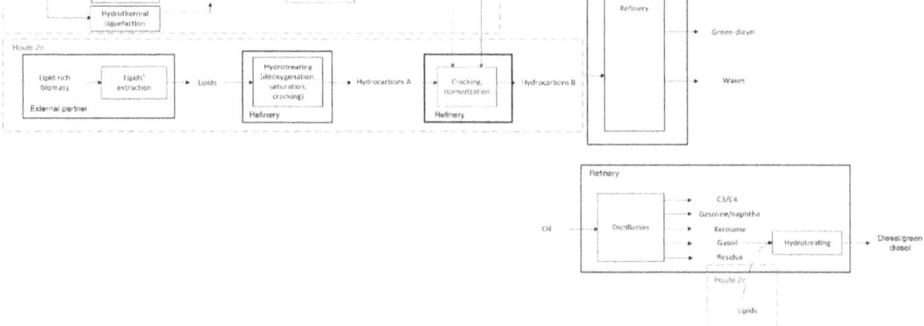

Figure 12.1 Options for processing biomass using the current infrastructure available in the refinery group.

Lignocellulosic (LG) biomass could potentially be processed by the biorefinery installations from the Refinery Group via the sugar routes (see Figure 12.1 route 1b) to produce ethanol. Ethanol could be dehydrated and oligomerized into hydrocarbons that could be processed by the oil refinery to obtain various commercial fuels. This strategy requires substantial investment in LG biomass pretreatment and ethanol dehydration and oligomerization, but it also establishes a synergy between the operations of the petroleum refinery and bie-ethanol operations of the group.

LG biomass could also be processed by the refinery via thermochemical pathways such as gasification-Fischer Tropsch, pyrolysis, and hydrothermal liquefaction (see Figure 12.1 route 2b). These three routes produce hydrocarbons that must be subjected to subsequent cracking and isomerization before being fed into the distillation columns or by repurposing existing cracking/isomerization units [14]. Any of these routes require potentially high investment but offer great flexibility in terms of the raw materials that can be fed into the process, as although these routes are presented from lignocellulosic biomass, other materials can also be processed by these.

The processing of lipids for their conversion into hydrocarbons can be accomplished via routes 2a and 2c in Figure 12.1. Route 2a in Figure 12.1 is the one adopted by Eni using the Ecofining™ process [15] and the one proposed by the Nesté Oil model [16]. Here, triglycerides are first converted into hydrocarbons through hydrotreatment, and then the chain length is adjusted by subsequent cracking and isomerization processes. This route may require new investments in hydrocracking and isomerization equipment, but the company had expertise in all of these operations. Finally, triglycerides can also be co-fed with fossil-based diesel cuts before their processing in the hydrocracker, as seen in Figure 12.1 route 2c. This route does not require investment and can be seen as a first step towards conversion to a bio-refinery.

12.3 Sustainable Aviation Fuel Process Design

The options discussed in Section 12.2 were presented to engineers of the Refinery Group, and based on their suggestion, the project focused on the design of a green hydrocarbon production process based on the hydrotreatment of triglycerides-rich material. Although green diesel, green naphtha, and SAFs are simultaneously produced by this route, the goal was to guarantee a jet fuel oil cut that could result in satisfaction with the specifications for alternative aviation fuels per norm ASTM D7566. The rationale behind this choice is that there is not yet a hydrocarbon-free solution for aviation fuels, and the production of biomass-derived fuels was already considered by the International Air

Transport Agency (IATA) as the most promising alternative for sustainable development of the aviation sector [17]. In all cases, it is assumed that vegetable oils, already extracted and with standard commercial qualities, are received as raw materials.

12.3.1 Base Process Overview

Figure 12.2 shows the flow diagram of the process we originally designed in this project. The diagram is divided into four areas: 100, 200, 300 and 400. The main equipment in area 100 is the hydrotreatment reactor, which is where the hydrogenation, depropanization and deoxygenation reactions take place. These reactions convert the triglycerides in the lipid-based material into hydrocarbons. This reactor has an oil stream (triglycerides) and a hydrogen stream at high pressure as inputs; the products are n-alkanes and iso-alkanes, as well as CO_2, CO, and H_2O. The specific reactions will be discussed in Section 3.2.2.

In area 200, gases and water are separated from hydrocarbons. The largest amount of water and gases are separated in the flash unit (F-200). The remaining water in the hydrocarbon stream is removed in the decanter unit (D-200). Removing water from this stream is important as it can affect the catalyst in the next area.

In area 300, hydrocracking and hydroisomerization reactions take place (R-300 reactor). These reactions convert heavier hydrocarbons into lighter hydrocarbons. At this point, the desired cut yield can be adjusted by varying the reaction conditions.

The hydrocarbons generated in area 300 are separated in area 400. This separation includes three distillation towers that allow for obtaining light gases (LPG), (green) naphtha, (bio) jet fuel oil, and (green) diesel as outputs. The column labeled D-400 separates light hydrocarbons (C3–C7) and heavier hydrocarbons (C8–C20); D-402 separates (bio) jet fuel oil (C8–C16) and (green) green diesel (C17 and higher); column D-401 separates LPG (C3 and C4) from (green) naphtha (C5–C7).

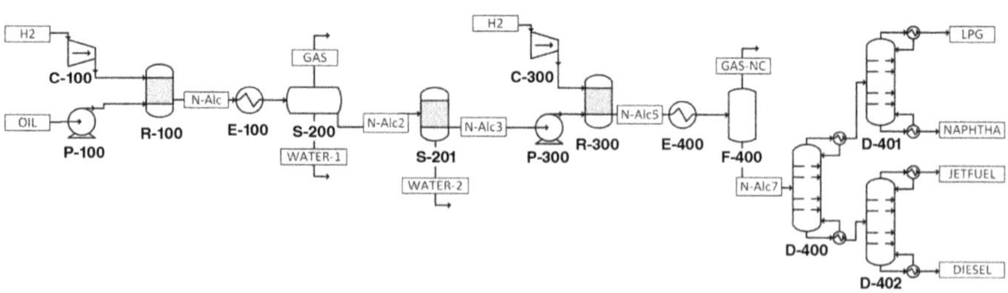

Figure 12.2 Process flow diagram – base process.

12.3.2 Modeling

12.3.2.1 Feedstock

Soybean, canola, sunflower, carinate oils, and tallow were selected as the potential feeds for the process. As observed in Table 12.1, these oils have different fatty acid compositions, so the distribution of products and hydrogen consumption in the reactor will vary.

The other feedstock required by the process is hydrogen, which is used in both area 100 and area 300 reactors. For the area 100 reactor, we calculated the theoretical hydrogen required for the following advances of reaction: depropanization 100%, hydrogenation 100%, deoxygenation through hydrodeoxygenation 70%, deoxygenation through decarboxylation 15%, and deoxygenation through decarbonylation 15%. An excess of hydrogen corresponding to three times the theoretical requirement was assumed. For the area 300 reactor, (hydrocracking and hydroisomerization reactions) the hydrocarbons/hydrogen ratio was taken from [23].

Current practice indicates that the hydrogen feedstock is obtained through naphtha reforming. For the base design, the hydrogen composition was assumed to be the one corresponding to the continuous catalytic reformer (CCR) reformer in [24] (see Table 12.2).

12.3.2.2 Hydrotreating Reactor (R-100)

Chemical Reactions Figure 12.3 summarizes the chemical reactions in the hydrotreating reactor. The first reaction is the hydrogenation of the triglycerides

Table 12.1 Fatty acid composition of oils (%).

Fatty acid	Soybean	Sunflower	Canola	Carinata	Tallow
C14:0	0.08	0.09	0.1	0	2.79
C16:0	10.98	6.33	4.8	5.3	26.78
C16:1	0.11	0	0.2	0	2.05
C18:0	4.39	3.45	1.9	0	34.85
C18:1	23.98	21.64	61.9	10	30.23
C18:2	52.56	67.78	19.8	24.6	0.74
C18:3	6.76	0.09	9.2	16.5	0
C20:0	0.43	0.23	0.6	0	0
C20:1	0.22	0.13	1.5	0	0
C22:0	0.49	0.22	0	0	0
C22:1	0	0	0	43.6	0
Ref.	[18]	[19]	[20]	[21]	[22]

Table 12.2 Typical CCR hydrogen quality.

Component	v (%)
Hydrogen	85–90
Methane	2–3
Ethane	2–3
Propane	1–2
i-Butane	1–2
n-Butane	1–2
C5-Plus	3
Chloride (as HCl)	(1–3 vppm)

Source: Treese et al. [24] / with permission of Springer Nature.

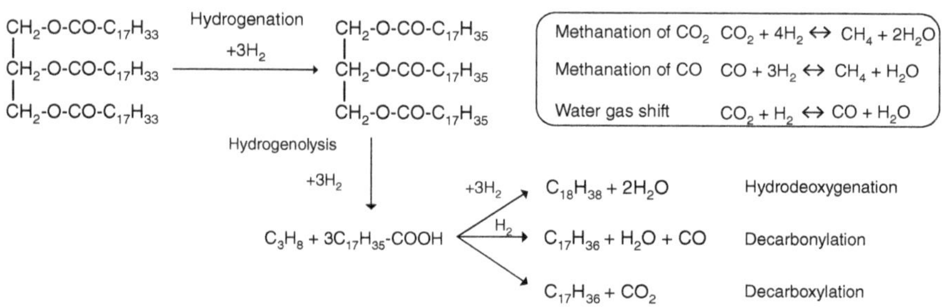

Figure 12.3 Reactions in the hydrotreating reactor.

in the oil feed. This stage is usually modeled by assuming that the glycerol is separated from the fatty acids, producing 3 moles of saturated free fatty acids and 1 mole of propane [25]. Then, saturated free fatty acids are transformed into hydrocarbons deoxygenation (oxygen removal) reactions: hydrodeoxygenation, decarbonylation, and decarboxylation. In decarboxylation, oxygen is removed in the form of CO_2, in decarbonylation as CO, and finally, in hydrodeoxygenation as H_2O [26]. The selectivity of fatty acids with respect to these reactions depends on the catalyst used and the operating conditions. The hydrodeoxygenation reaction is the preferred one since it maintains the number of carbons, and the secondary product is water [27]. However, it is the one that consumes the most hydrogen.

Formulation of an Optimization Problem to Estimate Component-Wise Reaction Yields
Open literature experimental data on the conversion of triacylglycerides (TAG) into hydrocarbons does not usually report the detailed composition of the hydrocarbons produced during the reaction. In general, data is reported in terms of:

1) Oil (TAG) mass in the feed (m_{oil}^{feed})
2) Fatty acid composition of the oil feed (weight %) ($wt_{FA}(i)$, where i indicates an individual fatty acid)
3) Weight percentage of water in the liquid phase at the end of the reaction (w_{aq})
4) Weight percentage of hydrocarbons in the liquid phase at the end of the reaction (w_{oil})
5) Molar percentage of methane in gas phase at the end of the reaction ($n_{CH_4}^{exp}$)
6) Molar percentage of CO_2 in the gas phase at the end of the reaction ($n_{CO_2}^{exp}$)
7) Molar percentage of CO in gas phase at the end of the reaction (n_{CO}^{exp})

However, for simulation purposes, we need to know the yields in terms of the percentage of each hydrocarbon (j) produced in the liquid phase ($w_{alk}(j)$).

Some studies (e.g. [18]) do report detailed hydrocarbon data $w_{alk}(j)$ for a given feedstock, a given catalyst, and reaction conditions. Data from these studies can be used to estimate the advance of each of the reactions in Figure 12.3 for such a feedstock, catalyst, and reaction conditions. Then, these estimates for the advance of reactions can be used to estimate the $w_{alk}(j)$ for a different feedstock on the same catalyst and under the same reaction conditions.

The procedure is as follows:

1) Starting from the experimental data for a given feedstock $F1$, we calculate the ratios A^{exp}, B^{exp}, and C^{exp}, the moles of triglycerides in the inlet stream (n_{TG}^{feed}), and the moles of individual fatty acids in the inlet stream ($n_{FA}(i)$) as follows

$$A^{exp} = nCO_2^{exp}/n_{CH_4}^{exp} \quad (12.1)$$

$$B^{exp} = nCO_2^{exp}/n_{CO}^{exp} \quad (12.2)$$

$$C^{exp} = w_{oil}/w_{aq} \quad (12.3)$$

$$n_{TG}^{feed} = m_{oil}^{feed}/(MW_{FA,mix} + MW_{glycerol}/3 - MW_{H_2O}) \quad (12.4)$$

$$n_{FA}(i) = n_{TG}^{feed} \cdot (wt_{FA}(i)/MW_{FA}(i))/\sum_i n_{FA,aux}(i). \quad (12.5)$$

2) We define the following variables:
   - DH: Reaction advance for the hydrodeoxygenation reaction
   - DCx: Reaction advance for the decarboxylation reaction
   - DC: Reaction advance for the decarbonylation reaction
   - DM1: Reaction advance for the CO_2 methanation reaction
   - DM2: Reaction advance for the CO methanation reaction
   - m_{water}^{teo}: Theoretical mass of water at the end of the reaction
   - m_{water}^{aq}: Theoretical mass of water at the end of the reaction in the liquid phase
   - m_{water}^{gas}: Theoretical mass of water at the end of the reaction in the gaseous phase

- $m_{\text{alk}}^{\text{teo}}(j)$: Theoretical mass of hydrocarbons at the end of the reaction
- $m_{\text{alk,total}}^{\text{teo}}$: Theoretical total mass of hydrocarbons at the end of the reaction
- $n_{\text{CO}_2}^{\text{teo}}$: Theoretical moles of CO_2 at the end of the reaction
- $n_{\text{CO}}^{\text{teo}}$: Theoretical moles of CO at the end of the reaction
- $n_{\text{CH}_4}^{\text{teo}}$: Theoretical moles of CH_4 at the end of the reaction
- A^{teo}: Ratio between the theoretical moles of CO_2 and CH_4 at the end of the reaction
- B^{teo}: Ratio between the theoretical moles of CO_2 and CO at the end of the reaction
- C^{teo}: Ratio between the theoretical moles of hydrocarbons and water at the end of the reaction
- HC^{teo}(j): Percentage of a hydrocarbon j within the total hydrocarbons
- A^{exp}: Ratio between the experimental moles of CO_2 and CH_4 at the end of the reaction
- B^{exp}: Ratio between the experimental moles of CO_2 and CO at the end of the reaction
- C^{exp}: Ratio between the experimental moles of hydrocarbons and water at the end of the reaction

3) The variables above are related through the following equations:
   - it is assumed that all triglycerides are converted to hydrocarbons, then the sum of the reaction progresses of the deoxygenation reactions is one:

$$\text{DH} + \text{DC}x + \text{DC} = 1, \tag{12.6}$$

   - water is produced by hydrodeoxygenation, decarbonylation, methanation, and Water Gas Shift reactions

$$m_{\text{water}}^{\text{teo}} = \left(2 \cdot \text{DH} \cdot \sum_{i=j} n_{\text{FA}}(i) + \text{DC} \cdot \sum_{i=j} n_{\text{FA}}(i) + 2 \cdot n_{\text{CH}_4}^{\text{teo}} + n_{\text{water}}^{\text{WGS}}\right) \cdot PM_{\text{H}_2\text{O}}, \tag{12.7}$$

   - the water produced is distributed in the aqueous and gaseous phases

$$m_{\text{water}}^{\text{teo}} = m_{\text{water}}^{\text{aq}} + m_{\text{water}}^{\text{gas}}, \tag{12.8}$$

   - hydrocarbons are formed from fatty acids by hydrodeoxygenation, decarbonylation, and decarboxylation reactions. Only the hydrodeoxygenation reaction maintains the number of carbons in the chain; the other two reduce it by 1.

$$m_{\text{alk}}^{\text{teo}}(j) = \text{DH} \cdot \sum_{i=j} n_{\text{FA}}(i) \cdot \text{MW}(i) + (\text{DC} + \text{DC}x) \cdot \sum_{i=j+1} n_{\text{FA}}(i) \cdot \text{MW}(i), \tag{12.9}$$

$$m_{\text{alk,total}}^{\text{teo}} = \sum_{j} m_{\text{alk}}^{\text{teo}}(j), \tag{12.10}$$

- CO_2 is assumed to be produced by decarboxylation (i.e. methanation and WGS reactions only consume it)

$$n_{CO_2}^{teo} = \mathrm{DC}x \cdot \sum_{i=j} n_{FA}(i), \tag{12.11}$$

- CO is formed by decarbonylation and WGS and is consumed by methanation

$$n_{CO}^{teo} = \mathrm{DC} \cdot \sum_{i=j} n_{FA}(i) - DM2 \cdot n_{CH_4}^{teo} + n_{CO}^{WGS}, \tag{12.12}$$

- all methane is produced by the methanation reactions

$$\mathrm{DM1} + \mathrm{DM2} = 1, \tag{12.13}$$

- stoichiometric ratios between the different species are taken into account

$$n_{CO_2}^{WGS} = -n_{CO}^{WGS}. \tag{12.14}$$

$$n_{CO_2}^{WGS} = -n_{H_2O}^{WGS}. \tag{12.15}$$

- A^{teo}, B^{teo}, and C^{teo} are calculated as follows

$$A^{teo} = n_{CO_2}^{teo}/n_{CH_4}^{teo}, \tag{12.16}$$

$$B^{teo} = n_{CO_2}^{teo}/n_{CO}^{teo}, \tag{12.17}$$

$$C^{teo} = \sum_{j} m_{alk}^{teo}(j)/m_{water}^{aq}, \tag{12.18}$$

where

$$m_{alk}^{teo}(j) = HC^{teo}(j) \cdot m_{alk,total}^{teo}. \tag{12.19}$$

4) Estimates for the advance of reactions DH, DCx, DC, DM1, and DM2 are obtained by solving the optimization problem in Eq. (12.20):

$$\min\ (A^{exp} - A^{teo})^2 + (B^{exp} - B^{teo})^2 + (C^{exp} - C^{teo})^2 + \sum_{i=13}^{20} (HC^{teo}(i) - HC^{exp}(i))^2 \tag{12.20}$$

s.t. Eq. (12.1)–(12.19).
Assuming that the experimental data is given in terms of moles of $CO_2(n_{CO_2}^{exp})$, CO (n_{CO}^{exp}), CH_4 ($n_{CH_4}^{exp}$) and mass weight of aqueous and oil phases after reaction (w_{oil}, w_{aq}) and that the initial composition in fatty acids ($\mathrm{wt}_{FA}(i)$) and total mass (m_{oil}^{feed}) are given, the decision variables of the problem in Eq. (12.20) are: $A^{teo}, B^{teo}, C^{teo}, HC^{teo}, A^{exp}, B^{exp}, C^{exp}, n_{FA}(i), n_{FA}(i)$, DH, DCx, DC, DM1, DM2, $m_{water}^{teo}, m_{water}^{aq}, m_{water}^{gas}, m_{alk}^{teo}(j), m_{alk,total}^{teo}, \mathrm{HC}^{teo}(j), n_{CO_2}^{teo}, n_{CO}^{teo}, n_{CH_4}^{teo}, n_{CO}^{WGS}, n_{CO_2}^{WGS}$, and $n_{H_2O}^{WGS}$.

5) The advance of reactions estimated by solving Eq. (12.20) indicates how fatty acids react in the presence of a certain catalyst and reaction conditions. Hence, they can be used to estimate the final hydrocarbon composition $w_{alk}(j)$ for any fatty acid mix, thus allowing the estimation of the reaction products for different triglyceride-based feeds.

In all cases, we assumed that cracking reactions did not occur. In addition, it was also found that two assumptions are usually made: (i) all the CO formed is by WGS, (ii) DCx and DC occur in parallel, and no WGS reaction occurs. Both cases were studied separately.

Estimation of Extent of Reactions for Conversion of TAG into Hydrocarbons on NiMo and CoMo Catalysts The optimization problem in Eq. (12.20) was solved using experimental data from [18], who reported detailed hydrocarbon compositions for soybean oil conversion on a NiMo catalyst at 92 bar and 400 °C. The results are shown in Table 12.3. Note that the assumptions listed in Table 12.3 were implemented by slightly modifying the Problem in Eq. (12.20) to account for them.

Estimation of Hydrocarbon Composition on NiMo and CoMo Catalysts for Different TAG-Based Feedstocks The advances of reaction in Table 12.3, obtained via experimental data for soybean, were used to predict the composition of hydrocarbons for the other oils in Table 12.1. In all cases, the hydrogen consumption is calculated as shown in Eq. (12.21). The results, expressed as weight fractions of the individual hydrocarbons yields, are presented in Table 12.4.

$$n_{H_2} = (3 \cdot \mathrm{DH} + \mathrm{DC}) \cdot \sum_{i=j} n_{FA}(i) + 4 \cdot \mathrm{DM1} \cdot n_{CH_4}^{teo} + 3 \cdot \mathrm{DM1} \cdot n_{CH_4}^{teo} + n_{H_2}^{WGS}. \tag{12.21}$$

Table 12.3 Advance of reaction estimates for conversion of TAG into hydrocarbons on NIMo and CoMo catalyst.

Catalyst	Assumptions	DH	DCx	DC	DM1	DM2
NiMo	DCx and DC occur in parallel and no WGS	0.304	0.557	0.139	0.807	0.193
NiMo	All CO formed is produced by the WGS reaction	0.304	0.696	0	1	0
CoMo	DCx and DC occur in parallel and no WGS	0.296	0.495	0.209	1	0
CoMo	All CO formed is produced by the WGS reaction	0.296	0.704	0	0.027	0.973

Source: [18] / with permission of Elsevier.

Table 12.4 Hydrocarbon yields estimated for sunflower, canola, and carinata oils and tallow.

Hydrocarbons	Sunflower	Canola	Carinata	Tallow
C13	0.001	0.001	0	0.016
C14	0	0	0	0.007
C15	0.036	0.029	0.031	0.169
C16	0.016	0.013	0.013	0.074
C17	0.533	0.532	0.296	0.385
C18	0.233	0.232	0.129	0.168
C19	0.002	0.012	0.000	0
C20	0.001	0.005	0.000	0
C21	0.001	0	0.252	0
C22	0.001	0	0.110	0
H_2O	0.044	0.044	0.041	0.045
CO	0.011	0.011	0.010	0.011
CO_2	0.068	0.067	0.063	0.069
CH_4	0.002	0.002	0.002	0.002
C3 (propane)	0.043	0.043	0.040	0.044
H_2	0.010	0.010	0.012	0.010

Cx expresses a hydrocarbon of chain length *x*.

Validation We validated the procedure above by comparing these estimates with the few available data in the literature on the same type of catalyst. The results are presented in Figure 12.4.

We consider the validation to be reasonably good since the operating conditions used in the references are slightly different from the ones used to derive the reaction extents (400 °C and 92 bar); hence, differences in yields are expected. In all cases, the major difference is in the percentage of C17.

12.3.2.3 Area 200: Separation

This area consists of two flash tanks (F-200 and F-201) and a decanter (D-200). The operating conditions of the equipment were selected by sensitivity analysis with the objective of maximizing the amount of hydrocarbons and minimizing the amount of CO_2, CO, propane, and water.

The F-200 unit was set to operate at the same pressure as the reactor (7 MPa) and 220 °C. The flow rates of n-C18 in the liquid and gaseous stream, as well as water and propane in the liquid stream, were monitored to select the temperature.

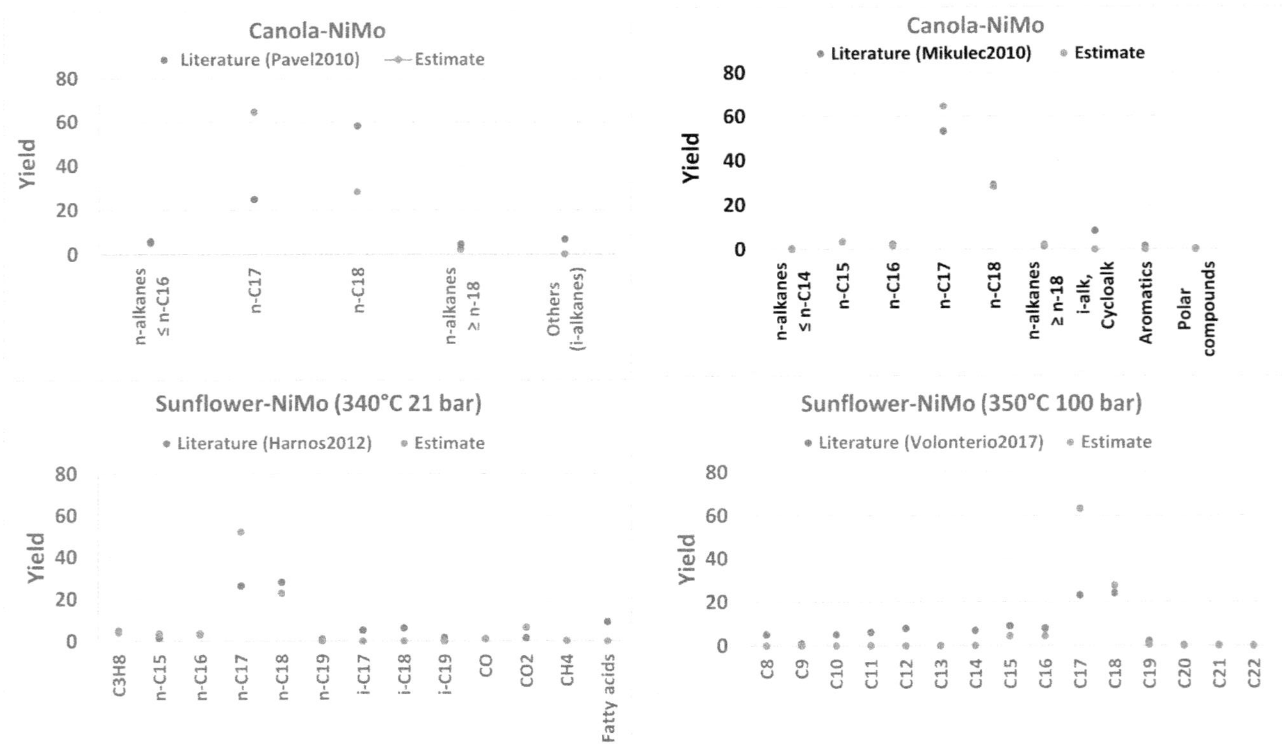

Figure 12.4 Comparison estimated and experimental yields. *Source:* Pavel2010 = Ref. [28]; Mikulec2010 = Ref. [29]; Harnos2012 = Ref. [30]; Volonterio2017 = Ref. [31] / MDPI / CC BY 4.0.

Lower temperatures resulted in a higher presence of water and propane in the hydrocarbon stream.

F-201 was set to operate at 1.1 MPa and 30 °C. The decrease in temperature and pressure is necessary for the separation of water from the other gases. The D-200 decanter was set to operate at 5 MPa and 25 °C. In this case, water separation is better at higher pressure and lower temperature.

12.3.2.4 Area 300: Hydrocracking and Hydroisomerization

These two reactions are required to meet the requirements of aviation fuels. Hydrocracking aims to produce hydrocarbons of the right chain length (C8–C16 for jet fuel oils), while hydroisomerization improves the fuel's cold flow properties and freezing point [27].

The R-300 reactor was modeled based on the data from [23]: 272 °C, 5.5 MPa, and zeolite Pt/US-Y catalyst. Hydrogen at the reactor inlet was considered at a ratio of 230 Nm^3/m^3 hydrocarbons. The yields were adapted from experimental data, as shown in Eq. (12.22):

$$\frac{m\,\mathrm{HC}_i \text{ feed in [23]}}{m\,\mathrm{HC}_i \text{ outlet in [23]}} = \frac{m\,\mathrm{HC}_i \text{ feed simulation}}{m\,\mathrm{HC}_i \text{ outlet in simulation}}. \tag{12.22}$$

As an example, we present the results for the canola oil reference in Table 12.5.

Further, the same ratio as in [23] was assumed to find the split into n-alkane and i-alkane. As an example, the final values for the canola oil simulation are shown in Table 12.6.

12.3.2.5 Area 400: Products Separation

The final separation steps aim to produce a jet fuel cut that meets the ASTM D7566 standard. The final design included three distillation columns that were rigorously modeled using RadFrac in Aspen Plus. The D-400 unit separates light (C3–C7) from heavy (C8–C20) hydrocarbons using 20 stages. In D-402, the jet fuel range (C8–C16) and the green diesel range (C17 onward) are separated in a 20 stage column. Finally, the D-401 tower is used to separate LPG (C3 and C4) from the green naphtha range (C5–C7) using a 6-stage column.

Distillate flows and refluxes required adjustment for the different simulations, which is compatible with the operation of a refinery.

12.3.3 Simulation Results

The previous design is able to produce sustainable aviation cuts that are by themselves close to meeting ASTM D7566 standards, as shown in Table 12.7 for

Table 12.5 Hydrocracking and isomerization yields based on data from [23].

	Data from [23]		Results used in the simulation from Eq. (12.22)	
Hydrocarbons	**Inlet weight (%)**	**Outlet weight (%)**	**Inlet weight (%)**	**Outlet weight (%)**
C4	0.00	0.81	0	0.94
C5	0.00	2.45	0	2.84
C6	0.01	3.97	0	4.60
C7	0.04	5.76	0	6.67
C8	0.12	7.50	0	8.69
C9	0.23	8.37	0	9.70
C10	1.03	9.71	0	11.25
C11	0.18	8.08	0	9.36
C12	0.17	7.03	0	8.14
C13	0.62	5.76	0.07	0.64
C14	2.93	4.77	0.03	0.05
C15	5.17	3.37	3.48	2.27
C16	6.08	4.30	1.52	1.08
C17	38.03	14.80	64.59	25.14
C18	43.38	12.89	28.21	8.38
C19	0.82	0.10	1.46	0.18
C20	0.51	0.06	0.64	0.08
C21	0.17	0.02	0	0
C22	0.10	0.02	0	0
C23	0.09	0.01	0	0

Simulation results are based on the canola oil feed reference.

the Canola feed, with the only exception being a small difference between the estimated and expected net heat of combustion. The properties listed in the table were retrieved from Aspen Plus by adding the request for extra properties to the Results Summary streams table.

In terms of yield, 75% (in mass) of the oil feed is recovered as jet fuel oil, naphtha, diesel, or gas. The product distribution is shown in Table 12.8, where it can be appreciated that diesel is also produced in a large amount.

Table 12.6 Product distribution in R-300 for the canola oil simulation.

Hydrocarbons	n-alkanes	i-alkanes	Others
C4	0.003	0.006	
C5	0.006	0.020	
C6	0.009	0.032	
C7	0.011	0.048	
C8	0.012	0.065	
C9	0.012	0.074	
C10	0.017	0.083	
C11	0.008	0.075	
C12	0.007	0.065	
C13	0.001	0.005	
C14	5.79E−5	0.0004	
C15	0.004	0.002	
C16	0.001	0.009	
C17	0.030	0.193	
C18	0.009	0.066	
C19	0.0002	0.001	
C20	0.0001	0.0006	
H_2			0.024
Propane			0.040
CO_2			0.046
CO			0.001
Water			0.0005
C1	0.0006		
C2	0.0006		

12.4 Sustainable Aviation Fuel Process Design: Adjustments in Design to Match Current Operations in the Refinery

Most refineries will need to modify the previous design to comply with internal operation restrictions or preferences. These modifications need to be discussed

Table 12.7 Selected properties of the jet fuel cut obtained by simulation of the base process.

	Canola oil	ASTM D7566
Distillation temperature (°C):		
10% recovery	167	205, max
50% recovery	184	15, min
90% recovery	244	40, min
Final boiling point (°C)	257	300, max
Flash point (°C)	48	38, min
Freezing point (°C)	−46	−47, max
Viscosity at −20 °C (mm^2/s)	3	8, max
Net heat of combustion (MJ/kg)	40	42.8, min
Density at 15 °C (kg/m^3)	741	775–840

Table 12.8 Product distribution and yield for base case design.

Feedstock	Jet fuel oil (%)	Naphta (%)	Diesel (%)	Gas (%)	Yield (Ljet/ton feed)
Canola oil	55.0	8.8	32.6	3.6	680

case by case. Figure 12.5 shows the final process reached after iterating with refinery technicians on the adjustments needed to comply with their operating preferences.

The main changes are:

Adjustment to Feed Hydrogen Composition

The refinery uses a technology that provides hydrogen at a higher composition than the one assumed in Table 12.2, and a different hydrogen excess/feed oil ratio. Also, hydrogen excess is recycled back to the hydrotreater reactor R-100 after separation in the decanter S-200. In addition, high purity hydrogen must be used in the reactor. Pressure swing adsorption (PSA) separation is used for hydrogen purification, and ANCAP already has this technology. However, the plant does not currently have idle capacity, meaning that the cost of a new PSA unit should be included in the economic evaluation. Hence, the PSA-100 unit was added to the simulation flowsheet to adjust the composition of the hydrogen fed to the reactor.

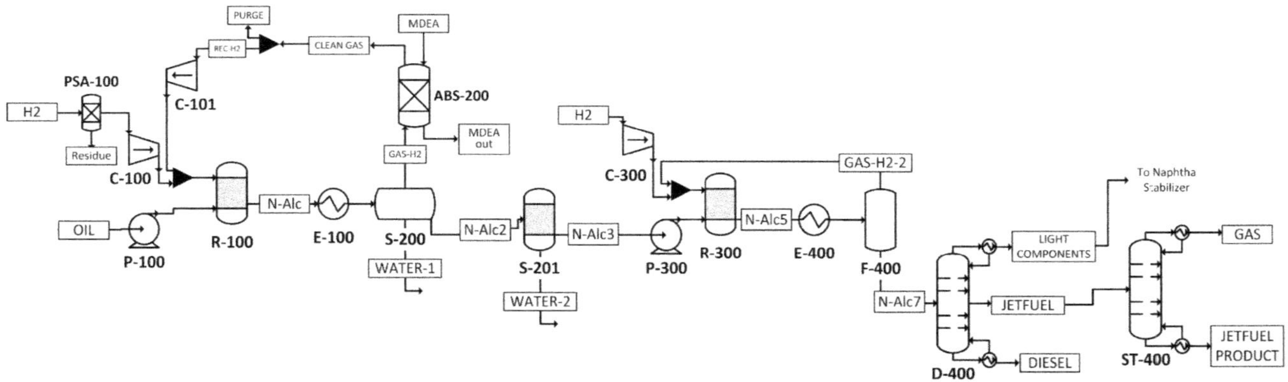

Figure 12.5 Process flow diagram – process adjusted to the refinery's existing operations.

Adjustments to Separation Sequence in Area 200

The objective in this area remains to be the maximization of the amount of hydrocarbons and the minimization of the amount of CO_2, CO, propane, and water in the hydrocarbons stream. However, the operation conditions in the three-phase separator (modeled as E-100 plus S-200) and the water separator (S-201) were changed to the temperatures and pressures used by the refinery in similar processes.

In addition, engineers from the group made us aware of the need to decrease the CO_2 and CO content in the H_2-rich GAS stream before recirculation to the hydrotreatment reactor R-100. Therefore, an absorption column with MDEA (methyldiethanolamine, the absorbent used in the refinery for other operations) is included to decrease the CO_2 content. This absorption column is rigorously modeled using the Radfrac unit in Aspen Plus. The CO content is decreased by purging.

Adjustments in Area 300

Here, adjustments were made to the hydrogen/ feed ratio to the hydrocracking and hydroisomerization reactor (R-300). Also, recirculation of the excess hydrogen was added; the MDEA column is not necessary in this case as this reaction does not generate CO_2.

Adjustments in Area 400

Several adjustments were made to this section following the recommendations from the refinery engineers. The aim is to reduce the number of operations and facilitate seamless connection with the rest of the plant. The changes are as follows:

- Operation conditions of the first flash column are changed to 50 °C and 5.5 MPa.
- The D-400 and D-402 towers are merged into one with more plates. Jet fuel is extracted in an intermediate cut, and light fuels are extracted from the head and diesel from the bottom. This unit is modeled as a rigorous column (Radfrac) with 30 plates, feed at plate 12, jetfuel outlet at plate 12, and a reflux ratio of 3.
- The reboiler of the new column operates at high temperatures (361 °C). A furnace already available at the refinery is used.
- Column D-401 is eliminated because light oils can be processed at an existing naphtha stabilizer.
- A steam stripper is added to improve jetfuel properties.

Table 12.9 Product distribution and yield for various feedstocks – modified process.

Feedstock	Jet fuel oil (%)	To naphtha stabilizer (%)	Diesel (%)	Gas (%)	Yield (Ljet/ton feed)
Canola oil	19.5	15.8	37.0	27.7	307
Sunflower oil	25.9	15.7	37.3	21.1	394
Carinata oil	31.4	15.6	25.8	27.2	481
Soybean oil	34.4	7.8	39.8	18.0	535
Tallow	35.5	15.8	25.1	23.6	541

12.4.1 Simulation Results

Table 12.9 compares jet fuel yields and product distributions for the different TAG-based feedstocks. Noticeably, the changes made to the process aimed at better adapting it to current operations negatively affected the jet fuel yields. However, hydrocarbons in the jet fuel range from the stream being sent to the naphtha stabilizer could potentially be recovered in downstream current refinery operations.

Another noticeable result is the difference in product distribution and yields for the different feedstocks, with tallow and soybean oil providing the largest yields. However, to make a fair comparison, this calculation should also consider the hydrocarbons in the jet fuel range that could potentially be recovered downstream of this process addition, as commented in the previous paragraph.

Table 12.10 shows the estimated properties of the jet fuel oil cut for the different feedstocks. As evidenced by the results for the canola oil feed, the changes made to the process negatively affected the closeness to the ASTM norm. However, it is important to mention again that (i) these properties should be satisfied upon blending; (ii) blending is anyways required as the aromatics requirement cannot be met by this pathway.

12.5 Environmental Assessment Using GREENSCOPE

Finally, an environmental assessment using the U.S. Environmental Protection Agency (EPA) GREENSCOPE methodology [10, 32–34] was performed to estimate the environmental impacts of the proposed process. The Refinery Group already possessed Life Cycle Assessment (LCA) data for the feedstocks considered in this study and their current processing of these feedstocks [35]. Hence, a gate-to-gate estimation of the environmental impacts associated with the design in Section 12.4

Table 12.10 Selected properties of the jet fuel cut obtained by simulation of the modified process.

	Canola oil	Sunflower oil	Carinata oil	Soybean oil	Tallow	ASTM D7566
Distillation temperature (°C):						
10% recovery	167	119	131	115	141	205, max
50% recovery	184	155	175	122	203	15, min
90% recovery	278	275	256	137	274	40, min
Final boiling point (°C)	257	290	267	140	288	300, max
Flash point (°C)	48	11	21	9	28	38, min
Freezing point (°C)	−46	−70	−59	−89	−49	−47, max
Viscosity a −20 °C (cSt)	3	2	2	2	3	8, max
Net heat of combustion (MJ/kg)	40	39	40	43	34	42.8, min
Density at 15 °C (kg/m^3)	741	713	723	702	729	775–840

was deemed to be the most relevant study as a complete Life Cycle Assessment (LCA) could be performed by the refinery group by combining the already existing LCAs with our estimates. The GREENSCOPE methodology includes 139 indicators in four areas: material efficiency, energy, economy, and environment, which are calculated from the mass and energy balances of a process flowsheet. Not all indicators are relevant to this case study. Below, we present a list of the categories and indicators that we considered.

12.5.1 Overview of Selected Indicators

12.5.1.1 Dangerous Materials

Hazardous materials are those listed by the U.S. Department of Transportation (DOT) [36]; they present a health or safety risk when transported by air, land, or sea. Indicators for this category include: *Number of hazardous materials input, Mass of hazardous materials input, and Specific hazardous raw materials input*. Chemical compounds present in our design that are classified as hazardous include: hydrogen, carbon dioxide, carbon monoxide, all the hydrocarbons, and MDEA.

12.5.1.2 Chemical Exposure Index (CEI)

The CEI is a measure of the potential health risk to people in the plant or surrounding communities if a chemical release incident occurs. Evaluation of CEI requires estimation of airborne quantity (*AQ*), which is the maximum amount in air, and *ERPG*, which is a value obtained from the AIHA (American Industrial Hygiene Association) emergency response guidelines [37]. Calculation of CEI is complex, and the reader is referred to GREENSCOPE's documentation for further details [10, 32]. From the chemicals used/produced in the process designed above, only carbon monoxide has a non-zero ERPG value.

12.5.1.3 Health Hazards in the Workplace

Two indicators are considered in this category. The *health hazard and irritation factor indicator* measures irritation to the skin and eyes of humans and animals. It calculates the volume of irritating substances in the workplace per mass of product. The *health hazard and chronic toxicity factor indicator* considers carcinogenicity, mutagenicity, and reproductive toxicity among others in the work environment, by calculating the volume of contaminated air. Values for these indicators are calculated from different properties, which are usually reported in the MSDS (Material Safety Data Sheet) of the compounds. GREENSCOPE prioritizes the estimation from the *Ecclass* property but if this property is not available for any of the compounds, the R-phrases, pH, or LD50 dermal lethality limits are used. *Ecclass* is a classification of hazardous substances from the European Parliament. If the substance is classified with *C*, it is corrosive, and if it is classified as *Xi*, it is irritant. *Rcode* refers to the risk phrases; codes 34, 35, 36, 37, and 38 are associated with irritant substances. The second indicator is calculated using a similar mechanism. From the chemicals used/produced in the process designed above: hydrogen and carbon dioxide are classified as very toxic (T+), carbon monoxide, ethane as toxic (T), propane, pentane, hexane, heptane, octane, isobutane, 2-methyl-butane, 2-methyl-pentane, 2-methyl-hexane, and 2-methyl-heptane as harmful (Xn), and MDEA as irritant (Xi).

12.5.1.4 Safety Hazards

Two indicators are considered in this category. The *safety hazard and fire/explosion* indicator is related to the probability of fire or explosion and is calculated from the enthalpy of combustion and an index calculated from the flash temperature, R phrase, or red zone of the National Fire Protection Association (NFPA) 704 diamond [38]. The *Safety hazard and acute toxicity* indicator estimates acute toxicity by calculating a volume of contaminated air at a hazardous level. Its calculation uses the Immediately Dangerous to Life or Health (IDLH) property, which represents a life-threatening atmosphere. This data was obtained from a list published by the US National Institute for Occupational Safety and Health [39]. From the chemicals used/produced in the process designed above, hydrogen, carbon monoxide, methane, ethane, propane, n-butane, n-pentane, and isobutane are classified with a rating NFPA-f severe; n-hexane, n-heptane, n-octane, n-nonane, 2-methyl-butane, 2-methyl-pentane, 2-methylhexane, 2-methylheptane, 2-methyloctane, 3-methylnonane, and 3-methyldecane are classified with a rating NFPA-f serious. All the other compounds are classified with a rating NFPA-f, either slight or moderate.

12.5.1.5 Substance Toxicity

The *specific toxic release, specific toxic release, and total mass of persistent, bio-accumulative, and toxic chemicals used* indicators are considered in this case. Toxic substances are those listed in the Toxics Release Inventory (TRI) Program [40] and include chemicals identified as causing cancer or other chronic health effects and adverse environmental effects. Persistent, Bioaccumulative and Toxic (PBT) is a subcategory of substances that have toxic, bioaccumulative, and persistent properties (remain in the environment for long periods of time). None of the chemicals used/produced in the design are in the TRI lists.

12.5.1.6 Enviromental Hazard

The *persistency of organic substances* indicator is the persistence of organic substances and is calculated by introducing the concept of organic mass. The organic mass is calculated from a property called half-life and is defined as the time it takes to halve the concentration of a compound due to degradation. This property was extracted from a computational tool available on the EPA website (EPI Suite [41]) that allows estimating degradability properties, among others, from the structure of a compound. As in other cases, GREENSCOPE allows the use of other data if the preferred one is not available. The *air hazard* indicator is similar to the health hazard indicator for substances with carcinogenic and mutagenic properties in the environment (HH chronic toxicity) explained above. For the calculation of the *water hazard* indicator, the LC50 property is used, which in this case is the concentration that causes 50% mortality in a species of fish (Fathead minnow) after 96 h. For this property, another EPA software called T.E.S.T. [42] was used, which estimates this value from the structure or obtains it from databases. The *bioaccumulation (in the food chain or soil)* indicator is calculated from the bioconcentration factor (BCF) parameter, which is the ratio of the concentration of a compound in an animal, for example a fish, compared to the concentration of the substance in the environment. The parameter Kow can also be used, which is the ratio of the partitioning of a substance in a two-phase system: one of octanol and one of water. Both properties were obtained from EPI Suite.

12.5.1.7 Indicators from Potency Factors: Global Warming, Smog, Acidification, Ozone Depletion, and Cancer

Global warming potential and intensity, photochemical oxidation (smog) potential and intensity, atmospheric acidification potential and intensity, stratospheric ozone-depletion potential and intensity, and human health burden, cancer effects indicators are calculated as the equivalent amount of a substance in reference to another substance that has a known effect using "potency factors." As an example, the global warming potential is expressed in CO_2 emission equivalents and can be positive even if CO_2 is not emitted per se. The indicators are always calculated as the mass of output per potential over the mass of product or product sales. Benzene is used to assess potential carcinogenic effects, while CFC-11 is used to assess the impact on the ozone layer. Ethylene is used to assess photochemical oxidation or smog, and sulfur dioxide is used to evaluate the potential acidification of the atmosphere. The factors for different compounds can be obtained from the book Sustainability Metrics by ICHEME [43]. From the chemicals used/produced in the process designed above, no one is listed to cause ozone depletion, be carcinogenic, or cause acidification (i.e. the potency factors for CFC-11, benzene, and sulfur dioxide are 0). Carbon dioxide, carbon monoxide, and methane have non-zero potency factors for global warming potential; while carbon monoxide, methane, ethane, propane, n-butane, n-pentane, n-hexane, n-heptane, n-octane, n-nonane, n-decane, n-undecane, n-dodecane, isobutane, 2-methyl-butane, 2-methyl-pentane, 2-methylhexane, 2-methylheptane, and 2-methyloctane have non-zero potency factors for smog.

12.5.1.8 Liquid Emissions

The indicators are *non-polluted liquid waste volume* and *polluted liquid waste volume*. The liquid emission from the process is, in all cases, the MDEA stream exiting the absorber. This stream was considered to be uncontaminated and can be recovered.

12.5.2 Results

The environmental indicators mentioned in Section 5.1 are shown in Figure 12.6. The results are expressed on a scale of 0–100% sustainability, as proposed by GREENSCOPE [33] and expressed in Eq. (12.23).

$$\text{Sustainability Score} = \frac{\text{Case Study} - \text{Worst case}}{\text{Best case} - \text{Worst case}} \cdot 100. \tag{12.23}$$

Overall, as expected, the results are similar for the different feedstocks, as they all involve relatively similar chemical substances (triglycerides, hydrocarbons, and hydrogen), which vary only slightly in the flows used/produced from each one.

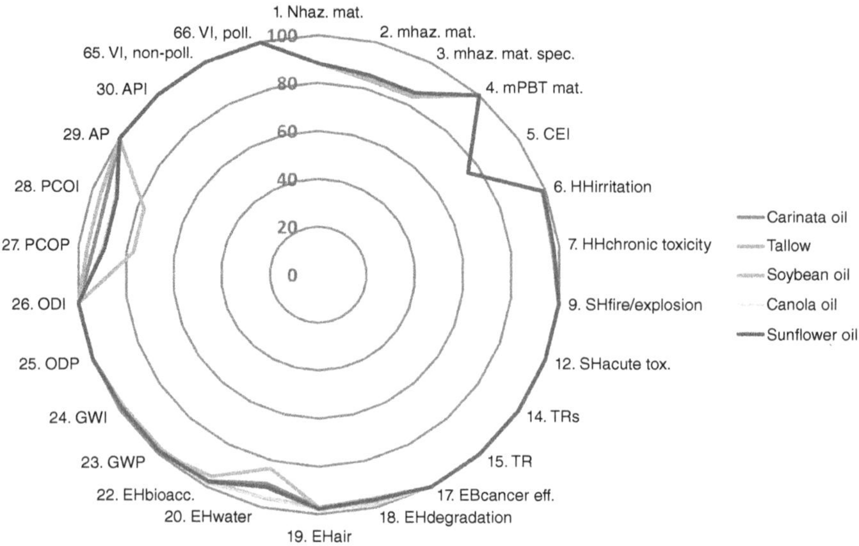

Figure 12.6 Environmental impacts for the process flowsheet in Section 12.4 and different feedstocks. Indicators are as follows: 1 = number of hazardous materials input; 2 = mass of hazardous materials input; 3 = specific hazardous raw materials input; 4 = total mass of persistent, bio-accumulative, and toxic chemicals used; 5 = chemical exposure index; 6 = health hazard, irritation factor; 7 = health hazard, chronic toxicity factor; 9 = safety hazard, fire/explosion; 12 = safety hazard, acute toxicity; 14 = specific toxic release; 15 = toxic release intensity; 17 = human health burden, cancer effects; 18 = environmental hazard, persistency of organic substances; 19 = environmental hazard, air hazard; 20= environmental hazard, water hazard; 22 = environmental hazard, bioaccumulation (the food chain or in soil); 23 = global warming potential; 24 = global warming intensity; 25 = stratospheric ozone-depletion potential; 26 = stratospheric ozone-depletion intensity; 27 = photochemical oxidation (smog) potential; 28 = photochemical oxidation (smog) intensity; 29 = atmospheric acidification potential; 30 = atmospheric acidification intensity; 65 = non-polluted liquid waste volume; 66 = polluted liquid waste volume.

In all cases, the indicators with the lowest sustainability scores are those associated with hazardous materials (indicators 1, 2, and 3) due to the presence of high-pressure hydrogen and MDEA and the CEI (indicator 5) due to the risk posed by a CO release. The photochemical oxidation potential or smog indicators (indicators 27 and 28) are not 100% due to the presence of hydrocarbons. As seen in Figure 12.7, the differential lower result for soybean oil responds to an increase in the production of methyl heptane and n-octane when processing this raw material. In the case of the water hazard indicator (indicator 20), the less than 100% score for tallow and carinata oil is because some hydrocarbons, such as methane, ethane, propane, as well as MDEA have high LC50 values, and these are disposed in a relatively larger amount for these feedstocks.

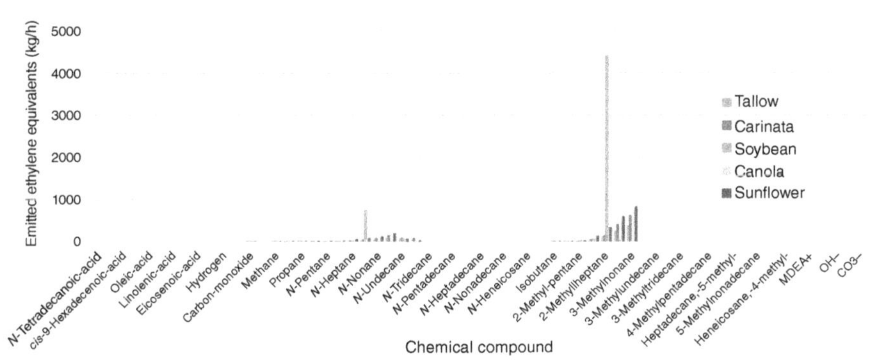

Figure 12.7 Emitted ethylene equivalents for the processing of different oils. These emissions affect indicators 27 (photochemical oxidation (smog) potential) and 28 (photochemical oxidation (smog) intensity).

12.6 Summary and Final Remarks

This work presents a practical and real example of collaboration between academia and an oil refinery to develop process flowsheets for the production of SAF and their techno-environmental evaluation.

The goal of the project was to propose pathways for converting an existing oil refinery into a bio-refinery, maximizing the use of existing infrastructure. After initial discussions with the industrial technical counterpart on different options, interest was directed toward SAFs via the hydrotreatment of local oil crops.

Pathways for oil crops to hydrocarbons have already been proposed in the literature. Catalysts that convert triacylglycerides into hydrocarbons were already commercialized. However, the product distribution for the different catalysts was not available for the feedstocks that we were considering. This data was needed to assess which of the already existing refinery operations could process the obtained hydrocarbons. Developing experimental capabilities to get this data was out of the scope of the project. We overcome this limitation by proposing an optimization problem that uses published experimental data for one catalyst and a specific oil-based feedstock to estimate the advance of the different reactions promoted by the catalyst and then use those advance of reactions to estimate the product distribution with the same catalyst and a different oil-based feedstock.

Two process flowsheets were then designed; one was adjusted to obtain a jet fuel cut as close as possible to meet the ASTM standard for renewable jet fuel; the other one was adjusted to meet the operational practices of the plant. It was found that yields and compliance with the ASTM standard deteriorated as a result of the changes made; however, we have not yet performed rigorous optimization that could potentially overcome this. In any case, jetfuel produced via these feedstocks and hydrotreatment will require mixing as the aromatic content can not be satisfied.

A gate to gate environmental analysis was performed using the Greenscope methodology. Most indicators have a sustainability score above 80%, with the only exceptions being the potential for smog in the case of processing soybean oil, and the Chemical Exposure Index for all feeds. Future studies could consider modifications aimed at improving these indicators.

Acknowledgments

The authors acknowledge funding from Agencia Nacional de Investigación e Innovación - Uruguay, Grant/Award Number: FSE_1_2011_1_152900.

Ana Inés Torres acknowledges support from the Department of Chemical Engineering at Carnegie Mellon University (start-up funds).

References

1 Chattopadhyay, S., Gandhi, R., Grossmann, I.E., and Torres, A.I. (2024). Optimization of retrofit decarbonization in oil refineries. *Systems and Control Transactions* 3: 426–433.
2 Greenhouse Gas Protocol Team (2011). *Corporate Value Chain (Scope 3) Accounting and Reporting Standard*. Washington, DC: World Resources Institute and World Business Council for Sustainable Development.
3 Shell Global. (2024). Scope 1, Scope 2 and Scope 3 emissions and related risks. https://reports.shell.com/annual-report/2023/strategic-report/our-journey-to-net-zero/climate-related-metrics-and-targets/emissions-and-related-risks.html (accessed September 2024).
4 Eni waste fuel. (2024). https://www.eni.com/enirewind/en-IT/water-waste/waste-to-fuel.html (accessed September 2024).
5 Eni Bioenergy. (2024). Energy from biogenic raw materials. https://www.eni.com/en-IT/actions/energy-sources/bioenergy.html (accessed September 2024).
6 Shell. https:www.shell.com/business-customers/catalysts-technologies/licensed-technologies/benefits-of-biofuels.html (accessed September 2020).
7 ExxonMobil. https:corporate.exxonmobil.com/News/Newsroom/News-releases/2020/0811_ExxonMobil-and-Global-Clean-Energy-Holdings-sign-agreement-for-renewable-diesel (accessed September 2020).
8 TotalEnergies. Biofuels. https://miningsolutions.totalenergies.com/renewables-and-low-carbon-energies/biofuels (accessed September 2024).

9 Aramco Aramco and ENOWA to develop first-of-its-kind e-fuel demonstration plant. https://www.aramco.com/en/news-media/news/2023/aramco-and-enowa-to-develop-first-of-its-kind-efuel-demonstration-plant (accessed September 2024).
10 U.S. Environmental Protection Agency. Assessing chemical process sustainability with GREENSCOPE. https://www.epa.gov/chemical-research/assessing-chemical-process-sustainability-greenscope (accessed September 2024).
11 Helal, A., Kreimerman, R., Gutiérrez, S., and Torres, A.I. (2019). A market-driven algorithm for the assessment of promising bio-based chemicals. *AIChE Journal* 65: e16775.
12 Torres, A., Gutiérrez, S., Kreimerman, R. et al. (2019). *Reporte Final Proyecto ANII FSE-1-2015-1-109976: Bio-refinerías en Uruguay: Evaluación tecnoeconómica de la producción de combustibles y químicos a partir de materia prima y residuos nacionales.*
13 International Air Transport Association (2015). iata.org. https://www.iata.org/contentassets/d13875e9ed784f75bac90f000760e998/safr-1-2015.pdf (accessed 08 August 2024).
14 Zhang, L., Torres, A.I., Chen, B. et al. (2024). Optimal retrofitting of conventional oil refinery into sustainable bio-refinery under uncertainty. *AIChE Journal* 70: e18371.
15 Eni Ecofining, a proprietary biofuel technology. https://www.eni.com/en-IT/actions/energy-transition-technologies/biofuels/biomass-ecofining.html (accessed September 2024).
16 Neste. https://www.neste.com/products/all-products/neste-my-sustainable-aviation-fuel#bb697708, https://www.neste.com/products-and-innovation/neste-my-renewable-diesel (accessed September 2024).
17 Gutiérrez-Antonio, C., Gomez-Castro, F., de Lira-Fores, J., and Hernandez, S. (2017). A review on the production processes of renewable jet fuel. *Renewable and Sustainable Energy Reviews* 79: 709–729.
18 Veriansyah, B., Han, J.Y., Kim, S.K. et al. (2012). Production of renewable diesel by hydroprocessing of soybean oil: effect of catalysts. *Fuel* 94: 578–585.
19 Krár, M., Kovács, S., Kalló, D., and Hancsók, J. (2010). Fuel purpose hydrotreating of sunflower oil on CoMo/Al_2O_3 catalyst. *Bioresource Technology* 101: 9287–9293.
20 Šimáček, P., Kubička, D., Šebor, G., and Pospíšil, M. (2009). Hydroprocessed rapeseed oil as a source of hydrocarbon-based biodiesel. *Fuel* 88: 456–460.
21 Dorado, M.P., Ballesteros, E., López, F.J., and Mittelbach, M. (2004). Optimization of alkali-catalyzed transesterification of *Brassica carinata* oil for biodiesel production. *Energy & Fuels* 18: 77–83.
22 Selvam, D.P. and Vadivel, K. (2012). Performance and emission analysis of DI diesel engine fuelled with methyl esters of beef tallow and diesel blends.

In: *International Conference on Modelling Optimization and Computing*, vol. 38 (ed. R. Rajesh, K. Ganesh, and S.C.L. Koh), 342–358. Elsevier.
23 Robota, H., Alger, J., and Shafer, L. (2013). Converting algal triglycerides to diesel and HEFA jet fuel fractions. *Energy & Fuels* 27: 985–996.
24 Treese, S.A., Jones, D.S., and Pujado, P.R. (2018). *Handbook of Petroleum Processing*, 1600. Springer Cham.
25 Wang, W.-C. (2019). Techno-economic analysis for evaluating the potential feedstocks for producing hydro-processed renewable jet fuel in Taiwan. *Energy* 179: 771–783.
26 Tao, L., Milbrandt, A., Zhang, Y., and Wang, W.-C. (2017). Techno-economic and resource analysis of hydroprocessed renewable jet fuel. *Biotechnology for Biofuels* 10: 261.
27 Why, E.S.K., Ong, H.C., Lee, H.V. et al. (2019). Renewable aviation fuel by advanced hydroprocessing of biomass: challenges and perspective. *Energy Conversion and Management* 199: 112015.
28 Šimáček, P., Kubička, D., Šebor, G., and Pospíšil, M. (2010). Fuel properties of hydroprocessed rapeseed oil. *Fuel* 89: 611–615.
29 Mikulec, J., Cvengroš, J., Joríková, Ľ. et al. (2010). Second generation diesel fuel from renewable sources. *Journal of Cleaner Production* 18: 917–926. Minimising emissions and energy wastage by improved industrial processes and integration of renewable energy.
30 Harnos, S., Onyestyák, G., and Kalló, D. (2012). Hydrocarbons from sunflower oil over partly reduced catalysts. *Reaction Kinetics, Mechanisms and Catalysis* 106: 99–111.
31 Volonterio, E., Bussi, J., Castiglioni, J. et al. (2017). Catalytic hydrotreating of vegetable oils for the production of liquid biofuels. *INNOTEC* 14: 37–43. Available https://www.redalyc.org/journal/6061/606164031004/html/.
32 Ruiz-Mercado, G., Smith, R., and Gonzalez, M. (2012). Sustainability indicators for chemical processes: I. taxonomy. *Industrial & Engineering Chemistry Research* 51: 2309–2328.
33 Ruiz-Mercado, G.J., Smith, R.L., and Gonzalez, M.A. (2014). *Appendix C: GREENSCOPE Technical User's Guide*. U.S. Environmental Protection Agency.
34 Ruiz-Mercado, G.J., Smith, R.L., and Gonzalez, M.A. (2012). Sustainability indicators for chemical processes: II. Data needs. *Industrial & Engineering Chemistry Research* 51: 2329–2353.
35 Herrera, I., de la Rúa, C., Caldés, N., and Lechón, Y. (2016). *Análisis de ciclo de vida del proceso de producción de biodiesel a partir de un mix de materias primas grasas en la empresa Alcoholes del Uruguay (ALUR)*. Technical Report. Available https://www.gub.uy/ministerio-industria-energia-mineria/sites/ministerio-industria-energia-mineria/files/documentos/publicaciones/2.1 - Ciclo de Vida Biodiésel ALUR.pdf

36 The Electronic Code of Federal Regulations eCFR Title 49 Subtitle B Chapter I Subchapter C Part 172 Subpart B, 172.101 Purpose and use of the hazardous materials table. https://www.ecfr.gov/current/title-49/subtitle-B/chapter-I/subchapter-C/part-172/subpart-B/section-172.101 (accessed September 2024).

37 AIHA. ERPG handbook and technical supporting documents (TSD). Available to purchase at https://www.aiha.org/get-involved/aiha-guideline-foundation/erpgs (accessed September 2024).

38 NFPA (2022). NFPA 704 Standard System for the identification of the Hazards of Materials for Emergency Response.

39 CDC The National Institute for Occupational Safety and Health (NIOSH). Immediately dangerous to life or health (IDLH) values. https://www.cdc.gov/niosh/idlh/intridl4.html (accessed September 2024).

40 US Environmental Protection Agency Toxics Release Inventory (TRI) Program. https://www.epa.gov/toxics-release-inventory-tri-program (accessed September 2024).

41 US Environmental Protection Agency EPI Suite-Estimation Program Interface. https://www.epa.gov/tsca-screening-tools/epi-suitetm-estimation-program-interface (accessed September 2024).

42 US Environmental Protection Agency. Toxicity Estimation Software Tool (TEST). https://www.epa.gov/comptox-tools/toxicity-estimation-software-tool-test (accessed September 2024).

43 Sustainability Working Party. (2010). The sustainability metrics Sustainable development progress metrics recommended for use in the process industries. https://www.icheme.org/media/16021/sustainability-metrics.pdf (accessed 16 December 25).

13

Uncertainty Quantification of Solid Sorbent-Based CO_2 Capture Processes

Ana Flávia Monteiro and Debangsu Bhattacharyya

Department of Chemical and Biomedical Engineering, West Virginia University, Morgantown, WV, USA

13.1 Introduction

It is desired that uncertainties in process models be quantified since they can lead to uncertainties in the prediction of performance, energy requirement, and operating behaviors of equipment items. These, in turn, can lead to uncertainties in the design and cost estimates of the plant. Furthermore, utilities and other ancillary equipment items might be inadequately designed due to uncertainties in the performance of the process equipment items. As process models often facilitate the analysis of systems that are costly or impractical to study experimentally by reducing the number of real experiments necessary [1–3], uncertainty quantification (UQ) is desired to make the results from these analyses realistic. Other than uncertainties in economics, energetics, and other performance-related issues, uncertainties in process models can lead to compromise in safety [4] and therefore UQ is highly desired for safety-critical applications.

Particularly for CO_2 capture systems, uncertainties in process models can lead to plant designs that may not satisfy the desired target CO_2 capture rates and/or handle the flue gas flowrate (e.g. lead to flooding or weeping in an absorber/stripper), lead to inadequate CO_2 purity, and/or higher or lower reboiler/heater duty than designed, and/or inaccurate design of the balance of plant (BOP) equipment items. Both for the solvent and sorbent systems, capture and regeneration systems/cycles are interconnected. For solvent-based capture systems, loaded solvent from the absorber goes to the regenerator and the regenerated solvent is recycled back to the absorber. For solid sorbent-based capture systems, there are varieties of solid–gas contactors [5]. The dominating contactor technologies are fixed beds [6, 7], moving beds [8–11], fluidized beds [12], and rotating packed beds. [13]. Fixed beds can be axial flow [14] or radial flow beds [15]. Irrespective of the type of contactor technology for the solid sorbent, solids undergo adsorption followed by desorption in the same equipment item (as in fixed beds) or in separate beds (such as for moving or fluidized beds). Thus, uncertainties in the adsorption model can lead to uncertainties in the desorption model results and vice versa as well as BOP model results.

Uncertainty is generally classified as aleatoric or epistemic [16]. Aleatoric uncertainty stems from natural stochasticity in the system and/or the measurement technology. Epistemic uncertainty stems from the lack of knowledge about a system and thus leads to uncertainties in the model, which in turn lead to uncertainties in model outputs. For the solvent/ sorbent-based CO_2 capture systems, examples of aleatoric uncertainties are uncertainties in the measurements of CO_2 loading of solvent/sorbent, solvent concentration, reboiler/heater duty, spatial temperature (e.g. there may be uncertainties in what temperature, liquid or gas, a temperature probe that is inserted in a tower measures at any instant of time), spatio-temporal evolution of solvent composition due to degradation, etc. Examples of epistemic uncertainties for solvent/sorbent-based CO_2 capture systems are lack of understanding of reaction kinetics including degradation reactions, if any, mass transfer and/or heat transfer mechanisms, hydraulic models (e.g., holdup and pressure drop models), physical/chemical equilibrium, physical properties, etc. Epistemic uncertainties can be reduced with improved understanding of various mechanisms/phenomena through additional research and investigation along with laboratory/bench-scale/pilot plant/commercial scale operations. For example, uncertainties in the degradation mechanism of a solvent can be reduced by collecting data of degradation products under various operating conditions for a solvent for sufficient time (like days or months) and developing reaction kinetic models and validating those with the experimental data. When there are unknown physics/chemistry for a process and/or it is difficult to improve understanding for a given mechanism given the current state-of-the-art in measurement technology, and/or it might require considerable resources (time, money, equipment, technology, etc.) for improving a process model, then it is desired that the uncertainties in process models are quantified.

Epistemic uncertainty in process models stems mainly from two sources – uncertainties in the process model and uncertainties in process model parameters. One of the most significant sources of model uncertainty is model parameters [17]. When uncertainties in model parameters are not necessarily Gaussian and it is desired that the user belief or prior knowledge be incorporated to inform posterior distribution of model parameters, Bayesian inference is the desired methodology. In addition, Bayesian inference can be applied to UQ of not only model parameters but also the model form, where the posterior probabilities naturally yield measure of uncertainty. Bayesian inference also seeks to avoid overfitting by rewarding goodness of fit while penalizing model complexity [18].

UQ of process models for solvent-based CO_2 capture systems has been undertaken by several researchers. Morgan et al. [19] presented a systematic approach for determining uncertainty in property models of a monoethanolamine (MEA)-CO_2-H_2O system and model parameters using a Bayesian inference approach. Morgan et al. [19] obtained a narrower posterior distribution in model parameters relative to the prior thus reducing uncertainty in results from the process model. Morgan et al. further expanded their approach and applied to the electrolyte NRTL (eNRTL)-based thermodynamic models of solvent-based CO_2 capture systems [20]. UQ of hydraulic and mass transfer model parameters for a MEA-based CO_2 capture system has been undertaken by Soares Chinen et al. [21, 22]. UQ of interfacial area model and kinetic model parameters for a membrane assisted chilled-ammonia capture system has been presented by Hughes et al. [23]. While the works cited earlier in this paragraph have mainly focused on epistemic uncertainties, Morgan et al. [24] have also investigated aleatoric uncertainties in the pilot plant data due to solvent concentration and loading measurements.

Compared to solvent systems, there is a lack of research in the area of UQ of solid sorbent-based CO_2 capture processes. In the area of CO_2 capture using solids sorbents, Bayesian uncertainty quantification for zeolite-based sorbents in axial flow beds has been undertaken [25, 26]. For solid sorbent-based capture systems, one of the greatest sources of uncertainty is the isotherm model and its parameters. Quite often a fundamental understanding of the actual mechanisms that finally lead to equilibrium loading is not available. Furthermore, it can be impractical to measure or difficult to understand the mechanisms that take place in functionalized sorbents. For example, while it can be of great interest to develop understanding of chemisorption mechanisms in a functionalized solid sorbent, the reaction mechanisms may be poorly understood. This is exacerbated by the difficulty in identifying the products formed, and measuring how their concentration evolves with time as a function of operating conditions in the interiors of the solids such as a functionalized metal organic framework (MOF) [8]. There are also often high uncertainties in thermo-physical, thermodynamic, and transport models for solid sorbents. For example, it is difficult to measure and develop a mechanistic model of thermo-physical properties such as thermal conductivity and heat capacity due to complex relationship of these properties with the CO_2 loading and temperature. Uncertainties in special variation in loading expected in many solid sorbents make it even more difficult to accurately measure and quantify these physical properties for model development and validation. In terms of thermodynamic properties, one of the important variables is excess enthalpy, which is again difficult, if not impossible, to measure locally. There are often several transport variables such as diffusivity that can play a key role in determining the performance of solid sorbents.

For example, measurement of diffusivity, especially for functionalized sorbents for CO_2 capture, can lead to additional complications due to unknown/poorly understood reaction mechanisms that can enhance mass transport flux, possible spatial nonuniformity in porosity/tortuosity, complex mechanisms and solid–gas interactions in micro-pores and macro-pores, etc. Furthermore, contactor type used in a solid sorbent-cased capture system can make it difficult to undertake UQ. For example, fixed beds are inherently dynamic; thus, UQ for the equipment model would need to consider dynamic data and dynamic simulation/optimization.

13.2 Methodology

Two major sources of epistemic uncertainty in models of solid sorbent capture systems are due to uncertainty in the model form (discrepancy) and/or uncertainty in the model parameters. In general, either the parameter uncertainty or discrepancy is addressed. Addressing both parameter uncertainty and discrepancy, together, is not very common [27]. When they are addressed together [26], it can lead to considerable computational challenges. In following texts, methodologies for UQ of these two major sources are described.

13.2.1 UQ of Model Parameters

A fully Bayesian approach to UQ of model parameters works on the principle that previous knowledge of the system can be updated and improved upon by using experimental data, based on Bayes' theorem:

$$\pi(\tilde{\theta}|Z) \propto P(\tilde{\theta})L(Z|\tilde{\theta}), \tag{13.1}$$

where $\pi(\tilde{\theta}|Z)$ denotes the posterior distribution that we are interested in obtaining for key parameters, $P(\tilde{\theta})$ denotes the prior distribution of the parameters based on the user knowledge or belief or literature information, and $L(Z|\tilde{\theta})$ denotes the likelihood function that is computed by considering the experimental data.

Equation (13.1) shows a generic process model where the model output (y) is a function of predictor variables (x), and a set of model parameters (θ):

$$y = f(x, \theta). \tag{13.2}$$

It is desired to quantify uncertainties in all or a subset of model parameters. For the Bayesian UQ, the first step is to consider prior distribution of parameters. Prior distributions can be uninformed (such as uniform distribution) or informed. For the latter, it can be based on the user belief or existing knowledge or based on the estimated uncertainty from solving the deterministic parameter estimation problem. For the last approach, one possible method has been outlined in the work of Morgan et al. [19] that estimates confidence intervals for the parameters based on the results from the deterministic parameter estimation problem and then the prior distributions of the parameters are determined by estimating their probability density functions (PDF) from a series of confidence intervals.

The Bayesian inference approach is computationally expensive even for moderate size problems since thousands of simulations are generally required to estimate the posterior distributions. This issue is of particular interest for CO_2 capture models that include large number of model equations and large parameter space, and there is frequent lack of data for adequately characterizing/estimating the model parameters. Two possible approaches for reducing computational expense are use of a surrogate model, which is significantly less computationally expensive than the actual process model, as the emulator and elimination of the parameters whose uncertainty estimate is unlikely to improve through Bayesian inference. For surrogate models to be used in Bayesian inference, it is not only desired that they reasonably accurately represent the existing process model but also that they are considerably less computationally expensive than the existing process model and well-behaved for Bayesian inference algorithms. Various advantages of the use of surrogate models for UQ and how inaccuracies in surrogate models would affect the outcome of UQ have been discussed in Giunta et al. [28]. Typically, a response surface model [29] given by Eq. (13.3), where output variables (e.g. CO_2 capture and sorbent loading) are represented as a function of inputs (parameters and predictor variables), are used as surrogate models:

$$y = f^*(\tilde{x}, \tilde{\theta}). \tag{13.3}$$

Surrogate response surface models have been widely used for Bayesian inference in varieties of applications [30, 31]. One of the popular response surface models used for Bayesian inference is multivariate adaptive regression splines (MARS) [32] given by Eq. (13.4):

$$y = \beta_0 + \sum_{j=1}^{M} \beta_j \, h_j(z), \tag{13.4}$$

where y is the response variable, β_0 is the intercept, β_j are the coefficients associated with the basis functions, $h_j(z)$ are the basis functions dependent on the values of the independent variables z that includes $\tilde{x}$, $\tilde{\theta}$, and M is the total number of basis functions.

For eliminating the parameters before undertaking Bayesian inference, that is, for downselecting the parameter space, sensitivity studies can be conducted. Various methodologies can be used for such sensitivity analysis such as a local method using partial derivatives [19] and variance-based global sensitivity computing Sobol indices [22]. Variance-based methods generally compute Sobol indices [33] indicating contributions of individual parameters (S_i) and interactions among

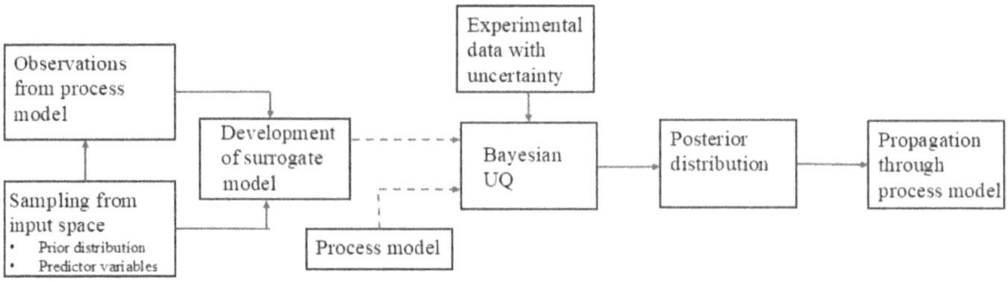

Figure 13.1 Methodology for parametric UQ and evaluation of its impact on results.

pairs (S_{ij}) or groups ($S_{1,2,\ldots,k}$) to the overall variability of the dependent variables of interest (e.g. CO_2 capture rate or utility requirement for a solid-sorbent-based capture process), while satisfying Eq. (13.5):

$$\sum_{i=1}^{k} S_i + \sum_{1 \le i < j \le k} S_{ij} + \cdots + S_{1,2,\ldots,k} = 1. \tag{13.5}$$

Values of Sobol indices lower than a threshold corresponding to a parameter would suggest that parameter can be neglected during Bayesian inference. Sobol indices can be calculated by using the surrogate model with the assumption that the surrogate model is a satisfactory representation of the process model.

For Bayesian UQ, random sampling from the posterior distribution is obtained by the Markov Chain Monte Carlo (MCMC) method [34–36]. Popular MCMC algorithms include Gibbs sampling and Metropolis Hastings (MH) algorithm. The Bayesian inference procedure yields joint posterior distributions of the parameters that can then be propagated through the process models to obtain uncertainties in the output variables of interest for various operating conditions.

The overall approach as described before is shown in Figure 13.1.

13.2.2 UQ of Model Form Discrepancy

When model form uncertainty or discrepancy is considered, a discrepancy function, $\delta\,(s, \beta)$, is considered as shown in

$$y = f(\mathrm{x}, \theta, \delta(s, \beta)), \tag{13.6}$$

where $\delta\,(s, \beta)$ is considered to be a function of variables s and parameters β. Variables s can include a subset of x and/or process variables. For example, in the work of Ostace et al. [26] and Bhat et al. [37], uncertainty in the Langmuir adsorption isotherm model of a solids sorbent for CO_2 capture is given by:

$$q^* = (q)(q_{\max}) = \frac{K_{\mathrm{uq}} P_{\mathrm{CO_2}} q_{\max}}{1 + K P_{\mathrm{CO_2}}}. \tag{13.7}$$

$$K_{\text{uq}} = \exp\left(\frac{\Delta S}{R_g}\right)\exp\left(\frac{-\Delta H}{R_g T_g} + \delta(q, T_g, \beta)\right)/P. \tag{13.8}$$

In Eq. (13.8), the departure from ideality in the adsorption enthalpy is represented by the discrepancy function δ, which is postulated to be a function of the fractional coverage q, the gas temperature T_g, and calibration parameters β.

The form of the discrepancy function should be chosen based on the specific model discrepancy it addresses but should be continuous and differentiable [25]. One potential discrepancy function can be a Bayesian Smoothing Spline Analysis of Variance (BSS-ANOVA) model, a Gaussian process (GP) with a special covariance function [37, 38] as shown in Eq. (13.9): This model decomposes the discrepancy function into main effects, two-way interactions, and higher-order interactions using functional ANOVA decomposition, with each functional component GP satisfying the ANOVA constraints [37–40]. Accordingly, and by letting $\delta(x, T_g, \beta)$ be noted as $\delta(\omega)$ for conciseness, the discrepancy function can be written as [39, 40]:

$$\delta(\alpha) = \beta_0 + \sum_{r=1}^{R} \delta_r(\alpha_r) + \sum_{r<p}^{R} \delta_{r,p}(\alpha_r, \alpha_p) + \cdots. \tag{13.9}$$

In Eq. (13.9), $\delta_r(\alpha)$ denotes main effect functional with distribution $\delta_r(\alpha) \sim GP\left(0, \varsigma_r^2 \Omega_1\right)$ and $\delta_{r,p}(\alpha_r,\ \alpha_p)$ denote two-way interaction functional with $\delta_{r,p}(\alpha_r, \alpha_p) \sim GP\left(0, \varsigma_{r,p}^2 \Omega_2\right)$, where ς_r^2 and $\varsigma_{r,p}^2$ are variance hyperparameters and Ω_1 and Ω_2 are BSS-ANOVA covariance functions [38]. Higher-order interaction functionals can also be added as needed to Eq. (13.9). Main effect functional can be represented as an infinite linear combination of orthogonal functions as shown in Eq. (13.10):

$$\delta_r(\alpha_r) = \sum_{l=1}^{\infty} \omega_{r,l} \zeta_l(\alpha_r), \tag{13.10}$$

where ζ_l are the eigenfunctions in the Karhunen–Loéve (KL) expansion. Two-way and higher interactions can be represented in a similar manner as the main effect functional with ζ_l representing pairwise (for two-way interactions) or multiple products (for higher interactions than two-way) of the respective main effect basis functions.

To select the optimal order for the discrepancy model, models of increasing complexity are generated by adding higher interaction functionals sequentially (this is okay since the terms in Eq. (13.9) are orthogonal and ordered) to the main effects. The model selection can be done through a suitable information-theoretic criterion such as Bayesian Information Criterion (BIC). Posterior distributions of parameters can be obtained as described in Section 13.2.1.

13.3 Example-UQ of a Solid-Based CO_2 Capture System in a Fixed Bed

13.3.1 UQ of the Isotherm Model

This example shows the application of Bayesian UQ for a solid sorbent-based CO_2 capture system. In particular, uncertainty in the isotherm model of diamine-appended metal organic framework (MOF), diamine 2,2-dimethyl-1,3-diaminopropane (dmpn) Mg_2(dobpdc) (dobpdc^{4-} = 4,4′-dioxidobiphenyl-3,3′-dicarboxylate) [dmpn-Mg_2(dobpdc)], that presents a steep step-shaped isotherm is evaluated [41]. This sorbent not only has a high equilibrium capacity even under low partial pressure of CO_2, but can be desorbed sufficiently at around 100 °C [6]. A dual-site Sips model given by Eqs. (13.11)–(13.16) from the literature is used, where both chemisorption and physisorption have been accounted for [6].

$$q^*_{CO_2} = q^*_{chem} + q^*_{phys} = n_{max,c}\left[\frac{(b_{chem}P)^{1/n_{chem}}}{1+(b_{chem}P)^{1/n_{chem}}}\right] + n_{max,p}\left[\frac{(b_{phys}P)^{1/n_{phys}}}{1+(b_{phys}P)^{1/n_{phys}}}\right]. \tag{13.11}$$

$$n_{max,p} = N_{phys}\left[\frac{\exp\left(K_c+\frac{K_d}{T}\right)}{1+\exp\left(K_c+\frac{K_d}{T}\right)}\right]. \tag{13.12}$$

$$b_{phys} = b_{phys,0}\exp\left[\frac{Q_{st,phys}}{RT_0}\left(\frac{T_0}{T}-1\right)\right]. \tag{13.13}$$

$$n_{chem} = n_{chem,0}\exp\left[\frac{E_{na}}{RT_0}\left(\frac{T_0}{T}-1\right)\right]. \tag{13.14}$$

$$n_{max,c} = N_{chem}\left[\frac{\exp\left(K_a+\frac{K_b}{T}\right)}{1+\exp\left(K_a+\frac{K_b}{T}\right)}\right]. \tag{13.15}$$

$$b_{chem} = b_{chem,0}\exp\left[\frac{Q_{st,chem}}{RT_0}\left(\frac{T_0}{T}-1\right)\right]. \tag{13.16}$$

In Eqs. (13.11)–(13.16), $q^*_{CO_2}$, q^*_{phys}, q^*_{chem}, are the total, physisorption, and chemisorption CO_2 equilibrium loadings, respectively, $n_{max,c}$ and $n_{max,p}$ are the maximum loading due to chemisorption and physisorption, respectively, b_{chem} and b_{phys} are equilibrium constants for chemisorption and physisorption, respectively, P is the equilibrium pressure, n_{chem} and n_{phys} are exponents used for computing chemisorption and physisorption loadings, respectively. Based on sensitivity studies, parameter downselection is undertaken leading to the following parameters used for Bayesian UQ: $b_{chem,0}$, $Q_{st,chem}$, N_{phys}, $n_{chem,0}$, K_a, K_b, E_{na}. Prior distributions for these parameters are generated based on the deterministic regression

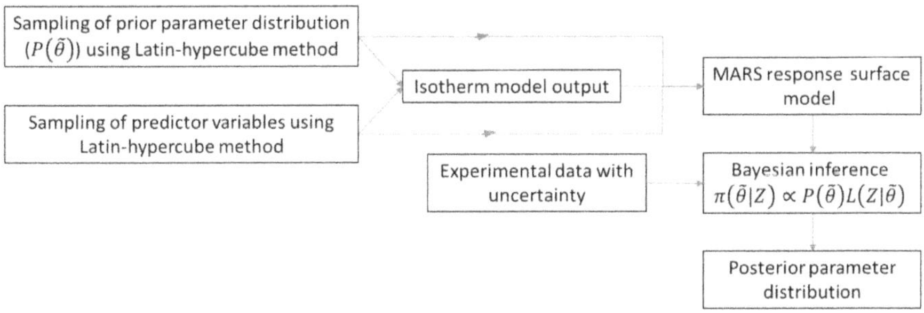

Figure 13.2 Bayesian inference schematic.

following the methodology described in Section 13.2.1. Latin hypercube sampling (LHS) is undertaken for the parameters and predictor variables. The resulting samples are propagated through the isotherm model. The outputs are then used to generate MARS response surface models that are used for Bayesian UQ using MCMC with Gibbs sampling.

The overall methodology is shown in Figure 13.2

Figure 13.3 displays the error and the predictions of the MARS surrogate model. The error histogram is centered around zero with an absolute mean error of 0.000032. Furthermore, the majority of the response surface's predicted values fall near the diagonal line in the parity plot, yielding a high R^2 value of 0.94. These performance measures indicate that the surrogate model is a satisfactory

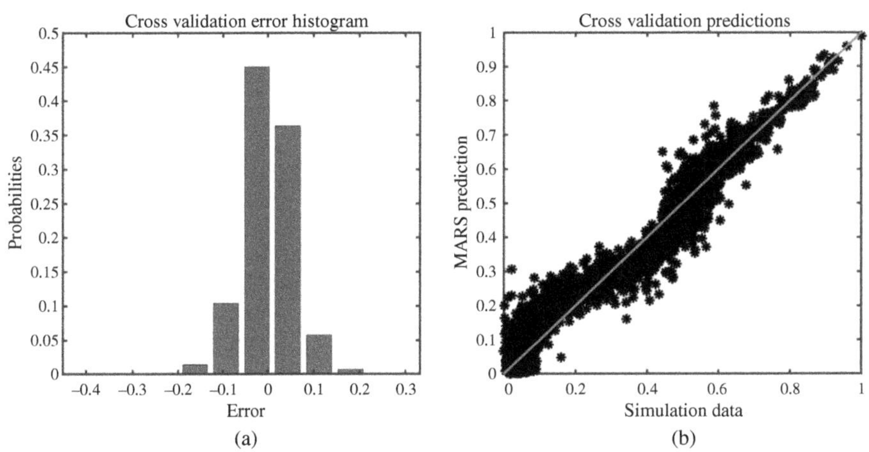

Figure 13.3 MARS response surface validation using 10-fold cross validation (R^2 = 0.94). Error histogram (a) and parity plot (b) comparing CO_2 loading predictions.

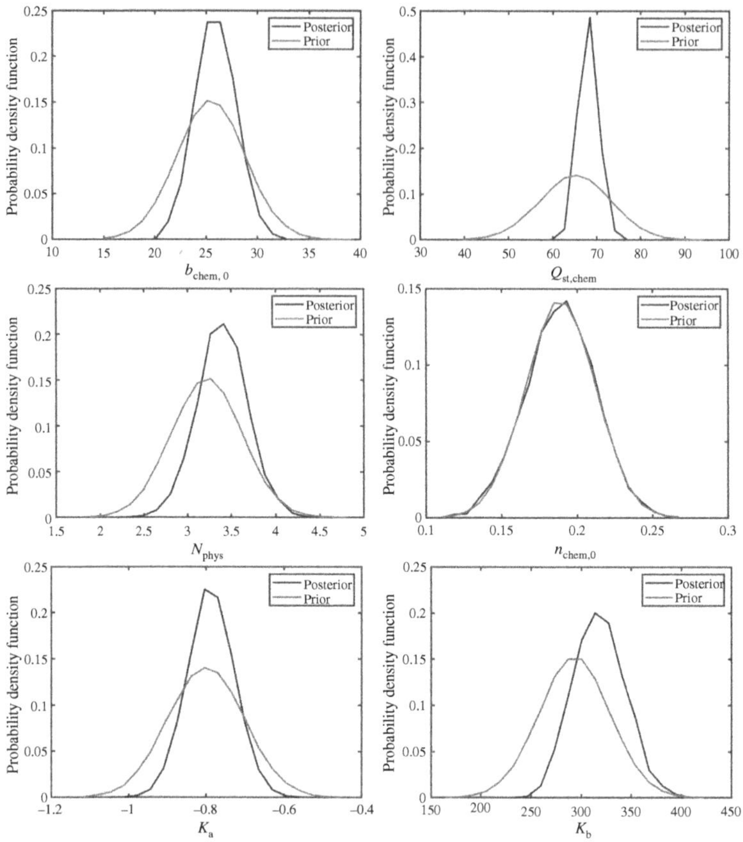

Figure 13.4 Single parameter marginal probability density functions for prior and posterior distributions.

representation of the isotherm model for the operating conditions and parameter space considered.

Five of the seven parameters considered showed a significant reduction in variance of 91%, 96%, 80%, 82%, and 84% for $b_{\text{chem},0}$, $Q_{\text{st,chem}}$, N_{phys}, K_a, and K_b, respectively. These shrinkages can be visually seen from the narrower posterior distributions shown in Figure 13.4 that shows the prior and posterior marginal probability density functions (PDF) for all parameters considered. The other two parameters, $n_{\text{chem},0}$ and E_{na}, showed practically no change, as can be seen by the overlapping prior and posterior PDFs $n_{\text{chem},0}$ in Figure 13.4 (PDFs of E_{na} are not shown for brevity) and therefore these two parameters are not considered further for uncertainty propagation.

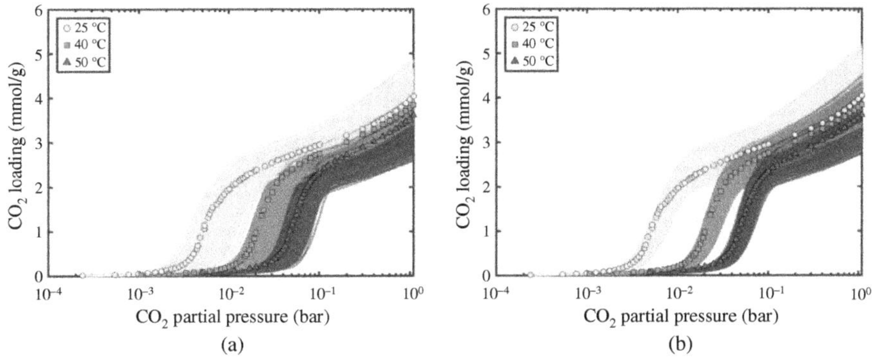

Figure 13.5 Results from the isotherm model obtained using prior (a) and posterior (b) parameter distribution for 25–50 °C temperature range.

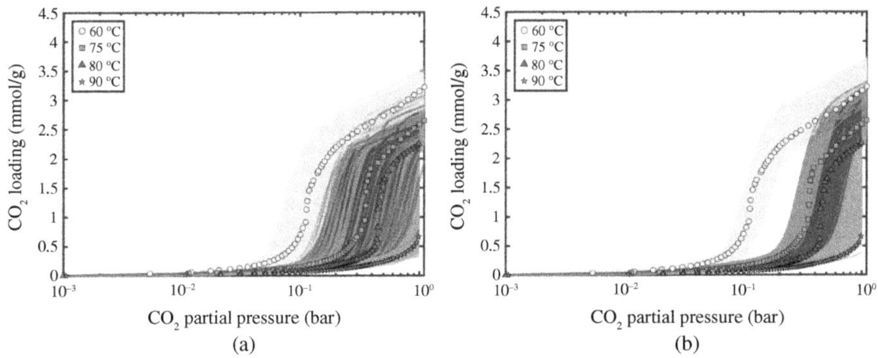

Figure 13.6 Results from the isotherm model obtained using prior (a) and posterior (b) parameter distribution for 60–90 °C temperature range.

Figures 13.5–13.7 show the results from the isotherm model obtained using prior and posterior parameter distribution in 25–50 °C temperature range (adsorption step), 60–90 °C temperature range (end of adsorption to beginning of desorption step), and 100–120 °C temperature range (desorption step), respectively. It can be observed that the uncertainty in the prediction of isotherm loading was greatly reduced by Bayesian inference, with the largest reductions observed for low CO_2 partial pressures and high temperatures.

13.3.2 Uncertainty Propagation

Uncertainties in the isotherm model parameters are propagated through a laboratory scale axial-flow-fixed bed model and a commercial-scale radial-flow-fixed bed model.

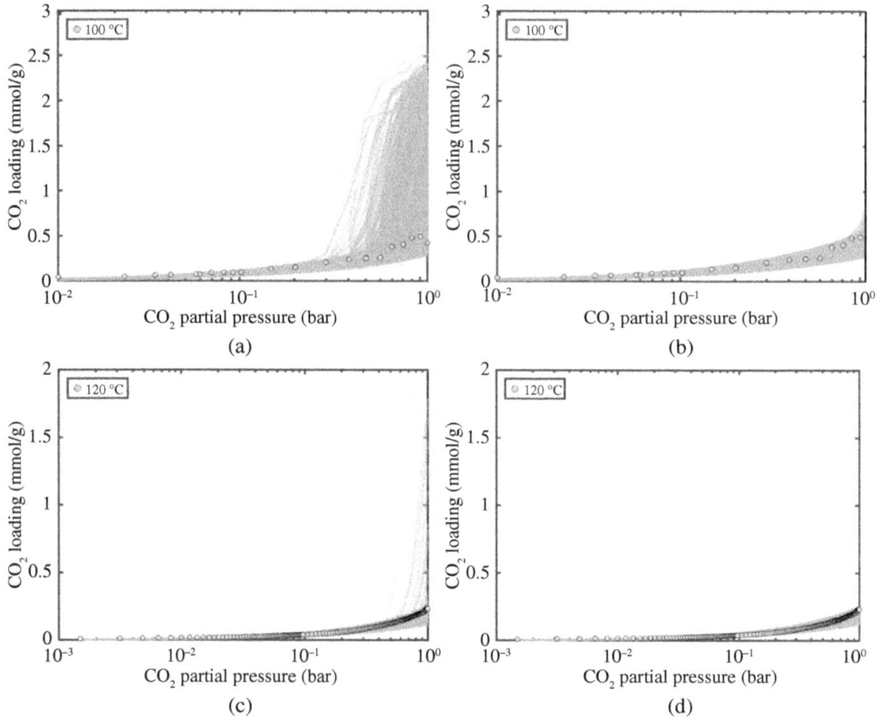

Figure 13.7 Results from the isotherm model obtained using prior (a, c) and posterior (b, d) parameter distribution for 100–120 °C temperature range.

13.3.2.1 Lab-Scale Axial Flow-Fixed Bed

The lab-scale axial flow-fixed bed model used for propagation of uncertainty in isotherm model parameters can be found in the work of Hughes et al. [6] and Monteiro & Bhattacharyya [15]. The prior and posterior parameter distributions are used to generate breakthrough curves for the experimental system detailed in Milner et al. [42] as shown in Figure 13.8. Here the breakthrough time is defined as the time when the concentration of CO_2 at the outlet of the bed surpasses 0.01. The distribution of breakthrough by using the prior and posterior parameter distributions is shown in Figure 13.9. The experimental breakthrough time for which only a single observation is available is also shown in Figure 13.10. It is observed that the distribution in the breakthrough time by using the posterior distribution is much narrower than that obtained from the prior distribution. It is also observed that the results from the posterior distribution yield high probability for the experimental observation as opposed to nearly zero probability from the prior distribution.

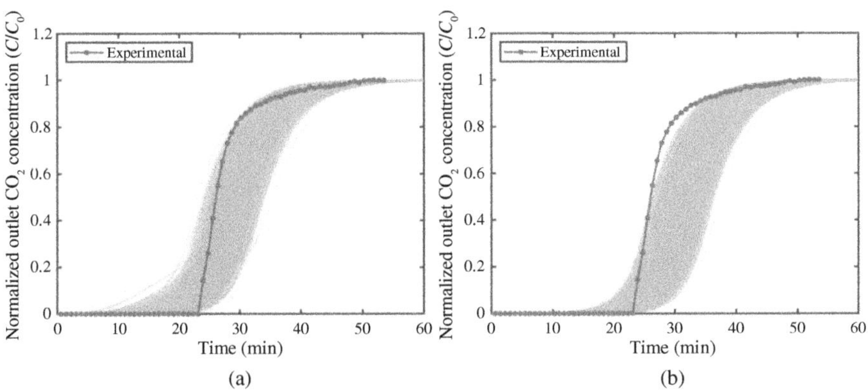

Figure 13.8 Axial flow-fixed bed model breakthrough curve predictions for prior (a) and posterior (b) sampled parameters.

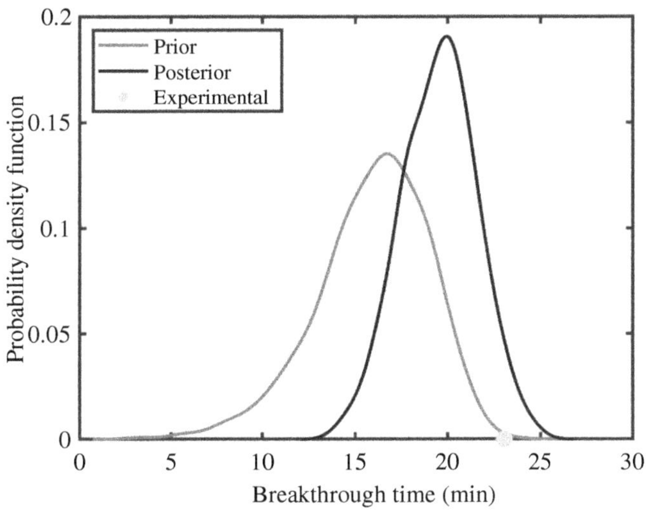

Figure 13.9 Distribution of axial flow-fixed bed breakthrough time for prior and posterior sampled parameters.

13.3.2.2 Commercial-Scale Radial Flow-Fixed Bed

The commercial-scale radial flow-fixed bed (RFFB) model used in this section for propagation of uncertainty in isotherm model parameters can be found in the work of Monteiro & Bhattacharyya [15] and Monteiro et al. [43]. Radial flow beds can result in much lower pressure drops than the axial flow beds. The RFFB model is multiscale where a particle scale model is coupled with a model of the bulk-scale. The bed height, inner diameter, outer diameter, and sorbent

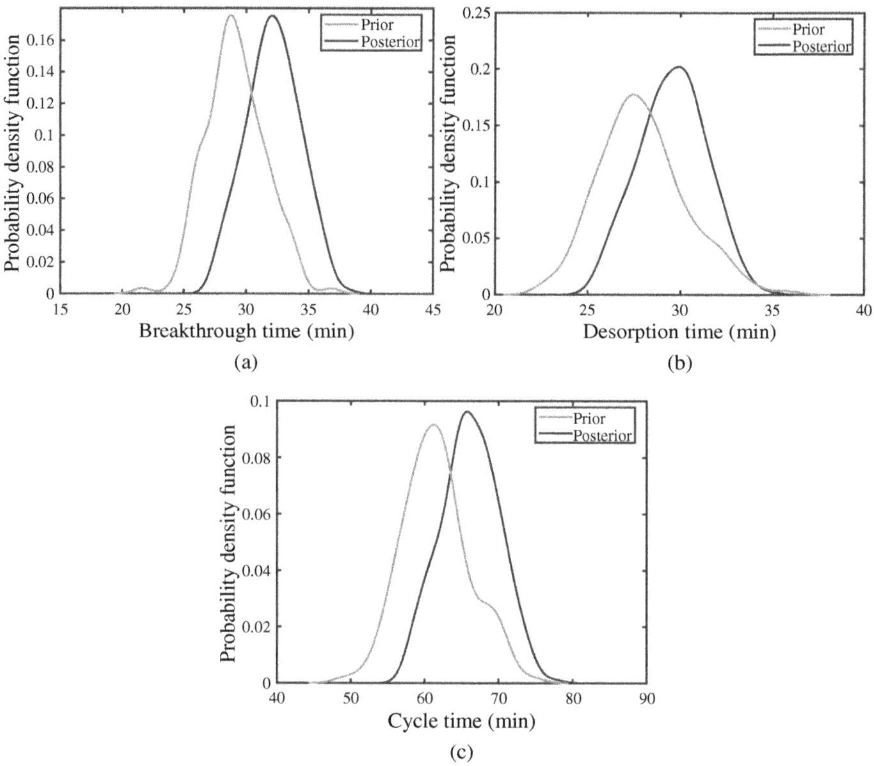

Figure 13.10 Distribution of RFFB breakthrough time (a), desorption time (b), and total cycle time (c) for prior and posterior sampled parameters.

volume are 5 m, 5 m, 15 m, and 251.3 m^3, respectively. Figure 13.10 shows that there is greater difference in desorption time obtained by using prior vs posterior distribution, with 19%, 40%, and 25% variance reduction for adsorption, desorption, and total cycle time, respectively by using the posterior distribution. This is expected as the posterior distribution led to much narrower uncertainty in the isotherm under desorption conditions (see Figure 13.7). The reduced uncertainty in desorption time has a direct effect in the process configuration/economics – for 90% capture from the flue gas corresponding to Case 11b of the National Energy Technology Laboratory baseline study [44], the results obtained from prior distribution results in requirement of 17–21 beds undergoing desorption at any time, while the results obtained from the posterior distribution suggests 17–19 beds undergoing desorption at any time. This also affects the utility requirement, as seen in Figure 13.11, where the reduction in the variance of the predicted amount of steam required in the desorption step is 23%. With respect to the

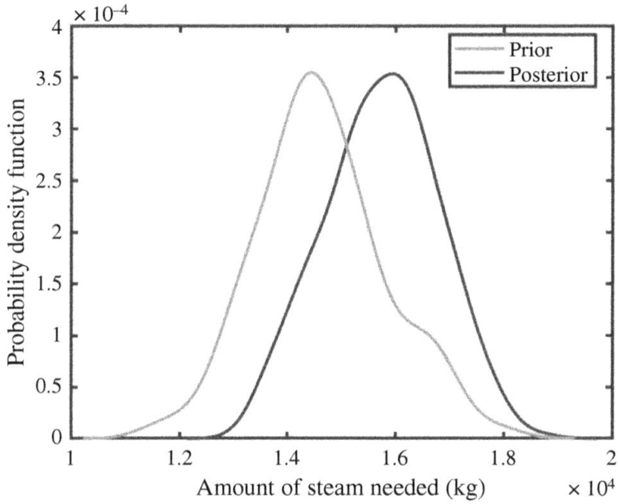

Figure 13.11 Distribution of RFFB requirement of steam during desorption step for prior and posterior sampled parameters.

adsorption step, a reduction of variance of 19% is observed for the average CO_2 loading in the bed at breakthrough time together with an increase in the mean loading, shown in Figure 13.12, which are expected given the reduction in breakthrough time uncertainty as well as its shift to higher meantime. Overall, there is considerable reduction in uncertainties in spatio-temporal CO_2 loading and bed temperature due to the Bayesian UQ. Reduction in uncertainties in CO_2 loading at the outer location and central anulus of the bed due to Bayesian UQ is shown in Figures 13.13 and 13.14, respectively. Corresponding reduction in uncertainties in temperature at the outer location and central anulus of the bed due to Bayesian UQ can be seen in Figures 13.15 and 13.16, respectively. Considerable reductions in uncertainties during desorption step are apparent from these figures.

13.4 Concluding Remarks

Bayesian UQ of solid sorbent-based CO_2 capture systems are discussed in this chapter. Quantification of epistemic uncertainty in the model form and parameters is discussed. For making the computational expense of Bayesian UQ tractable, strategies such as downselection of parameters and use of surrogate models as emulators are discussed. As an example, UQ of isotherm parameters of a functionalized MOF is undertaken. Bayesian UQ is found to lead to considerable reduction in uncertainties especially under desorption conditions. The breakthrough

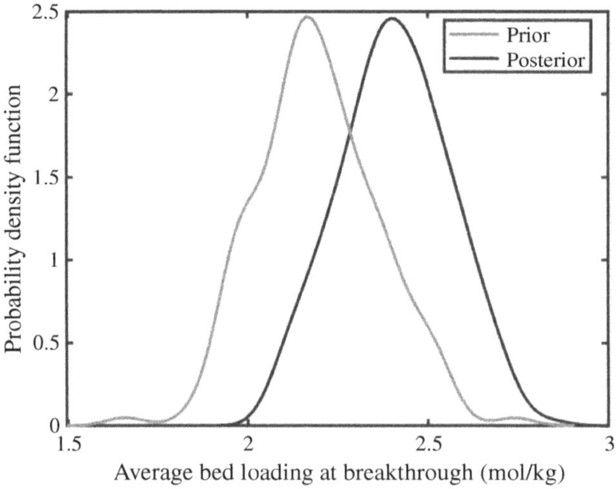

Figure 13.12 Distribution of average bed loading at breakthrough for prior and posterior sampled parameters.

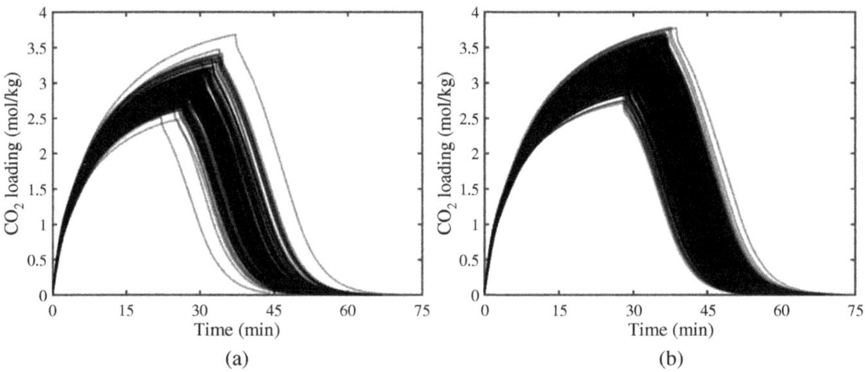

Figure 13.13 Uncertainty in the CO_2 loading at the outer location of the commercial RFFB bed for prior (a) and posterior (b) sampled parameters.

time of a lab-scale adsorber is analyzed. It is observed that Bayesian inference greatly reduces the variance of breakthrough time as well as shifts the mean breakthrough time toward the experimental value. Prior and posterior parameter distributions are also propagated through a commercial scale adsorption–desorption setup, where the Bayesian inference is found to reduce the uncertainty during both process steps, exhibited by a reduction in variance for breakthrough and desorption times of 19% and 40%, respectively.

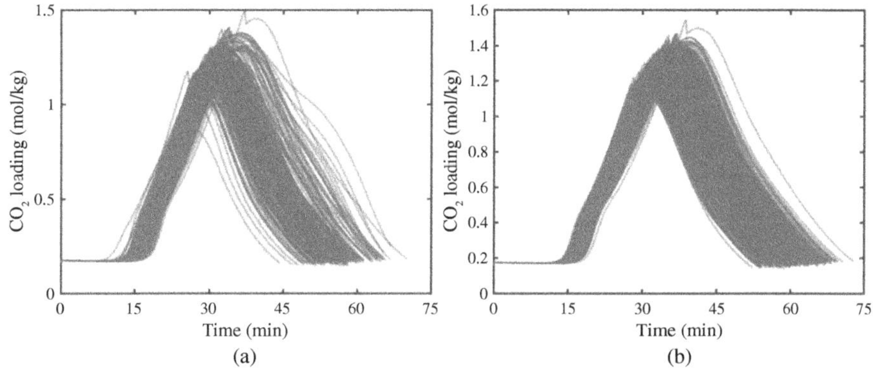

Figure 13.14 Uncertainty in the CO_2 loading at the center annulus location of the commercial RFFB bed for prior (a) and posterior (b) sampled parameters.

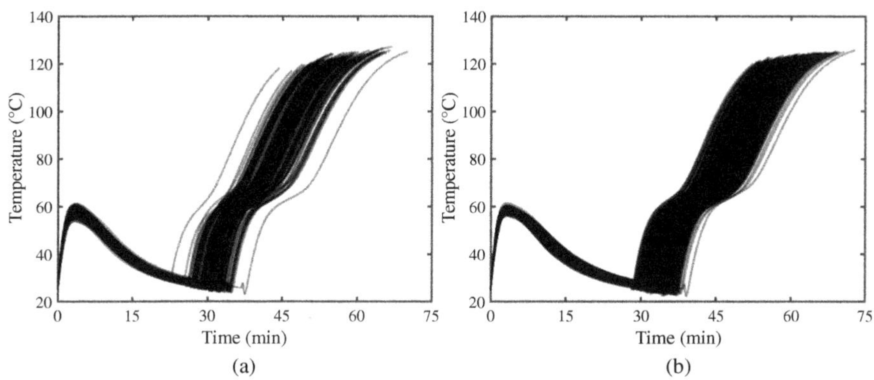

Figure 13.15 Uncertainty in the temperature at the outer location of the commercial RFFB bed for prior (a) and posterior (b) sampled parameters.

It is desired that Bayesian UQ for solid sorbent-based CO_2 capture processes be undertaken more often than currently present in the literature. Bayesian UQ of solid sorbents for CO_2 capture poses several challenges. One difficulty for solid sorbent processes is the difficulty in measurement and spatial variation of transport variables within the particles. Various contactor types and variability in their designs also make it difficult to develop Bayesian UQ methods with general applicability.

UQ of thermo-physical properties of loaded sorbent and transport mechanisms is highly desired due to complex interactions and mechanisms that take place in various sorbents, especially functionalized sorbents since inter- and intra-particle mass and heat transfer mechanisms can be complex based on the type of the

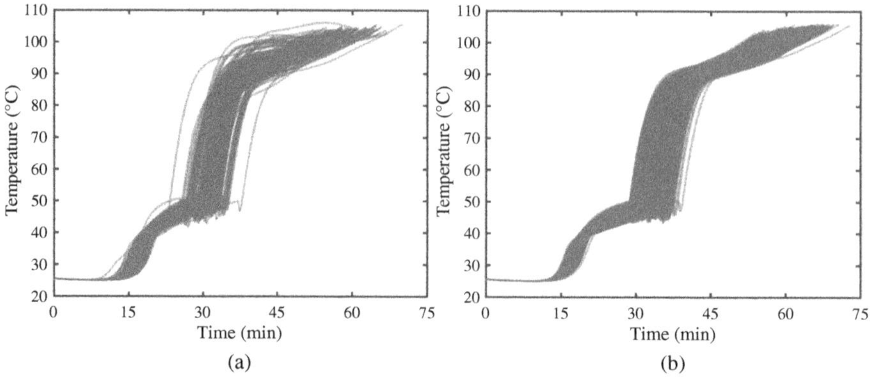

Figure 13.16 Uncertainty in the temperature at the center annulus location of the commercial RFFB bed for prior (a) and posterior (b) sampled parameters.

sorbent and type of beds. Inherent dynamics of fixed beds also make it far more difficult to undertake Bayesian UQ of transport kinetics in these beds. In the area of theoretical and algorithmic development, especially for Bayesian UQ for CO_2 capture applications using solid sorbents, it will be desired to investigate additional discrepancy functions with the desired properties for use in Bayesian UQ as discussed in Section 13.2.2. With the widespread use of machine learning models, their use for discrepancy function with quantified uncertainty will be a desired path to investigate.

References

1 Kuhnt, S. and Steinberg, D.M. (2010). Design and analysis of computer experiments. *AStA Advances in Statistical Analysis* 94: 307–309.

2 Levy, S. and Steinberg, D.M. (2010). Computer experiments: a review. *AStA Advances in Statistical Analysis* 94: 311–324.

3 Santner, T.J., Williams, B.J., and Notz, W.I. (2003). *The Design and Analysis of Computer Experiments*. New York: Springer.

4 Miller, D.C., Syamlal, M., Mebane, D.S. et al. (2014). Carbon capture simulation initiative: a case study in multiscale modeling and new challenges. *Annual Review of Chemical and Biomolecular Engineering* 5: 301–323.

5 Bhattacharyya, D. and Miller, D.C. (2017). Post-combustion CO_2 capture technologies—a review of processes for solvent-based and sorbent-based CO_2 capture. *Current Opinion in Chemical Engineering* 17: 78–92.

6 Hughes, R., Kotamreddy, G., Ostace, A. et al. (2021). Isotherm, kinetic, and process modeling and techno-economic analysis of a fixed bed temperature swing adsorption process for a diamine-appended metal–organic framework. *Energy & Fuels* 35: 6040–6055.

7 Kotamreddy, G., Hughes, R., Bhattacharyya, D. et al. (2019). Process modeling and techno-economic analysis of a CO_2 capture process using fixed bed reactors with a microencapsulated solvent. *Energy & Fuels* 33: 7534–7549.

8 Hughes, R., Kotamreddy, G., Bhattacharyya, D. et al. (2024a). Development of a chemistry-based isotherm model and techno-economic optimization of a moving bed process for CO_2 capture using a functionalized metal–organic framework. *Chemical Engineering Science* 287: 119679.

9 Kim, H., Miller, D.C., Srinivasarao, M. et al. (2016). Mathematical modeling of a moving bed reactor for post-combustion CO_2 capture. *AICHE Journal* 62: 3899–3914.

10 Kotamreddy, G., Hughes, R., Bhattacharyya, D. et al. (2023). Multi-scale dynamic modeling and techno-economic optimization of a CO_2 capture process using microencapsulated carbon sorbents in a moving bed contactor. *International Journal of Greenhouse Gas Control* 126: 103905.

11 Yancy-Caballero, D., Hughes, R., Zamarripa, M. et al. (2023). Isotherm modeling and techno-economic analysis of a TSA moving bed process using a tetraamine-appended MOF for NGCC applications. *International Journal of Greenhouse Gas Control* 128: n957.

12 Modekurti, S., Bhattacharyya, D., and Zitney, S. (2013). Dynamic modeling and control studies of a two-stage bubbling fluidized bed adsorber-reactor for solid sorbent CO_2 capture. *Industrial & Engineering Chemistry Research* 52: 10250–10260.

13 Bhattacharyya, D. (2022). Design and optimization of hybrid membrane-solvent- processes for post-combustion CO_2 capture. *Current Opinion in Chemical Engineering* 36: 100768.

14 Hughes, R., Yancy-Caballero, D., Zamarripa, M. et al. (2024b). Modeling and techno-economic optimization of a tetraamine-appended metal–organic framework for NGCC-based CO_2 capture using fixed bed contactors. *Energy & Fuels* 38: 2511–2524.

15 Monteiro, A.F. and Bhattacharyya, D. (2024). Multi-scale dynamic modeling of radial flow fixed bed contactors for CO_2 capture using a functionalized metal–organic framework. *Energy & Fuels* 38: 21052–21064.

16 Der Kiureghian, A. and Ditlevsen, O. (2009). Aleatory or epistemic? Does it matter? *Structural Safety* 31: 105–112.

17 Kennedy, M.C. and O'Hagan, A. (2001). Bayesian calibration of computer models. *Journal of the Royal Statistical Society, Series B (Statistical Methodology)* 63: 425–464.

18 Galagali, N. and Marzouk, Y.M. (2015). Bayesian inference of chemical kinetic models from proposed reactions. *Chemical Engineering Science* 123: 170–190.

19 Morgan, J.C., Tong, C., Miller, D.C., and Bhattacharyya, D. (2015). Uncertainty quantification of property models: methodology and its application to CO_2-loaded aqueous MEA solutions. *AICHE Journal* 61 (6): 1822–1839.

20 Morgan, J.C., Chinen, A.S., Omell, B. et al. (2017). Thermodynamic modeling and uncertainty quantification of CO_2-loaded aqueous MEA solutions. *Chemical Engineering Science* 168: 309–324.

21 Isenberg, N., Akula, P., Eslick, J. et al. (2021). A generalized cutting-set approach for nonlinear robust optimization in process systems engineering. *AICHE Journal* 67 (5): 1–21.

22 Soares Chinen, A., Morgan, J., Omell, B. et al. (2018). Development of a rigorous modeling framework for solvent-based CO_2 capture. Part 1: Hydraulic and mass transfer models and their uncertainty quantification. *Industrial and Engineering Chemistry Research* 57: 10448–10463.

23 Hughes, R., Kotamreddy, G., Bhattacharyya, D. et al. (2022). Modeling and Bayesian uncertainty quantification of a membrane-assisted chilled ammonia process for CO_2 capture. *Industrial & Engineering Chemistry Research* 61: 4001–4016.

24 Morgan, J.C., Soares Chinen, A., Omell, B. et al. (2018). Development of a rigorous modeling framework for solvent-based CO_2 capture. Part 2: Steady-state validation and uncertainty quantification with pilot plant data. *Industrial and Engineering Chemistry Research* 57: 10464–10481.

25 Mebane, D.S., Bhat, K.S., Kress, J.D. et al. (2013). Bayesian calibration of thermodynamic models for the uptake of CO_2 in supported amine sorbents using ab initio priors. *Physical Chemistry Chemical Physics* 15: 4355–4366.

26 Ostace, A., Kocan, K., Mebane, D. et al. (2020). Probabilistic model building with uncertainty quantification and propagation for a dynamic fixed bed CO_2 capture process. *Energy & Fuels* 34: 2516–2532.

27 Arendt, P.D., Apley, D.W., and Chen, W. (2012). Quantification of model uncertainty: calibration, model discrepancy, and identifiability. *Journal of Mechanical Design* 134 (10): 100908.

28 Giunta, A.A., McFarland, J.M., Swiler, L.P., and Eldred, M.S. (2006). The promise and peril of uncertainty quantification using response surface approximations. *Structure and Infrastructure Engineering* 2 (3–4): 175–189.

29 Khuri, A. and Mukhopadhyay, S. (2010). Response Surface Methodology. *Wiley Interdisciplinary Reviews: Computational Statistics* 2: 128–149.

30 Bontinck, Z., De Gersem, H., and Schöps, S. (2016). Response surface models for the uncertainty quantification of eccentric permanent magnet synchronous machines. *IEEE Transactions on Magnetics* 52 (3): 72030404.

31 Ma, Z. and Zabaras, N. (2009). An efficient Bayesian inference approach to inverse problems based on adaptive sparse grid collection method. *Inverse Problems* 25 (3): 035013.

32 Friedman, J.H. (1991). Multivariate adaptive regression splines. *The Annals of Statistics* 19 (1): 1–141.
33 Sobol, I.M. (1993). Sensitivity estimates for nonlinear mathematical models. *Mathematical Modelling and Computational Experimentation* 1: 407–414.
34 Andrieu, C., DeFreitas, N., Doucet, A., and Jordan, M.I. (2003). An introduction to MCMC for machine learning. *Machine Learning* 50: 5–43.
35 Beers, K.J. (2007). Probability theory and stochastic simulation. In: *Numerical Methods for Chemical Engineering: Applications in MATLAB®*, 372–435. New York: Cambridge University Press.
36 Tierney, L. (1994). Markov chains for exploring posterior distributions. *The Annals of Statistics* 22 (4): 1701–1728.
37 Bhat, K.S., Mebane, D.S., Mahapatra, P., and Storlie, C.B. (2017). Upscaling uncertainty with dynamic discrepancy for a multi-scale carbon capture system. *Journal of the American Statistical Association* 112 (520): 1453–1467.
38 Reich, B.J., Storlie, C.B., and Bondell, H.D. (2009). Variable selection in Bayesian smoothing spline ANOVA models: application to deterministic computer codes. *Technometrics. A Journal of Statistics for the Physical, Chemical, and Engineering Sciences* 51 (2): 110–120.
39 Gu, C. (2013). *Smoothing Spline ANOVA Models*, vol. 297. New York: Springer.
40 Storlie, C.B., Lane, W.A., Ryan, E.M. et al. (2015). Calibration of computational models with categorical parameters and correlated outputs via Bayesian smoothing spline ANOVA. *Journal of the American Statistical Association* 110 (509): 68–82.
41 Forse, A.C., Milner, P.J., Lee, J. et al. (2018). Elucidating CO_2 chemisorption in diamine-appended metal–organic frameworks. *Journal of the American Chemical Society* 140 (51): 18016–18031.
42 Milner, P.J., Siegelman, R.L., Forse, A.C. et al. (2017). Diaminopropane-appended metal–organic framework enabling efficient CO_2 capture from coal flue gas via a mixed adsorption mechanism. *Journal of the American Chemical Society* 139 (38): 13541–13553.
43 Monteiro, A., Gribble, D., Jayaraman, A. et al. (2025). Multi-scale dynamic modeling and validation of radial flow fixed bed contactors for post-combustion CO_2 capture using bench scale and pilot plant data. *International Journal of Greenhouse Gas Control* 147: 104481.
44 Fout, T., Zoelle, A., Keairns, D. et al. (2015). *Cost and Performance Baseline for Fossil Energy Plants Volume 1a: Bituminous Coal (PC) and Natural Gas to Electricity Revision 3. DOE/NETL-2015/1723, 1480987*. U.S. Department of Energy Office of Scientific and Technical Information.

Index

www.ingramcontent.com/pod-product-compliance
Ingram Content Group UK Ltd.
Pitfield, Milton Keynes, MK11 3LW, UK
UKHW020222290926
15470UKWH00001B/324

9 781394 205578